D0203043

VISION AND ACQUISITION

VISION AND ACQUISITION

Fundamentals of Human Visual Performance, Environmental Influences and Applications in Instrumental Optics

Ian Overington, *B.Sc., A.R.P.S., M.Inst.P.*

Chief Optics Engineer,
British Aircraft Corporation
(Guided Weapons Division)

Pentech Press
London

Crane, Russak & Company, Inc.
New York

First Published 1976
by Pentech Press Limited
London 8 John Street, WC1N 2HY

ISBN 0 7273 2201 X

Published in the United States of America
by Crane, Russak & Company, Inc.
347 Madison Avenue
New York, N.Y. 10017

Library of Congress Catalog Card No. 75-45872

ISBN 0-8448-0917-9

Set by Preface Limited, Salisbury, Wiltshire, England
Printed in England by Redwood Burn Limited,
Trowbridge & Esher, England

Preface

During a period of some fourteen years of direct involvement in the field of practical visual acquisition I have become acutely aware of the diversity of disciplines with which it is necessary to become familiar in order to appreciate the problem as a whole. Although a mass of relevant literature exists, the important sources remain widely scattered. Furthermore, while a great deal of academic work has been carried out on facets of the problem, our frequent experience at British Aircraft Corporation is that a further and necessary step must be taken to translate the ideal into the practical.

In the course of work which has required my colleagues and myself to search widely through the literature and carry out a number of pieces of supporting research, it became clear that there was a place for a book which, while not going into any particular discipline in extreme depth, would provide, in one volume, a survey of the whole field of visual acquisition and copious references for more detailed study. The layout of the book is aimed at introducing the reader progressively to the basic workings of the eye, threshold behaviour in simple acquisition situations, the effects of imperfections in image quality and of complex stimulus structure, and finally external influences due to the atmosphere and surface reflectance properties. The final chapter is then devoted to a discussion of the acquisition problem in typical practical field situations and the possibilities of providing adequate simulation in the laboratory. In keeping with modern practice the units used throughout the book are the SI system. This has inevitably meant that many equations and figures familiar to established workers in the field have taken on a new look. It is hoped that this is beneficial to new workers. As a further aid, a conversion table of relevant SI units and older units is included.

In several areas the subject is far from completely understood and widescale research is currently proceeding. Whilst some attempt has been made to take account of new data becoming available during the preparation of the book, it is inevitable that some of what is reported will need to be modified as time goes on. This is particularly true of the fields of image evaluation and understanding of the visual process itself where, had this book been written a year or two later, I would expect the emphasis to have been changed considerably.

In compiling this book my thanks are due to my employer, British Aircraft Corporation Limited (Guided Weapons Division), for sponsoring the work and to a number of colleagues, past and present, who have carried out portions of the work reported. I should like particularly to acknowledge the collaboration in our early vision modelling work of Edward P. Lavin, the collaboration in our early image evaluation work of Mrs. Sheila A. Gullick and the survey work on atmospheric optics and surface reflectance of Alan G. Crowther & Gerald E.

Titmuss, on which chapter 14–16 are largely based. My thanks are also due to my colleagues John Ackroyd and Dr. Ray W. Williams, and to Geoffrey P. Owen of the Defence and Operational Analysis Establishment, for reading the original draft manuscript and offering many helpful criticisms and suggestions. I acknowledge gratefully the permission granted by the Ministry of Defence (Procurement Executive) for publication of considerable amounts of data which were collected under contract to them and their predecessors and last, but by no means least, I am indebted to my wife, Dorothy, for supporting me throughout this venture and for typing all the draft material.

Ian Overington.

Contents

1 The Basic Problem of Acquisition

Of man's 5 senses vision is claimed to provide some 75% of the total input to the brain about his environment. This means both that the amount of data being received through two small optical sensors is truly enormous and that it is important to know how this information is received and processed. Not surprisingly a very great deal of research effort has gone into discovering more about man's visual system over many, many years. Researchers are hampered, however, by the fact that the visual receiver – the retina at the back of the eye – has numerous individual detectors implanted in it which are known to couple through a very complicated array of neural networks to the cortex.[1] It is hardly practical to probe deep into these networks with living subjects and, although some work has been carried out on eyes bequeathed to medical research by deceased persons, there is a severe limit on what can be learnt from such studies. This has led many scientists to carrry out extensive studies on various forms of lower animals, from which a great deal has been learnt as to what *might* be facets of human vision. However, in the author's opinion care should always be taken in drawing too close parallels between the established behavioural characteristics of the visual systems of lower animals and that of man. It seems highly probable that evolutionary processes will have been largely responsible for roughly optimising a given animal's visual system for its environment. Having said this, it is difficult to justify an assumption that, say, a cat's visual system is the same as man's, except perhaps in spectral sensitivity (which might be expected in most animals to be optimised about the peak emission of the sun). The foregoing comments are not intended to imply that research on animals is of no use. Far from it – a lot of very valuable data has come out of such studies. All that must be stressed is that it is not *enough* – it must be supplemented by data obtained about man by whatever method possible.

Much has been written on the anatomy and neurophysiology of the eye. The reader wishing to study these in depth is referred, for instance, to Pirenne[1] and Brindley.[2] It is not the purpose of this book to go into these subjects in any more detail than necessary to develop the subject of the title – Vision and Acquisition. Instead the emphasis will be on a *general* consideration of everything, both external and internal to the eye, which goes towards the ability of man to see things in practice. An attempt will be made to show that at the present time, whilst there are still a lot of unanswered questions, a real possibility exists of estimating when many classes of visual stimuli can be 'acquired'.

1.1 THE MEANING OF ACQUISITION

Before going further let us define what we mean by acquisition. The word has to many people a military flavour – and indeed our involvement in visual research was predominently centred for a long time around the ability of man to see military targets. As we have delved deeper into the subject, however, we have come to look upon this word *acquisition* in a very much broader sense. We now prefer to consider it to relate to any object which we are interested in looking at.

The word *acquisition* does not only cover one specific visual process, but rather a series of processes, which range from the first awareness of some local difference in energy at a specific point in the visual field through a progressive awareness of the detail structure of an object. The various stages of this progressive process of acquisition are variously defined by groups of workers. For the purposes of this book we shall choose to define three specific facets. Firstly the awareness of existence of local difference energy we shall refer to as *detection*. Secondly we shall refer to awareness that an object is of a particular class – for instance a vehicle rather than a bush – as *recognition*. Finally the ability to specify that an object is a particular one of a class – for instance a particular model of car – we shall term *identification*. It will be seen as we proceed to study the subject that there are no watertight compartments of detection, recognition and identification, but rather a continuum. Even so, it is believed that the three definitions can be related to different aspects of the acquisition process.

Detection, which is usually defined as the ability to detect the presence of something (unknown) in an otherwise uniformly illuminated field, is frequently associated with military situations (although, of course, it is equally applicable to other fields such as astronomy). However, recognition, which, in a military sense, is the next stage in the acquisition process, and involves the detection of the presence of some characteristic structure (or movement perhaps), is really a word which can be applied to each and every visual task we carry out, and for whatever purpose. Looked at in this way the whole of the understanding of the visual process is tied up with an understanding of ability to *detect* details (without at this stage defining what we mean by 'details'). It is for this reason that we feel justified in spending a large amount of time in this book discussing the process of visual detection, and what might appear to be relatively little time discussing other facets of acquisition.

1.2 FACTORS AFFECTING 'SEEING'

With this preamble let us look a little further into the factors which potentially affect our ability to see.

1.2.1 Available energy

Firstly, there must be sufficient of the right form of energy available. The human visual system is only sensitive to a very small portion of the electromagnetic spectrum – wavelengths from about 0.4 μm to 0.7 μm with a predominent peak sensitivity at around 0.55 μm. In terms of evolutionary optimisation it is interesting to note that the band of visual sensitivity coincides very roughly with the peak of radiation from the sun (around 0.5 μm), thus tending to optimise the efficiency of the visual system to natural illumination. Because of the above it is no use trying to see an an object which only emits radiation around 0.3 μm (ultraviolet) or, perhaps, at 10 μm (well out into the infra-red) without first transducing the energy by some means. It is perhaps unfortunate that until very recently it has not been possible to visualise radiation at 10 μm, since this is a valuable spectral region for viewing through the atmosphere (due to relative freedom from atmospheric attenuation coupled with the fact that it coincides with the peak thermal radiation from objects at ambient temperature[3,4]). Recent advances in electro-optics seem likely to make such a visualisation become much more common in the not too far distant future.[5,6,7]

As with any detector system, the more energy that is available to the visual system the less serious the internal system noise is, and the more sensitive and reliable the detector becomes. That this is so for the visual system can be readily appreciated when one realises how relatively poor visual performance is at night compared to that in good daylight. In order to cope with the tremendous range of luminance levels met with in natural viewing (from about 5000 cd/m^2 down to 10^{-6} cd/m^2) whilst retaining a high sensitivity, the human visual system has developed two distinct forms of receptors – the cones and the rods.[1] The cones, of which there are several types with different spectral sensitivities, look after our colour vision and, it is believed, have individual connections directly to the brain. They are most concentrated in a tiny area of the retina known as the fovea, which covers a circular portion of the visual field subtending between 10 and 20 mrad diameter. As we move further and further from the fovea into the peripheral regions of the retina the cone density falls off sharply.[8] This means that colour vision, which is very acute in the fovea, becomes progressively less acute at angles away from the fovea. In contrast the rods, which have predominently blue sensitivity and appear to be grouped together in the primary neural networks, are completely absent in the fovea and progressively increase in number as one moves to the periphery. These rods are primarily responsible for night vision at luminance levels where the cones are inoperative. Night vision is thus 'colour blind', and with a maximum sharpness at a position several degrees from the fovea.[1] The above properties of the eye will be covered in rather more detail in Chapter 2.

1.2.2 Stimulus characteristics

Other factors which must be considered in discussing visual performance are several associated with the object being viewed (which we shall refer to as the stimulus). Some of them may appear obvious in general terms but can be difficult to define in absolute terms. For the moment suffice it to say that size, shape and form, contrast against surroundings, texture, edge sharpness and interaction with surroundings are all stimulus parameters which may affect ability to see. In this book contrast will always be assumed to be defined as the psychometric contrast $(B/B' - 1)$ where B and B' are the luminances of the stimulus and immediate background respectively. In many structured situations it will be found that such a concept cannot be used. In such cases local luminance structure will be described in terms of local luminance differences (ΔB). Some of the many effects due to stimulus and background will be discussed in detail in Chapters 4, 5, 11, 12 and 13.

1.2.3 Other factors

In addition to the available energy and the stimulus characteristics, other factors which may affect visual performance are effective exposure time, any search requirements, stimulus motion, atmospheric effects and scene structure. Let us take these in order.

The effective exposure time is the total time available for inspection of any particular part of a stimulus. It may be limited by physical constraints external to the eye, or may be an implied limit due to visual search. Search itself may be present due to uncertainty as to where a stimulus is in the visual field – as for instance when searching the sky for an aircraft – or it may be an imposed detail search within a local area of a visual scene, or even within local parts of a complex stimulus. In each case the search strategy will yield progressive bits of data on which to build a brain 'picture' of the fine details of the scene being studied, but at the expense of available time to study any one elemental area. If a stimulus to be studied has relative motion with respect to the observer then the effect on the ability to see it, or detail within it, depends very much on the rate of motion, its size and contrast and the position within the visual field at which it is viewed. These various facets of the presentation situations are covered in detail in Chapters 4, 5, 8 and 13.

If there is any significant distance between the observer and the stimulus it may be necessary to consider the effects of the atmosphere on the appearance of the stimulus. The main atmospheric effects are of two kinds. Firstly, and usually most important, there are the effects due to the significant scattering properties of most real atmospheres. These properties are the predominent cause of mists and fogs, and result in the light from parts of a scene being attenuated (extinction) at the same time as additional veiling light is added into the viewing path from elsewhere (air-light). The second effect, most usually seen as shimmer from hot surfaces, is atmospheric turbulence. This is the result of eddying of air due to thermal effects, is a refraction phenomenon, and usually leads to time variant detail image motion or blurring, dependent on the viewing conditions. The above facets of atmospheric optics will be considered in detail in Chapters 15 and 16.

Finally, in this list of factors which can affect visual performance we come to scene structure. This can have several effects on performance, most of them ill-defined at present. It may include similar objects to the stimulus studied, thus producing conflicting input data. It may upset the adaptation level at which the local visual inspection is taking place. It may introduce local veiling glare effects which detract from visual performance. These are considered in more detail in Chapter 13.

1.3 AIDED VISION

In order to improve viewing conditions, it is often the practice to employ one of a number of forms of visual aid. For instance a microscope, telescope or binoculars may be used to produce a magnified image of the stimulus to be studied. Alternatively television may be used to translate an image to a more accessible spot, or to give enhancement of luminance in difficult lighting situations. Then again, photographic processes are frequently used to record visual data for later study. Finally, for very low luminance situations, image intensifiers are becoming fairly common instrumental visual aids.

If these visual aids were perfect the appraisal of their effects on visual performance would be easy, but unfortunately they each have quality limitations which must cause complex interactions with the structure of the visual system itself. Because of the importance of this aspect of vision, and the number of difficulties associated with it, the whole of Chapters 9 and 10 are devoted specifically to the problems of aided vision.

REFERENCES

1. Pirenne, M. H. (1967). *Vision and the Eye*, Chapman & Hall, London
2. Brindley, G. S. (1970). *Physiology of the Retina and Visual Pathway*, 2nd edition, Edward Arnold, London
3. Taylor, J. H. and Yates, H. W. (1957). 'Atmospheric Transmission in the Infra red', *J. Opt. Soc. Am.*, **47**, 223
4. Benford, F. (1939), 'Laws and Corollaries of the Black Body', *J. Opt. Soc. Am.*, **29**, 92

5. Kruse, P. W. (1972). 'Optics at Honeywell', *Applied Optics*, **11**, 2129
6. Barnes, R. B. and Gershon-Cohen, J. (1963). 'Thermomastography', *J. Albert Einstein Med. Center*, **11**, 107
7. Bivans, E. W. (1965). 'Scanning Radiation Pyrometer', *Inst. & Control Systems*, July 1965, 115
8. Østerberg, G. (1935). 'Topography of the Layer of Rods and Cones' *Acta Ophthal.*, **13**, Suppl. 6

2 Some Basic Properties of the Human Visual System

In order that the reader may more readily understand the implications of the following chapters, this chapter is devoted to a discussion of the known physical and physiological properties of the visual system which may contribute to the human visual capabilities. Such factors include the actual form of the eye, the basic components of the neural networks coupling the eye to the brain, the quality of the refraction optics of the eye, the structure and spectral response of the retina upon which the final optical image is formed, the relationship between the diameter of the eye pupil and the prevailing light level, the effects of sudden changes of scene luminance, the form and extent of involuntary eye movements which are always present, interactions between the two eyes in binocular vision and the phenomenon of after-images.

2.1 THE BASIC STRUCTURE OF THE VISUAL SYSTEM

Figure 2.1 shows a stylised representation of a cross-section through an eye and the associated neural networks. Light entering the eye from the top of the figure first encounters the cornea, which is a hard transparent layer, after which it shortly reaches the eye lens. Successive refraction of the light at the various surfaces of the cornea and the eye lens together serve to focus an image (ideally) on the curved rear surface of the eyeball known as the retina. The lens, which is readily deformable, can be made more or less bulbous, thus changing its focal length, by means of the ciliary muscles, which hold it in a continual state of tension[1]. By this means closer or more distant objects may be focussed on the retina, a process known as accommodation. For a fuller description of the eye the reader is referred to Pirenne[2].

As a means of rapid control of the amount of light falling on the retina the eye lens is fitted with an iris diaphragm, by which the pupil diameter can be varied from less than 1.5 mm to around 8 mm in the normal eye[3]. The retina contains a large number of small photo receptors of varying shape and spectral response. Broadly speaking these receptors fall into two classes, the cones and the rods (so called because of their appearances). The cones are further subdivided into a number of categories – variously believed to be between two and four or five – these various categories having different spectral sensitivities and providing the basis of colour vision[4–7]. Rod vision, which is more sensitive, provides the basis for night vision and is of only one spectral sensitivity[6,7]. The cones are predominently clustered around a small circular area of the retina known as the fovea – of between 10 and 20 mrad angular subtense – which provides the facility for critical vision. This area is completely free from rods,

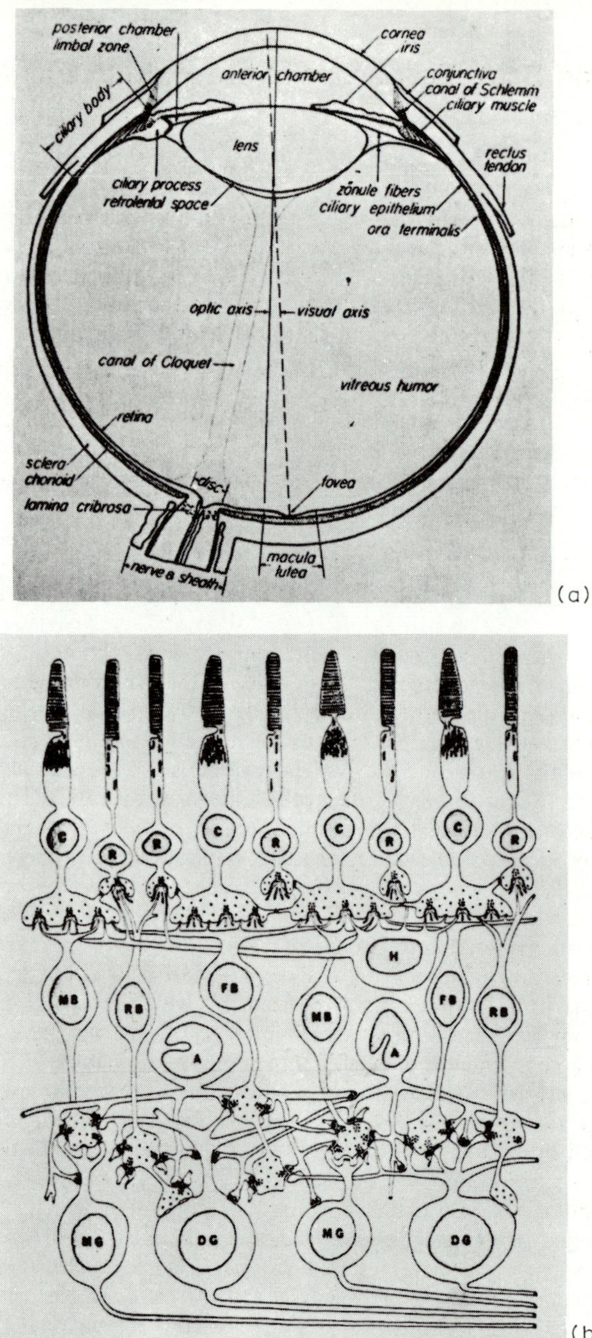
posterior chamber
limbal zone
ciliary body
ciliary process
retrolental space
cornea
iris
anterior chamber
conjunctiva
canal of Schlemm
ciliary muscle
lens
rectus
tendon
zonule fibers
ciliary epithelium
ora terminalis
optic axis
visual axis
canal of Cloquet
vitreous humor
retina
sclera
choroid
disc
fovea
lamina cribrosa
nerve & sheath
macula
lutea
(a)
C
R
H
MB
RB
FB
A
MG
DG
(b)

which are progressively more concentrated out to an angle around 0.35 rad from the fovea[8]. All the foveal cones are believed to be individually served by nerve fibres, whilst peripheral cones are believed to be grouped in small numbers. The rods are believed to be grouped in large numbers (in clusters of up to 10 000)[9] – presumably to provide sensitivity for low light viewing at the expense of resolution. All nerve fibres from the whole of the retina are grouped together and pass from the eyeball along the optic nerve at a position of the retina some 0.25 rad to the nasal side of the fovea. In this region of the retina is an approximately 50 mrad diameter patch which is devoid of receptors and is known as the blind spot[2]. Simple experiments may readily be devised to show the existence of this blind area. The optic nerves from each eye are subsequently split in two and are fed to the two lobes of the visual cortex. At the cortex a visual map is believed to be reconstructed of the images formed on the two retinae[7], binocular depth information being obtained by comparisons.

Between the retinal receptors and the optic nerve are a series of neural networks comprising neurons, axions and ganglion cells. Data is transmitted through these various networks – which contain complex interconnections and feedback loops – by a series of pulse discharges[2,7].

2.2 VARIATIONS OF PUPIL DIAMETER

In conditions of stable viewing – that is, when viewing a fairly uniformly illuminated scene for a prolonged period – the pupil diameter tends to take up a mean value which is related to the scene luminance[3]. Figure 2.2 shows collected data from various sources on the relationship between pupil diameter and scene luminance. It will be seen that, broadly speaking, the pupil diameter is an inverse function of scene luminance. It might at first be thought that this is the obvious way in which the eye compensates for changes of scene luminance. However, since the variation of transmission possible by varying the eye pupil from 1.5 mm to 8 mm is only of the order of 30:1, whilst the eye is capable of adapting over a range of some 10^7:1 in scene luminance, this assumption is obviously wrong (as will be discussed further in Section 2.5). Rather it would appear that the pupil sets itself to a diameter where it is able to compensate partially and rapidly for the most likely sudden changes of scene luminance. For instance, if the eye is in dark surroundings it is most likely that it will be required to cope with a sudden large increase in luminance. Hence the pupil is

---

Fig. 2.1. (a) Horizontal section of the human eye (Reproduced from Walls, G. L. (1942), The Vertebrate Eye, *by courtesy of the Cranbrook Institute of Science). (b) Diagrammatic representation of the synaptic connections of the primate retina, R, rod; C, cone; MB, midget bipolar; FB, flat bipolar; RB, rod bipolar; H, horizontal cell; A, amacrine cell; MG, midget ganglion cell; DG, diffuse ganglion cell. (Reproduced from Dowling and Boycott[33] by courtesy of the Royal Society).*

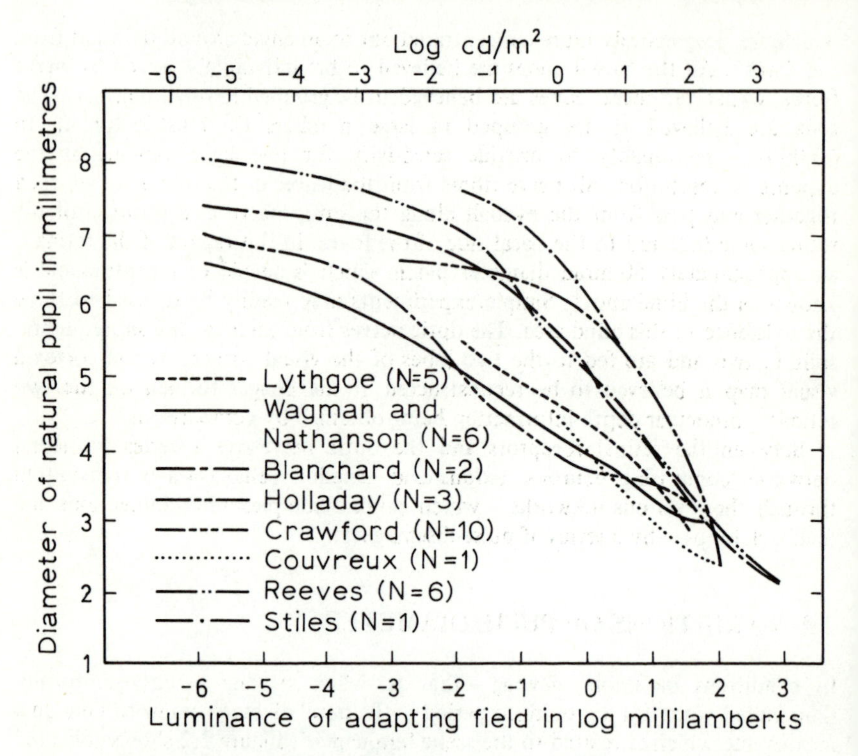

Fig. 2.2. Diameter of the natural pupil due to the luminance of large adapting fields. Separate measurements from eight studies (N = number of subjects studied). (Reproduced from DeGroot and Gebhard[3] by courtesy of the Journal of the Optical Society of America*).*

wide open. Conversely, in conditions of high luminance the most likely change is a sudden darkening of the scene. Hence a fully closed down pupil. Yet again, if the luminance is moderate – say deep shadow or indoors in daylight – the possibilities of sudden increase or decrease are roughly matched. Hence an intermediate pupil diameter.

Pupil diameters are not only controlled by prevailing light levels. Other factors which may affect them are of nervous origin. Emotions such as anger tend to close the pupils, whilst the opposite emotions such as adoration tend to open them[10]. Similarly certain drugs can have major biassing effects on pupil diameters, either through the general nervous system or by local nerve reaction[10]. The latter effects are made use of considerably by ophthalmologists and physiologists in studying the behaviour of the eye. For such studies certain drugs known as cycloplegics are particularly useful, since they fully dilate the eye pupil for extended periods (mydriasis), and allow controlled studies to be carried out using artificial pupils of various diameters[1,11].

2.3 THE OPTICAL QUALITY OF THE EYE

Since the eye consists of a few very simple optical components it is unrealistic to expect that the retinal image formed will be 'perfect' (that is, diffraction limited). For several decades studies have been made of the extent of various primary aberrations present in typical eyes. Of recent years these have been supplemented by measurements of the quality of the refraction optics as a function of pupil diameter and spectral composition of the image forming light carried out by several workers.

Some of the more extensive work on the various aberrations has been carried out by Ames and Proctor[12], Bahr[13] and Ivanoff[14,15], who between them have provided a considerable body of data. In addition several other workers have carried out extensive work on particular aberrations of the eye. Nevertheless the data are still specific to relatively few eyes, and show very considerable variations from eye to eye. There are also in evidence marked asymmetries in

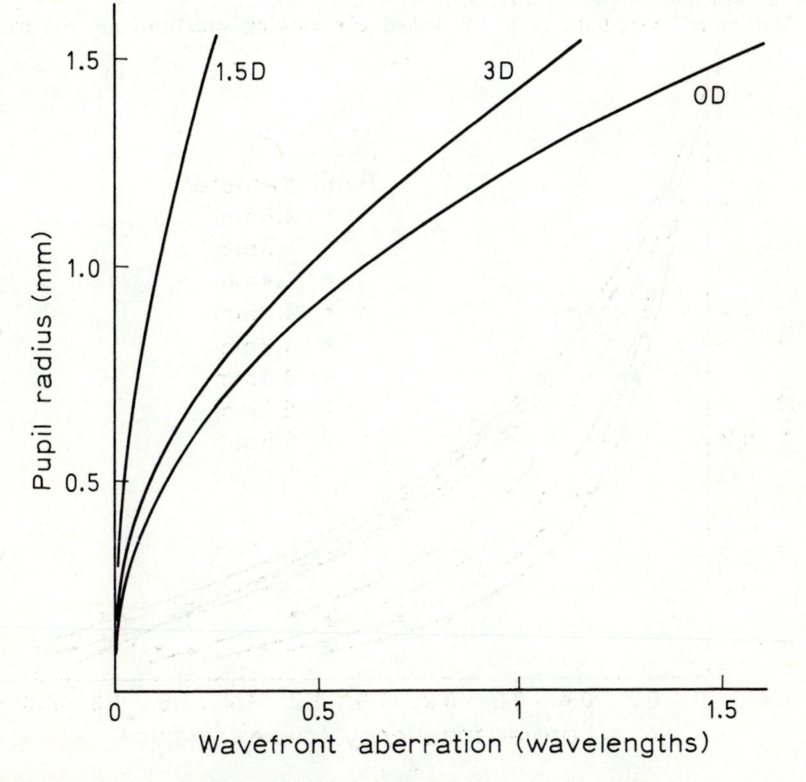

Fig. 2.3. Average wavefront aberrations of the human eye for accommodations of 0, 1.5 and 3 dioptres as computed from spherical aberration data. (after Overington[19]).

any particular eye, making it difficult to define an 'average' amount of any particular aberration (e.g. Ivanoff[16]). However, from this body of data a few important general trends do emerge. For instance, it is fairly conclusively shown that the eye has a predominent chromatic aberration resulting in a progressive shift of focus from in front of the retina to behind the retina as one changes spectral band from blue through to red (e.g. Ames and Proctor[12]). The extent of this trend is large, amounting to some 2 dioptres change of focus from extreme blue to extreme red, but its effect has been shown to be relatively small when viewing neutral shades in white light[17], due presumably to the relative sensitivity of the retinal receptors to various spectral regions (see Section 2.4). It has also been shown that the eye has a residual spherical aberration which varies as a function of viewing distance (state of accommodation), although the magnitude of the spherical aberration under many viewing conditions appears to be masked by residual defocus effects of up to 0.75 dioptres[16,18]. The combined effect of spherical aberration and defocus may be represented in terms of a wavefront aberration (Fig. 2.3)[19].

Of more immediate value for naked eye viewing situations are the data

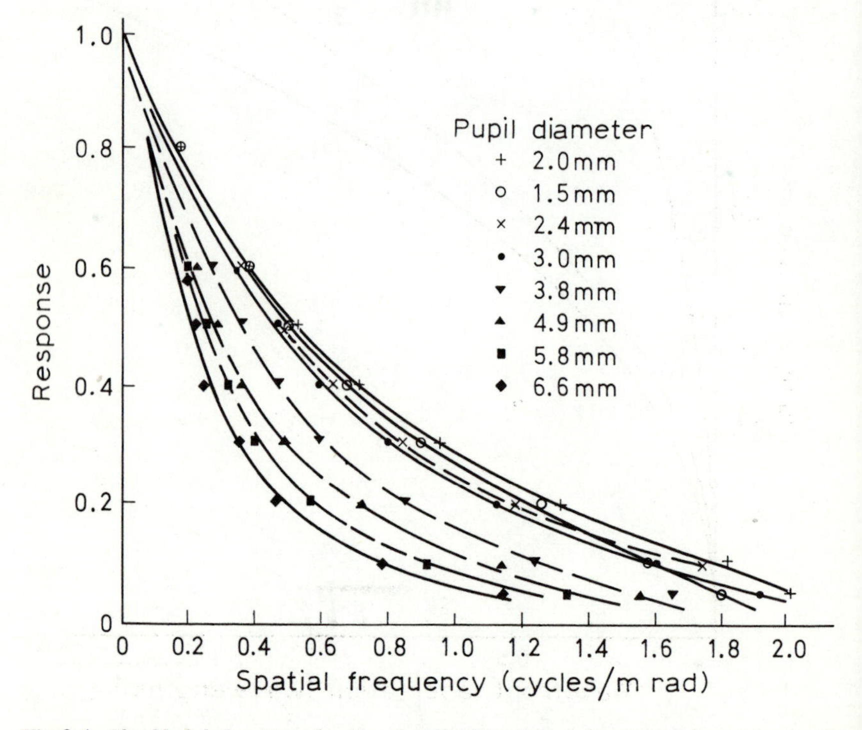

Fig. 2.4. The Modulation Transfer Function (MTF) of the refraction optics of the human eye as a function of pupil diameter (prepared from data in Campbell and Gubisch[17]).

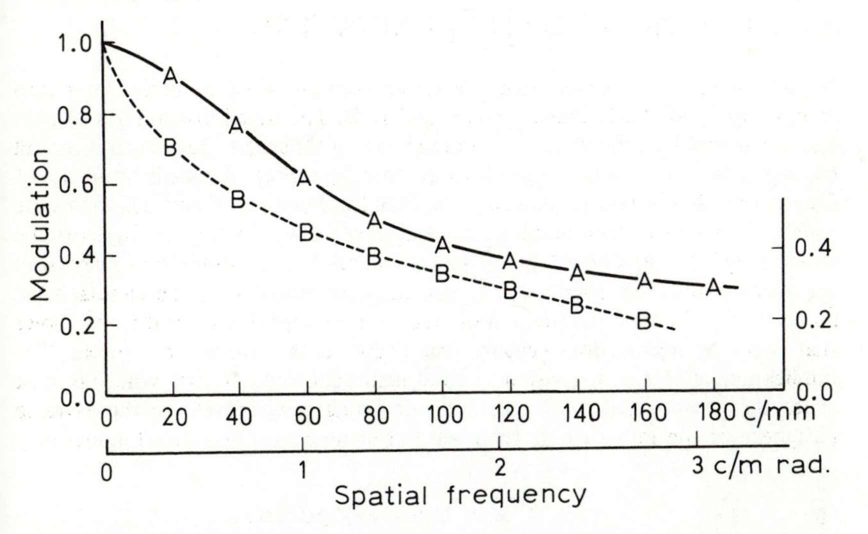

Fig. 2.5. The MTF of the human foveal retina (Curve A) compared to that of the refraction optics (Curve B). (Reproduced from Ohzu and Enoch[25] by courtesy of Pergamon Press).

of Flamant[20], Arnulf and Dupuy[21] and Campbell *et al*[17,22,23], which define the linespread function or the modulation transfer function* of the refraction optics of the eye as a function of pupil diameter. However, since such studies require that accommodation be paralysed (cycloplegia) they have usually been limited to conditions of best focus or controlled defocus. A typical set of MTF curves are shown in Fig. 2.4, where it will be seen that the frequency response is relatively independent of pupil diameter at pupil diameters below about 3 mm for an optimally refracted eye, whilst falling off rapidly for pupil diameters greater than 3 or 4 mm.

In addition to the image degradations due to the optical limitations of the refraction optics, further degradation has recently been found to occur due to diffusion in the retina itself[24,25]. This occurs due to the construction of the retina, which is such that the retinal receptors are at the back of the retina, with the neural cells and linkages overlaying them[26]. Whilst the overlay is thin near the centre of the fovea, it still produces diffusion effects which are comparable to the quality limitations of the refraction optics (Fig. 2.5). As one moves away from the fovea the neural layers thicken substantially and it is believed[25] (although not reliably measured at this time) that the resultant diffusion MTF for the peripheral retina may well be markedly worse than that for the fovea.

*The modulation transfer function (MTF) is the percentage response to a sinusoidally modulated spatial bar pattern as a function of spatial frequency of the modulation. It is the Fourier transform of the linespread function and is a basic measure of optical performance of growing popularity (see Chapter 10.3).

2.4 THE RETINA AND NEURAL NETWORKS

It has already been stated that the retina contains a large number of photo receptors of two basic kinds – cones and rods. The detail distribution of rods and cones usually referenced is from the work of Østerberg[8] (although it should be noted that this was obtained from only one eye). A useful summary of Østerberg's distribution patterns is to be found in Pirenne[2]. Figure 2.6 shows the usually accepted distribution of rods and cones as a function of angle off the fovea. It will be seen how rapidly the cone concentration diminishes away from the fovea, whilst the rods reach a maximum concentration in an annular zone about 0.35 rads from the fovea. More recent work reported by O'Brien[27] shows that cone concentrations change markedly even within the fovea. The implication of this is that vision in good light (photopic or cone vision) is most acute in the very centre of the fovea, declining progressively to the extreme periphery of the retina where there are not many cones. In contrast, under very

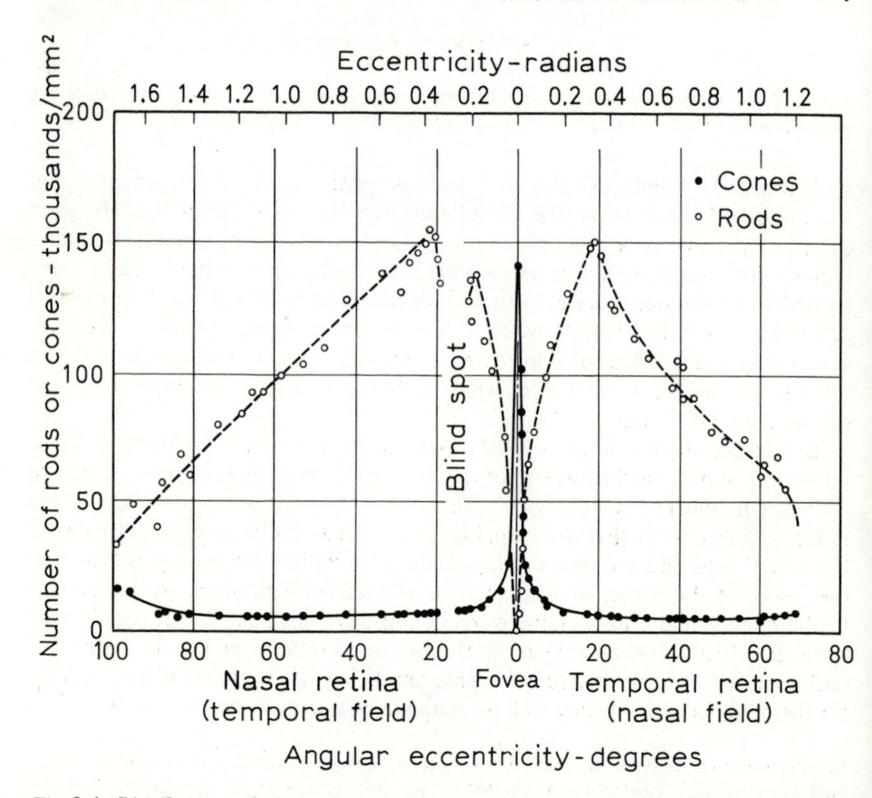

Fig. 2.6. Distribution of the rods and cones across the human retina in the horizontal meridian. (Reproduced from Østerberg[8] courtesy of Munksgaard, Copenhagen).

low light conditions, where vision is predominently by rods (scotopic), the maximum acuity is many tens of milliradians from the fovea, and the fovea is virtually blind.

There are known to be a variety of neural cells and linkages immediately associated with the retina (see Fig. 2.1). Some of the cells are linked horizontally, some in direct line to the optic nerve. Some can transmit electrical signals forwards, some backwards, some in both directions[7]. A very great deal of discussion has ensued up to the present day as to the exact functions which these various cells and linkages perform[7,26,28-34]. Some have been claimed to be associated with complex local field interactions, as apparently evidenced in 'receptive field' studies (see Chapter 13). Other functions proposed have been to provide various forms of colour vision mechanisms, feedback control of adaptation state and rod grouping for enhancement of scotopic sensitivity at the expense of quality. One general property which is widely claimed for neural units is that they are characterised by a positive response to a central excitatory

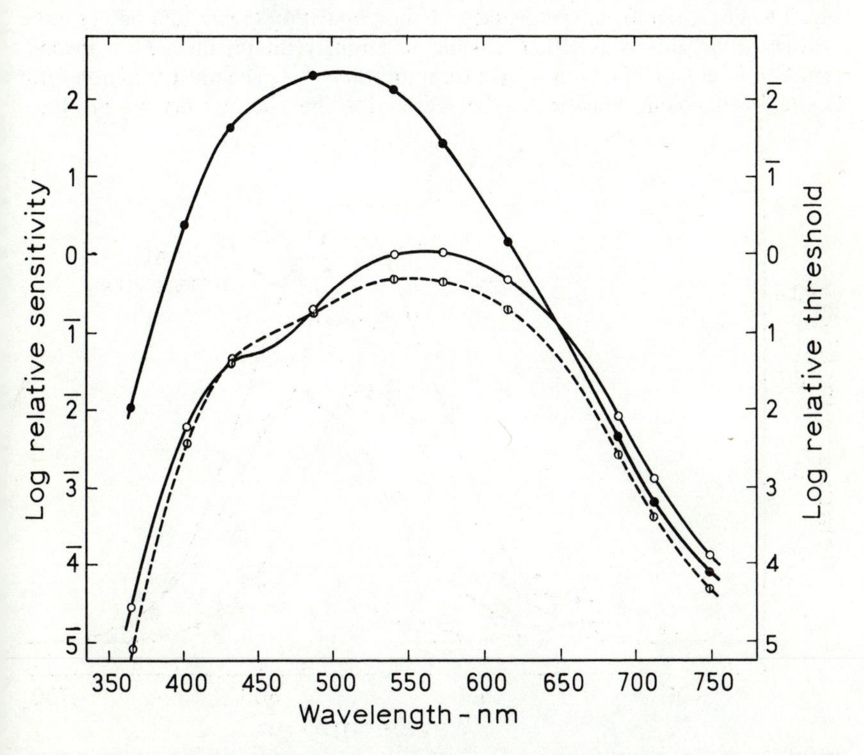

Fig. 2.7. Basic photopic and scotopic spectral sensitivity curves for the human eye. Rods (black circles, full line), peripheral cones (broken line), foveal cones (open circles, full line), (Reproduced from Wald, G. (1945), Science, **101**, *653, by courtesy of the American Association for the Advancement of Science).*

field, with an inhibition of response being caused by receipt of energy from a surrounding annular field (e.g. Brindley[7] and Cornsweet[35]). The size of these two fields appears to be very variable from neural unit to neural unit and from species to species. However, any unit with such antagonistic response fields has a basic ability to act both as an integrator of signals received from a central area and as a differentiator by considering the central and surround fields as units. It will be seen in later chapters how various authors have interpreted this dual property of neural units.

Overall spectral sensitivity curves for cone and rod vision have been established and internationally agreed[36,37]. These are shown in Fig. 2.7. The spectral sensitivities of several types of cone receptor are claimed to have been isolated by various workers, by one of several methods[4,5,6,28,31,38]. Taking a concensus of these various 'findings', it seems most likely that, in an observer with normal colour vision, the daylight mechanisms are based, at receptor level, on three classes of receptor with spectral sensitivities broadly as shown in Fig. 2.8. For a significant percentage of the population (some 8%) one or more of these mechanisms is either missing or strongly interacting with a second, resulting in colour blindness of one form or other[4,6,7]. The most common form is red/green colour blindness – the reason for the classic story of Sir Isaac

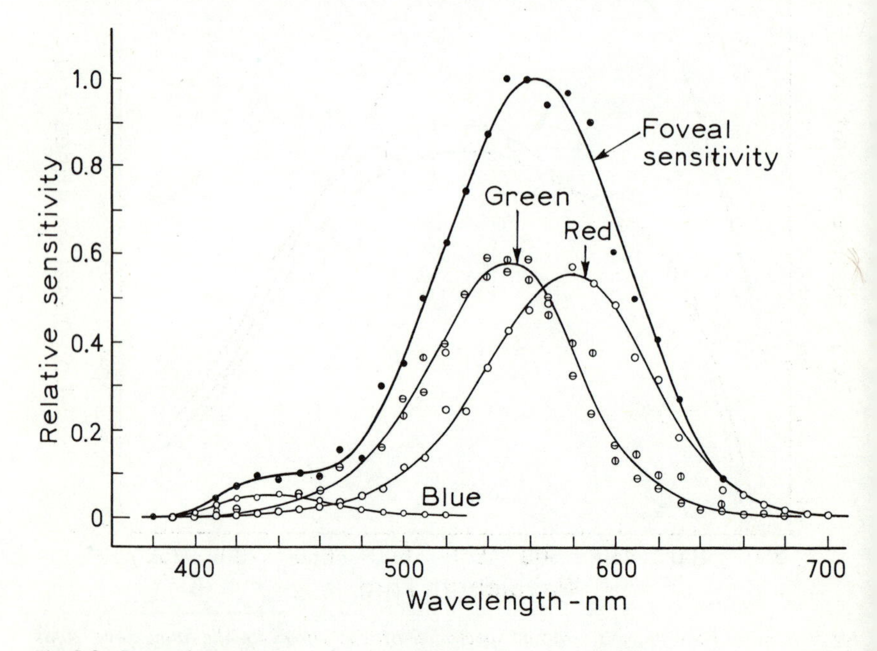

Fig. 2.8. Spectral sensitivities of individual human cone receptors, and the total foveal sensitivity. (Reprinted by permission from Wald, G. (1964), Science, **145**, *1007. Copyright (1964) by the American Association for Advancement of Science).*

Newton throwing his scarlet riding tunic onto the lawn and then not being able to find it. Other, much rarer forms are blue/yellow colour blindness and total colour blindness, where a person can only see shades of grey.

The specific nature of the forms of colour blindness, and certain limitations of a simple trichromatic theory of colour vision, have led some workers to suggest that colour vision should really comprise three opponent mechanisms, black versus white, red versus green and blue versus yellow. This was first

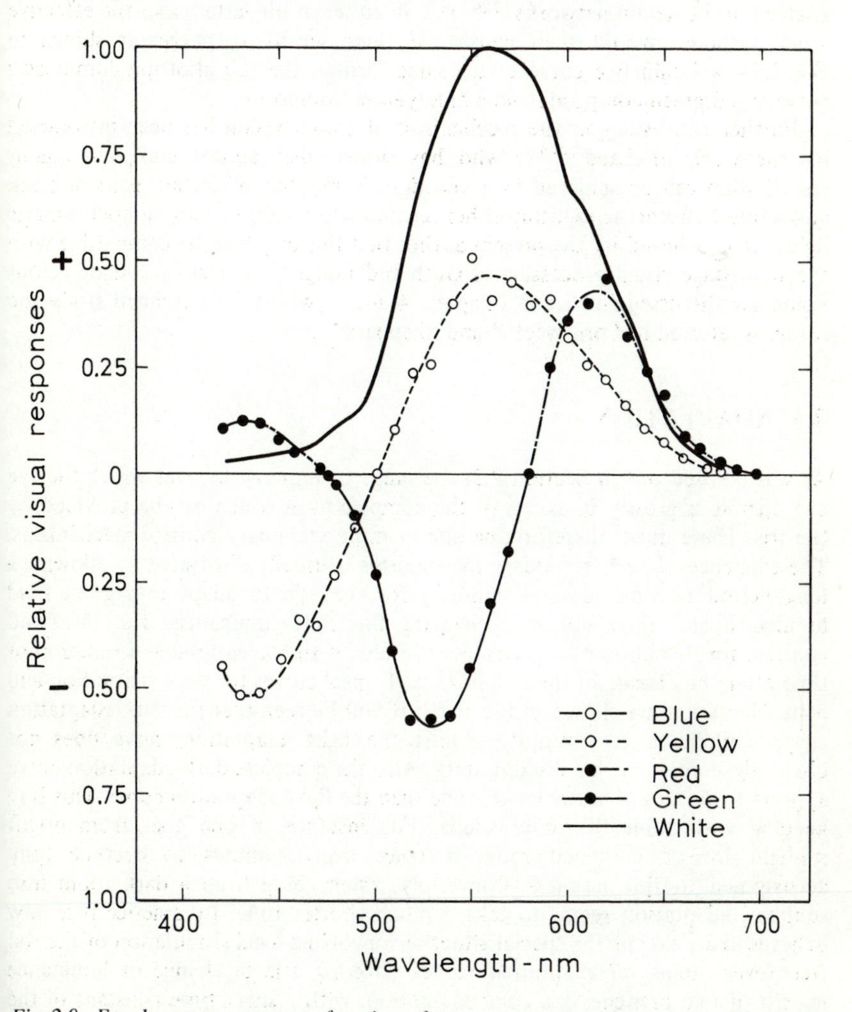

Fig. 2.9. Equal energy response functions for an opponent–colours vision mechanism. (Reprinted by permission from Hurvich and Jameson[40]. Copyright (1957) by the American Psychological Association).

proposed almost a century ago by Hering[39] but has only been quantified relatively recently by Hurvich and Jameson[40]. Talbot[31] claimed in 1951 that such a process in the retina was proven. The opponent colours system, the supporters argue, would permit explanation of certain phenomena which do not fit in with the simple trichromatic theory. Other workers have suggested a multistage process, with trichromatic response at the receptor level followed by intercoupling to yield an opponent-colours system and possible polychromatic analysis in the neural networks[41,42,43]. In either of the latter cases the effective visual response would then consist of three sensitivity curves as shown in Fig. 2.9 – a luminance curve of the same form as the full photopic luminosity curve, a red/green composite and a blue/yellow composite.

Further confusion on the mechanisms of colour vision has been introduced by the work of Land[44-46], who has shown that almost complete colour visualisation can be achieved by projection, in register, of certain pairs of black and white transparancies, using either red and white light or two monochromatic lights. It is believed by the present author that this may well be compatible with the multistage visual processing of Guth and Lodge[43]. Certain facets of colour vision are discussed further in Chapters 4 and 7, whilst for extended study the reader is referred to Cornsweet[35] and Sheppard[47].

2.5 ADAPTATION

As was pointed out in Section 2.2, the range of luminances over which the eye can operate is grossly in excess of the compensation which can be provided by the iris. There must, therefore, be one or more secondary control mechanisms. The existence of such secondary mechanisms is usually illustrated by allowing a long period of time (several minutes) for the eye to adapt to a given field luminance and then suddenly changing this field luminance. The threshold contrast for detection of a given test stimulus is then measured as a function of time after the change of field luminance. Typical curves for dark adaptation and light adaptation are shown in Fig. 2.10. It will be seen that the dark adaptation curve contains a discontinuity, whilst the light adaptation curve does not obviously contain such a discontinuity. Also the complete dark adaptation curve appears to occupy a much longer time than the light adaptation curve. This is in keeping with subjective impressions. For instance, if one goes from bright sunlight into a darkened room it takes many minutes to become fully accustomed to the darkness. Conversely, when going from a dark room into sunlight, adaptation seems to take a much shorter time. In practice it is now believed that, except for special situations involving local stimulation of the rod free fovea alone, *all* adaptation curves covering a large change of luminance consist of two branches – a cone adaptation, with a short time constant of the order of 2 min, and a rod adaptation, with a time constant of the order of 7 or 8 min[9,48]. Since, at high photopic levels, the rods are believed to be relatively

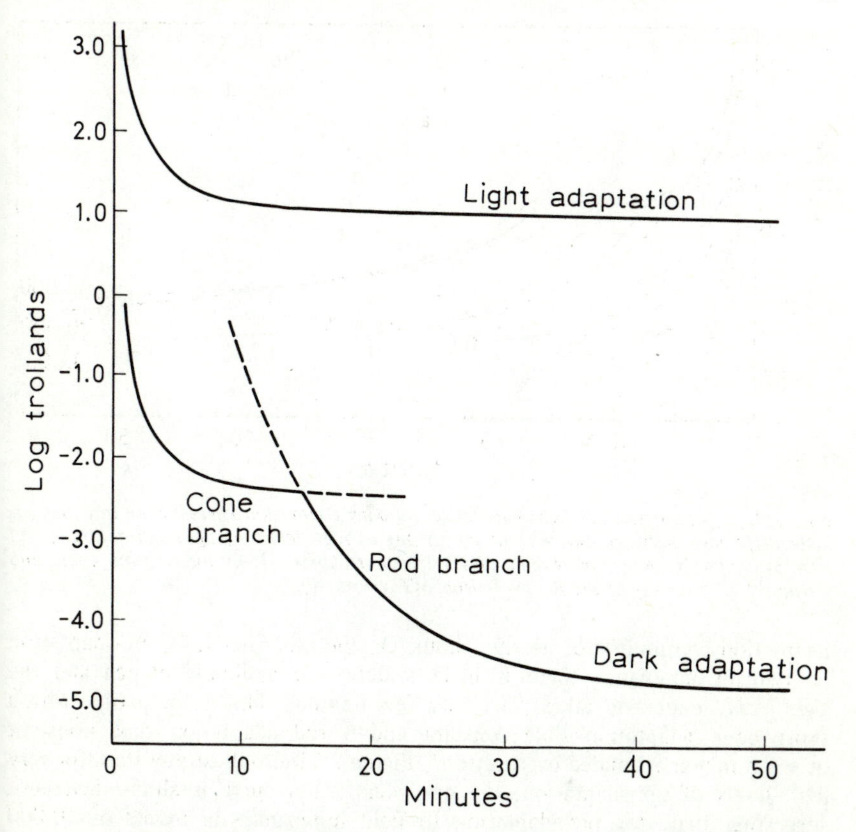

Fig. 2.10. Typical dark and light adaptation curves. In the dark adaptation curve the upper branch is due to cone adaptation, the lower branch being due to rod adaptation.

insensitive, during dark adaptation from a high photopic level the first phase of adaptation will be controlled by the cones. When the cone adaptation is almost complete the rods, with their longer adaptation time constant but superior low light sensitivity, will eventually take over, thus giving the adaptation curve a second 'branch'. The two separate adaptation curves are continued with dotted lines in Fig. 2.10. For light adaptation, since the final state is predominently controlled by cone vision, the lag of rod adaptation is almost unnoticeable – although in a critical experiment Wright[48] *has* shown a long term improvement in threshold, of the order of 2, after effective completion of cone adaptation.

The variation of the overall rate of adaptation to absolute darkness as a function of level of prior light adaptation is well illustrated by Fig. 2.11 (taken from Craik and Vernon[49]). It will be seen that, for pre-adaptation to very high luminances, there is a very prominent discontinuity and a total *significant*

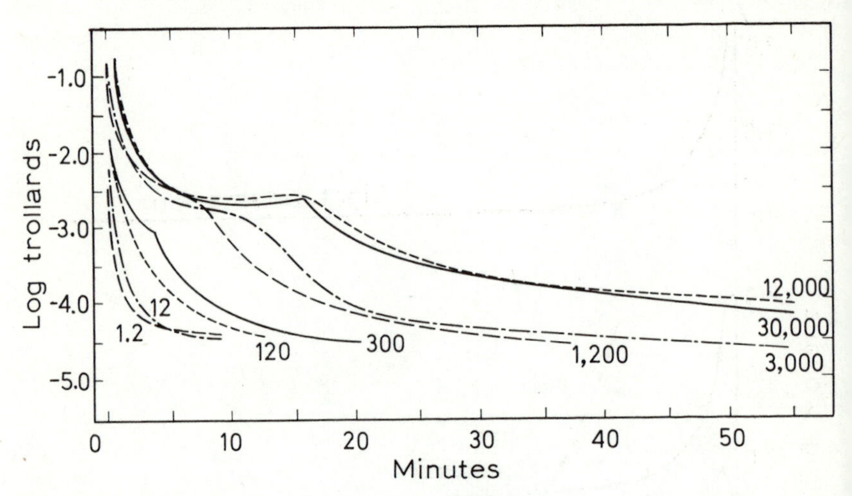

Fig. 2.11. Effect of state of light adaptation on dark adaptation curves. Note how there is little difference between various curves starting at high levels of light adaptation. (light adaptation levels against individual curves are in trollands) (Reproduced from Craik and Vernon[49] by courtesy of the British Journal of Psychology).

adaptation continuing for nearly 1 hour. On the other hand, for pre-adaptation to twilight luminance there is little evidence of a double branch and the *significant* adaptation takes place in a few minutes. This is compatible with a short cone adaptation time constant and a rod adaptation time constant of ≈ 7.5 min as estimated by Rushton[9]. Figure 2.11 also illustrates that, for very high levels of pre-adaptation, the dark adaptation curve is almost invariant, suggesting that, for pre-adaptation to field luminances in excess of 10 000 trollands, the adaptation processes are 'saturated'.

It is generally believed that one main mechanism determining state of adaptation is photochemical in origin. The existance of a light sensitive chemical substance called rhodopsin (visual purple) associated with the rods has been known for many years[2]. This substance is bleached progressively by light action and regenerates in the dark. Its state of bleach at any instant is thus dependent on both the instantaneous level and the recent time history of retinal illumination. It has been shown by Rushton[9] that there is a striking similarity between the state of regeneration of rhodopsin and the increment threshold. He achieved this by establishing the form of dark adaptation curves for detection of various diameters of stimuli after flash adaptation and by comparing these curves with increment threshold curves for the same stimuli against various background luminance levels (Fig. 2.12). Now the state of rhodopsin regeneration after flash adaptation is directly related to time after adaptation, and a given steady background luminance level is also related to a given rhodopsin level. Therefore times can be equated to background luminances for equal rhodopsin bleach

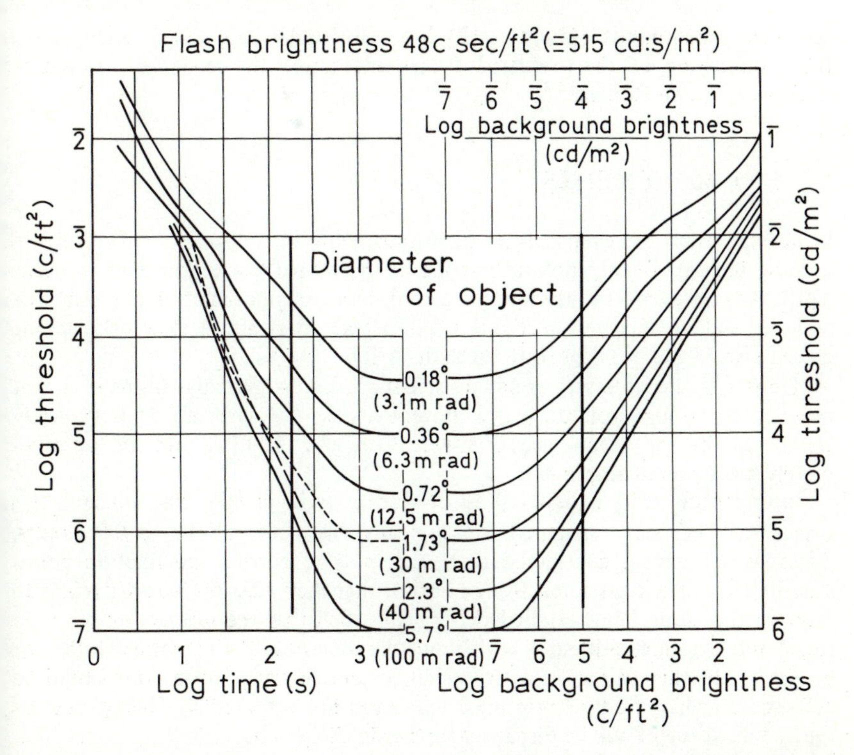

Fig. 2.12. Comparison between a family of dark adaptation curves (implying state of rhodopsin regeneration) after exposure to stimuli of the diameter shown (left-hand curves) and a family of increment thresholds for the same stimuli (right-hand curves). (Reproduced from Rushton[9] by courtesy of the Royal Society).

states. A pair of equated conditions are shown by the two thick vertical lines in Fig. 2.12. In recent years other photochemical substances associated with the various cone mechanisms have been isolated and similar relationships demonstrated between regeneration rates and increment thresholds[9] for cone vision.

Despite the fairly conclusive evidence that a major factor in adaptation is the state of bleach of photopigments, other workers claim that, in addition, changes take place in lateral neural connections, again as a function of recent time history of illumination (e.g. Arden and Weale[50] and Hecht *et al*[51]). Yet others claim a close relationship between some dark adaptation curves and the decay of after-images (e.g. Craik and Vernon[49] and Barlow and Sparrock[52]) whilst Rushton and Powell have recently shown conclusive evidence of the existence of *some* secondary mechanism controlling the early phase of rod dark adaptation[53]. It is felt by the author that the complete story may well be clarified when the reasons for behaviour of the eye when viewing stimuli against complex backgrounds (Chapter 13) are fully understood. Meanwhile, for a fuller discussion of the problem of dark adaptation the reader is referred to Cornsweet[35] and Rodieck[54].

2.6 EYE MOVEMENTS

In normal vision the eyeball is in a continual state of motion[55]. This results in minute, but significant, motion blurring of the retinal image. The main motions which it is necessary to consider are a continuous small oscillation (tremor), the periodic sudden changes of fixation point (saccades) and a slow drift taking place between the saccades (intersaccadic drift).

Tremor is the necessary residual oscillation due to muscular balance so that muscles controlling voluntary eye movements do not seize up. It is typically found to be of frequencies between 10 and 100 Hz and of approximately ± 0.15 mrad amplitude.

Intersaccadic drift is believed to be due to residual muscular inbalance (as typical of D.C. servo systems) and typically has a rate of 0.7 to 0.9 mrad/s. Saccades are widely considered to be attempts to correct the fixation point. Alternatively it is considered by the author that they may also be an attempt to reimage the scene being studied periodically such that the interaction between signal and the inevitable noise which must be present in the neural networks can be varied. In this way an effective enhancement of signal/noise ratio would be achieved, as when multiple printing is employed in photography. This concept of signal enhancement will be discussed further in Chapter 7.

2.7 ACCOMMODATION

In a state of rest it is believed that the average eye will focus an object at a distance of approximately 0.8 m onto the retina[1]. A young person with good eyesight will usually be able to bring into sharp focus objects at distances ranging from a small fraction of a metre to infinity by use of the ciliary muscles. The shortest distance which can be brought into sharp focus is known as the near point. With advancing years the accommodative power reduces (presbyopia), making it more difficult to bring close objects into sharp focus and, by inference, reducing the clarity of distant objects, although this latter fact is probably not of great importance.

Many people, of course, have other than 'average' eyes, which most frequently results in loss of long range clarity with accentuated short range performance (short sight or myopia) or loss of short range performance with accentuated long range performance (long sight or hypermetropia)[2]. Both the foregoing defects are due to the resting focal length of the eye being other than optimally related to the diameter of the eyeball.

In practical viewing the eyes are continually attempting to optimise image quality on the retina but, as with any servo system, for that is what the accommodative mechanism must essentially be, there will always be some state of inbalance. Thus in practice the overall imagery will always contain some element of defocus (see Section 2.3, Ivanoff[16] and Westheimer[18]). This in turn means that the eye quality as measured by Campbell and others[17,20-23] (see Section 2.3) is better than will ever be achieved in practical viewing. To the author's knowledge no directly comparable quality data exist for a naturally accommodated eye and paralysed accommodation.

2.8 OVERALL QUALITY OF THE VISUAL SYSTEM

It should be clear from the preceeding sections that the overall quality of the visual system is dependent on several factors. The basic quality of the refraction optics, retinal diffusion, the state of accommodation (controlling residual defocus), involuntary eye movements, the distribution of receptors in the retina and neural coupling must all play a part in defining the quality of the whole system.

For very many years (from Lord Rayleigh's time) it was common practice to define visual performance in terms of one of several forms of resolution criteria[56]. It was found that the eye could usually just detect a black disc of angular subtense about 0.15 mrad against a plain background in good light. Similarly it was found possible to detect the presence of twin points of light as twin points when they were separated by about 0.15 mrad, again in good light. If lines or simple patterns were studied different resolution limits were found, as also was the case if the objects were of low contrast[57]. Of recent years attempts have been made to measure the spatial frequency response of the complete visual system, thus producing a performance measure of the complete system which is similar in form to the MTF already described for study of the refraction optics (Section 2.3). To achieve this, it is usually arranged for a spatial bar pattern of variable frequency and sinusoidal spatial modulation to be presented to an observer, the modulation depth then being altered (for a given frequency) until the bar pattern can no longer be seen (e.g. Campbell and Green[22]). If this procedure is repeated at various frequencies a frequency response function is produced. It is conventional to present such functions with an ordinate of the reciprocal of threshold contrast, thus retaining absolute information about threshold performance as well as relative response as a function of frequency. Such a function is then known as a *contrast sensitivity function*. Since such an experimental measurement does not necessarily require any special treatment of the eye, it is possible to carry out measurements either with the eye accommodation paralysed – thus comparing with the quality measurements of the refraction optics alone[22] or with natural accommodation (e.g. Campbell and Gubisch[58]). Typical contrast sensitivity functions from the two foregoing

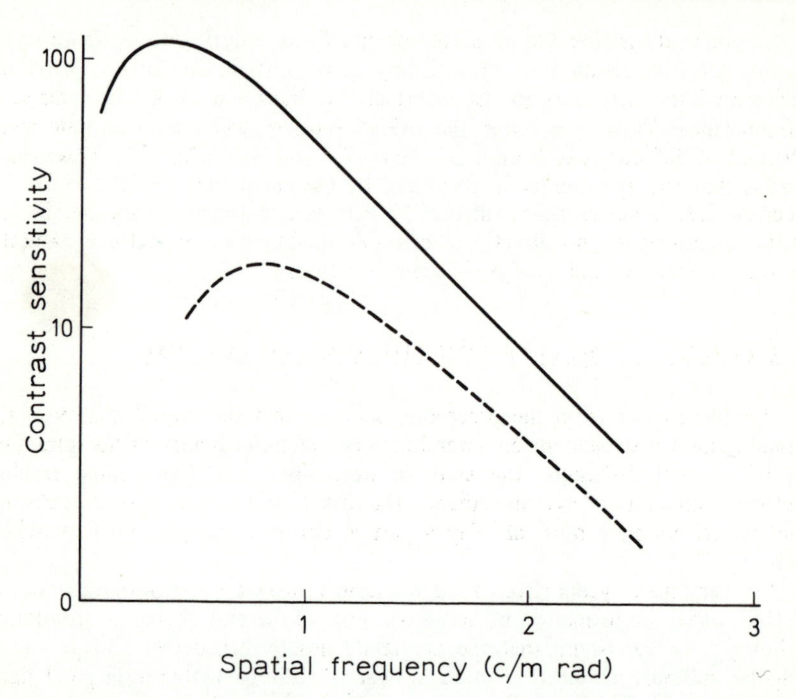

Fig. 2.13. Comparison of contrast sensitivity functions for large and small patterned field sizes. (Data from Campbell et al.[22,58]*).*

references are shown in Fig. 2.13. It will be seen that, although they both exhibit the same overall shape, they are very dissimilar. It is tempting to explain this difference in terms of the residual defocus of the naturally accommodating eye. However, work by Hoekstra *et al*[59] shows that the absolute form of the contrast sensitivity function is very considerably sensitive to the display field of the bar pattern, whilst other work (e.g. Green and Campbell[11]) has shown contrast sensitivity to be relatively insensitive to the small amounts of defocus which might be expected to exist in natural accommodation. Since Campbell and Gubisch[58] only used a very small presentation field (about 3 mrad diameter) compared to a 35 mrad field used by Campbell and Green[22], it must thus be expected that a considerable part of the difference between the curves in Fig. 2.13 is due to display field size.

Contrast sensitivity and variants will be further discussed in Chapters 4, 10 and 12. In particular Chapters 10 and 12 include discussion of the surprising fall-off of response at low frequencies which is characteristic of all total visual system response testing with sinusoidal bar patterns and is particularly forcibly illustrated by a demonstration figure due to Campbell (see Ratliff[60]). The relationship between the contrast sensitivity and the retinal image MTF is also discussed in Chapter 10.

2.9 BORDER ENHANCEMENT – THE MACH EFFECT

It is now over 100 years since Ernst Mach[61] first observed the phenomenon which now bears his name (the Mach effect). This phenomenon is that, if the eye is presented with an illuminated field containing a high contrast sharp discontinuity of luminance, a lightening of the field adjacent to the light side of the border and a darkening to the dark side will be observed. The magnitude of the lightening and darkening is found to be considerably dependent on the sharpness of the luminance discontinuity presented[62]. There has been speculation about the reason for the phenomenon. Some workers have held that it is a subjective effect, implying non-linear processing in the neural networks. Others (e.g. Campbell and Gubisch[17]) have suggested that such assumptions are unnecessary, it being possible to explain the phenomenon in conjunction with other experimentally measured performance of the visual system. Yet others have elaborated on Mach's own studies and suggested several forms of local interaction in the neural networks to explain the observed distortions of apparent luminance[61]. Many of these are surveyed in Chapter 6, whilst the author's own suggested explanation of the phenomena are discussed further in Chapter 12.

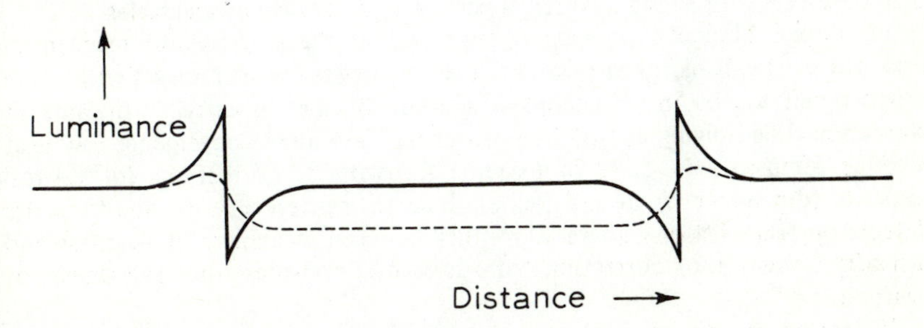

Fig. 2.14. The form of luminance distribution across borders (solid curve) necessary to promote the Craik–O'Brien illusion. The perceived brightness is shown dotted.

Similar visual phenomena to the Mach effect are to be found in a whole range of 'brightness illusions' which have been made use of by artists down the ages but have to this present day caused puzzlement to scientists[35,60]. Of particular surprise are the effects of various forms of line contours having asymmetric luminance profiles applied to otherwise substantially uniform fields. A circular line contour of luminance profile as in Fig. 2.14 drawn on a plain field, for instance, will cause the field to appear to be of two considerably different luminances – as shown dotted in Fig. 2.14 (see Ratliff[60]). Such illusions will be discussed further in Section 12.5.

2.10 BINOCULAR/MONOCULAR EFFECTS

Thus far little note has been taken of the fact that most people have two good eyes. What additional properties of vision are associated with the combined use of two eyes? The answer is that, if a person has two equally good eyes, and if they are used as nature intended (that is, with the same stimulus presented to both), visual threshold performance for simple tasks is improved by a factor of $\sqrt{2}$[63,64], whilst at the same time additional information associated with object distance is provided, due to the slightly different view points of the two eyes (stereopsis). However, many people do not possess two equally good eyes, in which case threshold performance for binocular vision will not improve over monocular performance with the preferred (or master) eye by the optimum factor of $\sqrt{2}$. Furthermore, since optimal performance depends on the ability to fuse two retinal images at the cortex by means of the basic binocular functions convergence and cyclorotation, any differential eye defect between the two eyes may result in a conflict between the accommodative and convergence mechanisms, resulting in other than optimal overall performance. In this sense observers with significant eye defects, but having optimal refractive correction by means of spectacles, are more likely to function in an optimal binocular sense than observers with eyesight which appears adequate without spectacles.

In normal binocular viewing it appears that the convergence mechanism responds to the disparity in position of two associated retinal images with a rate proportional to the instantaneous magnitude of the disparity[65], the rate of correction thus slowing as fusion approaches. The rate of convergence can be as high as 50 mrad/s in cases of extreme disparity[66]. Corrections for vertical disparity (dip vergence) are less flexible than for convergence. If, due to ocular defects or other reasons, there is conflict between awareness of nearness and disparity, then the correction of disparity overrides the awareness of nearness[67,68].

If, instead of a threshold detection task, the observer is required to carry out a typical visual inspection or recognition task, it appears likely that a simple relationship between binocular and monocular performance no longer exists. Experiments by Spicer[69], using high and low contrast Snellen letters, have shown that ability may be only marginally improved by binocular viewing over that with *either* eye monocularly. The observers did, however, comment on the greater 'comfort' of the binocular viewing – believed by the author to be evidence of the interelationship between retinal image quality and system noise. This will be further discussed in Chapters 7 and 12.

If, instead of using the eyes in their natural state, any form of visual aid is used, then a large range of distortions to visual performance can occur. For instance, disparity or stereoscopic parallax can be introduced (an imposed error in vertical alignment or convergence related to a particular presentation distance). This will cause degradation of binocular performance and, if the disparity is too great, can even cause binocular performance to fall below optimal monocular performance[70]. Equally, differential illumination of the two eyes, intentionally or accidentally, can have serious effects on perception of depth (e.g. Gogel[71], Enright[72] and Harker[73]). In particular this gives rise to the classical Pulfrich effect (e.g. Lit[74]) where a swinging pendulum is observed to describe an elliptical path. Some of the effects associated with binocular vision using visual aids are further discussed in Chapter 9.

2.11 NEURAL SIGNALS AND BASIC RHYTHMS

In an endeavour to find out more about the visual processes beyond the retina, a great deal of research work has been carried out on anaesthetised lower animals, in the hope that such neural behavioural patterns as could be determined could give a lead to what happens in man (e.g. Brindley[7], Hubel and Weisel[75] and Campbell *et al*[76]). These studies have shown nerve fibres at deep neural levels in animals such as the cat, the squirrel and certain monkeys which respond to all manner of specific stimuli. Some respond only to onset of light, others to switching off of light, others to motion in one particular direction and so on. A number of interpretations may be put on such findings – for instance they may be specific or they may be part of the total processing which must go on in order to analyse the enormous amount of data input to the visual channel. It is not the intention of this book to study the inner workings of the neural networks and brain in any detail – the reader is referred to Pirenne[2], Brindley[7] and Rodieck[77] for such information – but it is wise for anyone studying vision to be aware that such complex processing must be going on somewhere beyond the retina. In man it is obviously impractical to probe into the optic nerve and the visual cortex. However, modern electroencephalographical (EEG) techniques have permitted some experimenters to isolate minute electrical signals in the head which can be related to certain forms of visual processing (e.g. Walter[78]). Of special interest is the ability of some researchers (for instance, Walter[79]) to isolate particular brain rhythms and associate them to the build up of confidence that a stimulus is present, even before the observer is conscious of any build up of confidence. The same waves may also be used to monitor the instant of making a decision much more accurately than by means of physical responses.

2.12 AFTER–IMAGES

In a completely general treatment of vision processes a phenomenon which cannot be ignored is that of after-images (e.g. Brindley[80] and Padgham[81]). This is the name given to any residual imagery by the visual system which is not related to immediate retinal imagery. After-images take the form of either negative or positive representations of a scene imaged on the retina up to several seconds earlier, the form and intensity being largely dependent on the intensity of the scene previously imaged. Strong, obvious after-images are thus related to excessive stimulation of the retina, and can be considered to be associated with temporary local blindness in the limit. An obvious example is the temporary degradation of vision after the eyes have been subjected to a very strong flash of light.

The importance of after-images in visual acquisition is somewhat difficult to assess, but it must be assumed that *any* local stimulation of the retina is producing some temporary local adaptation which will interact with subsequent imagery at that point on the retina, whether or not an obvious after-image is seen. Hence, whilst single glimpse detection in a plain field is unlikely to be affected by after-images, this is almost certainly not true for extended viewing, whilst for acquisition in a highly structured field strong interactions may be expected. In such situations the previous time history of local illumination of the retina may well play an important part. Some factors which may be associated with after-images are discussed in Chapter 13.

2.13 STABILISED VISION

In Section 2.6 the existence of several forms of involuntary eye movements were discussed. An interesting question which had to remain academic until a few years ago is 'What would the visual process be like if these eye movements could be removed?' In 1959 Ditchburn *et al* reported an equipment comprising a miniature mirror attached to a contact lens[82] which made it possible to present to the eye an image which was almost perfectly compensated in retinal position as the eye moved. It was found that, under such viewing situations, the image gradually faded and after some 10 s disappeared entirely (e.g. Ditchburn and Fender[83]). This finding is extremely important in attempting to understand the visual processes, since it tends to imply that it is not the absolute level of local retinal illuminance, but periodic variations of it, which permit the visual process to operate. As such it must have a major influence on modelling of the visual process, as will be seen in later chapters.

REFERENCES

1. Westheimer, G. (1963). 'Optical and Motor Factors in the Formation of the Retinal Image', *J. Opt. Soc. Am.*, **53**, 86
2. Pirenne, M. H. (1967). *Vision and the Eye*, Chapman and Hall, London
3. Groot, S. G. de and Gebhard, J. W. (1952). 'Pupil Size as determined by Adapting Luminance', *J. Opt. Soc. Am.*, **42**, 492

4. Lobanova, N. V. (1971). 'Possible forms of Colour Vision', in *Problems of Physiological Optics, Vol. 15*, (Ed. V. G. Samsonova *et al*), NASA Translation TTF–650, p. 43
5. Stiles, W. S. (1959). 'Colour Vision: the Approach through Increment Threshold Sensitivity', *Proc. Nat. Acad. Sci., U.S.A.*, **45**, 100
6. Wald, G. and Brown, P. K. (1965). 'Human Colour Vision and Colour Blindness', *Symposia on Quantitative Biology*, Cold Spr. Harb., New York, **30**, 345
7. Brindley, G. S. (1970). *Physiology of the Retina and Visual Pathway*, 2nd edn, Edward Arnold, London
8. Østerberg, G. (1935). 'Topography of the Layer of Rods and Cones', *Acta Ophthal.*, **13**, Suppl. 6
9. Rushton, W. A. H. (1965). 'Visual Adaptation' *The Ferrier Lecture, Proc. R. Soc. B.*, **162**, 20
10. Duke-Elder, W. S. (1944). *Textbook of Ophthalmology, Vol. 1*, C. V. Mosby Co., St. Louis
11. Green, D. G. and Campbell, F. W. (1965). 'Effect of Focus on the Visual Response to a Sinusoidally Modulated Spatial Stimulus', *J. Opt. Soc. Am.*, **55**, 1154
12. Ames, A. (Jr.) and Proctor, C. A. (1921). 'Dioptrics of the Eye', *J. Opt. Soc. Am.*, **5**, 22
13. Bahr, G. V. (1945). 'Investigations into the Spherical and Chromatic Aberrations of the Eye and their Influence on its Refraction', *Acta Ophthal.*, **23**, 1
14. Ivanoff, A. (1947). 'The Chromatic and Spherical Aberrations of the Eye – their Role in Night Vision' (in French), *Revue d'Optique*, **26**, 145
15. Ivanoff, A. (1950). 'On the Subject of Asymmetry of the Eye' (in French), *Comptes Rendus, Academie des Sciences*, **231**, 373
16. Ivanoff, A. (1956). 'About the Spherical Aberration of the Eye', *J. Opt. Soc. Am.*, **46**, 901
17. Campbell, F. W. and Gubisch R. W. (1966). 'Optical Quality of the Human Eye', *J. Physiol.*, **186**, 558
18. Westheimer, G. (1955). 'Spherical Aberration of the Eye', *Optica Acta.*, **2**, 151
19. Overington, I., (1975). 'On the Interaction between State of Accommodation and Retinal Image Quality', *Atti della Fondazione Giorgio Ronchi*, **29**, 909
20. Flamant, F. (1955). 'Study of the Distribution of Light in the Retinal Image of a Slit', (in French), *Revue d'Optique*, **34**, 433
21. Arnulf, A. and Dupoy, O., (1960). 'The Transmission of Contrast by the Optical System of the Eye and the Thresholds of Retinal Contrast' (in French), *C.r. hebd. Séanc. Acad. Sci.*, Paris, **250**, 2757
22. Campbell, F. W. and Green, D. G. (1965). 'Optical and Retinal Factors affecting Visual Resolution', *J. Physiol.*, **181**, 576
23. Westheimer, G. and Campbell, F. W. (1962). 'Light Distribution in the Image Formed by the Living Human Eye', *J. Opt. Soc. Am.*, **52**, 1040
24. Ohzu, H., Enoch J. M. and O'Hair, J. C. (1972). 'Optical Modulation by the Isolated Retina and Retinal Receptors', *Vision Research*, **12**, 231
25. Ohzu, H. and Enoch, J. M. (1972). 'Optical Modulation by the Isolated Human Fovea', *Vision Research*, **12**, 245
26. Polyak, S. (1957). *The Vertebrate Visual System*, University of Chicago Press, Chicago
27. O'Brien, B. (1951). 'Vision and Resolution in the Central Retina', *J. Opt. Soc. Am.*, **41**, 882
28. Valois R. L. de (1965). 'Analysis and Coding of Colour Vision in the Primate Visual System', *Cold Spr. Harb. Symp.*, **30**, 567
29. Guth, S. L., Donley, N. J. and Marrocco, R. T. (1969). 'On Luminance Additivity and Related Topics', *Vision Research*, **9**, 537
30. Granit, R. (1945). 'The Colour Receptors of the Mammalian Retina', *J. Neurophysiol.*, **8**, 195
31. Talbot, S. A. (1951). 'Recent Concepts of Retinal Colour Mechanisms. I. Contributions from Psychophysics', *J. Opt. Soc. Am.*, **41**, 895

32. Hartline, H. K. (1942). 'The Neural Mechanisms of Vision', *The Harvey Lectures,* **37**, 39
33. Dowling, J. E. and Boycott, B. B. (1966). 'Organisation of the Primate Retina: Electron Microscopy', *Proc. R. Soc. B.,* **166**, 80
34. Gallego, A. (1971). 'Horizontal and Amacrine Cells in the Mammal's Retina', *Vision Research Supplement No. 3*, 33
35. Cornsweet, T. N. (1970). *Visual Perception*, Academic Press, New York and London
36. Gibson, K. S. and Tyndall, E. D. T. (1923), 'Visibility of Radiant Energy', *U.S. Bur. of Stand., Scientific Paper No. 475,* **19**, 131
37. Judd, D. B. (1951), 'Report of the U.S. Secretariat Commission of Colorimetry and Artificial Daylight'. *C.I.E. Proc. I. Pt 7,* p. 11. (Stockholm). Paris, Bureau Central C.I.E., 57, rue Cuvier
38. Stiles, W. S. (1939). 'The Directional Sensitivity of the Retina and the Spectral Sensitivities of the Rods and Cones', *Proc. R. Soc. B.,* **127**, 64
39. Hering, E. (1880). 'An Explanation of Colour Blindness from the Opponent-Colour Theory', *Lotos, Jb. f. Naturwiss.,* **1**, 76 (in German)
40. Hurvich, L. M. and Jameson, D. (1957). 'An Opponent-process Theory of Colour Vision', *Psych. Review,* **64**, 384
41. Judd, D. B. (1949). 'Current Views on Colour Blindness', *Doc. Ophthal.,* **3**, 251
42. Luria, S. M. (1966). 'Colour Vision', *Physics Today*, 34
43. Guth, S. L. and Lodge, H. R. (1973). 'Heterochromatic Additivity, Foveal Spectral Sensitivity, and a new Colour Model', *J. Opt. Soc. Am.,* **63**, 450
44. Land, E. H. (1959). 'Experiments in Colour Vision', *Scientific American*, May
45. Land, E. H. (1959). 'Colour Vision and the Natural Image, Part I', *Proc. Nat. Acad. Sci.,* **45**, 115
46. Land, E. H. (1959). 'Colour Vision and the Natural Image, Part II.' *Proc. Nat. Acad. Sci.,* **45**, 636
47 Sheppard, J. J. (1968). *Human Color Perception*, North Holland Publishing Co., Amsterdam
48. Wright, W. D. (1937). 'The Foveal Light Adaptation Process', *Proc. R. Soc. B.* **122**, 220
49. Craik, K. J. W. and Vernon, M. D. (1941). 'The Nature of Dark Adaptation', *Brit. J. Psychol.,* **32**, 62
50. Arden, G. B. and Weale, R. A. (1954). 'Nervous Mechanisms and Dark Adaptation', *J. Physiol,* **125**, 417
51. Hecht, S., Haig, C. and Wald, G. (1935). 'The Dark Adaptation of Retinal Fields of Different Size and Location', *J. Gen. Physiol.,* **19**, 321
52. Barlow, H. B. and Sparrock, J. M. B. (1964). 'The Role of After-images in Dark Adaptation', *Science,* **144**, 1309
53. Rushton, W. A. H. and Powell, D. S. (1972). 'The Early Phase of Dark Adaptation', *Vision Research,* **12**, 1083
54. Rodieck, R. W. (1973). *The Vertebrate Retina – Principles of Structure and Function*, (Chapter 22.) W. H. Freeman & Co., San Francisco
55. Ditchburn, R. W. and Foley-Fisher, J. A. (1967). 'Assembled Data on Eye Movements', *Optica Acta,* **14**, 113
56. Hartridge, H. (1922). 'Visual Acuity and the Resolving Power of the Eye', *J. Physiol.,* **57**, 52
57. Ronchi, V. (1970). 'The Energetic Theory of Resolving Power', *Atti della Fondazione Giorgio Ronchi,* **25**, 565
58. Campbell, F. W. and Gubisch, R. W. (1967). 'The Effect of Chromatic Aberration on Visual Acuity', *J. Physiol.* **192**, 345
59. Hoekstra, J. J., van der Goot, D. P. J., van den Brink, G. and Bilsen, F. A. (1974). 'The Influence of the Number of Cycles upon the Visual Contrast Threshold for Spatial Sine Wave Patterns', *Vision Research,* **14**, 365
60. Ratliff, F. (1972). 'Contour and Contrast', *Scientific American,* 91

61. Ratliff, F. (1965). *Mach Bands*, Holden-Day Inc., San Francisco, London and Amsterdam.
62. Remole, A. (1974). 'Relation between Border Enhancement Extent and Retinal Image Blur' *Vision Research*, **14**, 989
63. Campbell F. W. and Green D. G. (1965). 'Monocular versus Binocular Visual Acuity', *Nature*, **208**, 191
64. Ronchi, L. and Longobardi, G. (1971). 'Luminance-Time Relationships in Binocular Vision', *Atti della Fondazione Giorgio Ronchi*, **26**, 239
65. Rashbass, C. and Westheimer, G. (1961), 'Disjunctive Eye Movements', *J. Physiol.*, **159**, 339
66. Stewart, C. R. (1961). 'Jump Vergence Responses', *Am. J. Optom.*, **38**, 57
67. Jones, R. (1972). *Psychophysical and Oculomotor Responses of Normal and Stereo-anomalous Observers to Disparate Retinal Stimulation*, Ph.D. Dissertation, Ohio State University.
68. Stewart, C. R. (1963). 'Convergence Response to Sinusoidal Stimulation', *Am. J. Optom*, **40**, 447
69. Spicer, P. J. (1972). 'The Measurement of Visual Acuity', App. 3, Section 1 of *Research into Factors affecting the Detection of Aircraft through Optical Sights*, April 1972, (B.A.C.(GW) Ref. L50/186/1449)
70. Joshua, D. and Bishop, P. O. (1970). 'Binocular Single Vision and Depth Discrimination. Receptive Field Disparities for Central and Peripheral Vision and Binocular Interaction on Peripheral Single Units in the Cat Striate Cortex', *Experimental Brain Research*, **10**, 389
71. Gogel, W. C. (1968). 'The Measurement of Perceived Size and Distance', in *Contributions to Sensory Physiology*, Vol. 3, (Ed. W. D. Neff) Academic Press, New York. 125
72. Enright, J. T. (1970). 'Distortions of Apparent Velocity: a new Optical Illusion', *Science*, **168**, 464
73. Harker, G. S. (1967). 'A Saccadic Suppression Explanation of the Pulfrich Phenomenon', *Percept. and Psychophys.* **2**, 423
74. Lit, A. (1949). 'The Magnitude of the Pulfrich Stereophenomenon as a Function of Binocular Differences of Intensity at Various Levels of Illumination', *Am. J. Physchol.*, **62**, 159
75. Hubel, D. H. and Wiesel, T. N. (1968). 'Receptive Fields and Functional Architecture of Monkey Striate Cortex', *J. Physiol.*, **195**, 215
76. Campbell, F. W., Cooper, G. F. and Enroth-Cugell, C. (1969). 'The Spatial Selectivity of the Visual Cells of the Cat', *J. Physiol.*, **203**, 223
77. Rodieck, R. W. (1973). *The Vertebrate Retina – Principles of Structure and Function*, W. H. Freeman & Co., San Francisco
78. Walter, W. G. (1953). *The Living Brain*, Duckworth and Co., London
79. Walter, W. G. (1967). 'Slow Potential Changes in the Human Brain associated with Expectancy, Decision and Intention', in *The Evoked Potentials, Electroenceph. Clin. Neurophysiol. Suppl. 26*, (W. Cobb and C. Morocutti, Eds.), Elsevier Publishing Co., Amsterdam, 123
80. Brindley, G. S. (1959). 'The Discrimination of After-Images', *J. Physiol.*, **147**, 194
81. Padgham, C. A. (1957). 'Further Studies of the Positive Visual After-image', *Optica Acta.*, **4**, 102
82. Ditchburn, R. W. (1955). 'Eye-movements in Relation to Retinal Action', *Optica Acta*, **1**, 171
83. Ditchburn, R. W. and Fender, D. H. (1955). 'The Stabilised Retinal Image', *Optica Acta*, **2**, 128

3 Observer Variability

If the contrast of a stimulus against its surroundings is progressively reduced there will naturally come a time when it can no longer be seen. However, this 'threshold' point will vary from observation to observation. If many observations are made, and if the number of times that the threshold is above given values is plotted as a percentage of the total number of observations, a cumulative probability curve will result (see for instance Blackwell[1]). Such a curve is conventionally known as a 'frequency of seeing' curve. One may equally expect a form of frequency of seeing curve if presentation time is varied for a fixed stimulus, if size or scene luminance is progressively varied for fixed contrast and presentation time, or if a stimulus is caused to move at various rates during presentation. However, the most common form of frequency of seeing curve to be found in literature is, without a doubt, that as a function of stimulus contrast. This is reasonable since contrast is, by definition, a measure of differential energy, and hence should be related to concepts of signal to noise ratio.

In studying threshold performance, then, it is possible to study the probability of acquisition as a function of one or more basic properties of the viewed scene, or to determine the trend of the 50% probability of acquisition as a function of scene parameters. The majority of threshold studies to be discussed in this book will be concerned with the trends in 50% probability of acquisition. However, before proceeding to such studies it is necessary to survey the factors which contribute to threshold variability and their extent. For convenience these may be divided into 'within observer' variability, 'between observer' variability and influence of environment.

3.1 WITHIN OBSERVER VARIABILITY

3.1.1 The frequency of seeing curve

One of the largest sets of experimental data from which frequency of seeing curves have been extracted is that reported by Blackwell,[1] who was involved in a major detection experiment using disc stimuli in the early 1940's. This experiment employed 20 female observers over a period of between six months and 2½ years, and involved some two million observations covering stimulus diameters from 0.15 mrad to 100 mrad and scene luminance levels from moderate daylight (350 cd/m^2) down to starlight (3×10^{-6} cd/m^2). At each presentation the observers were forced to make a decision about the stimulus. Such experimentation is known as 'forced choice' experimentation. The results reported – based on statistical analysis of some 450 000 observations – claimed

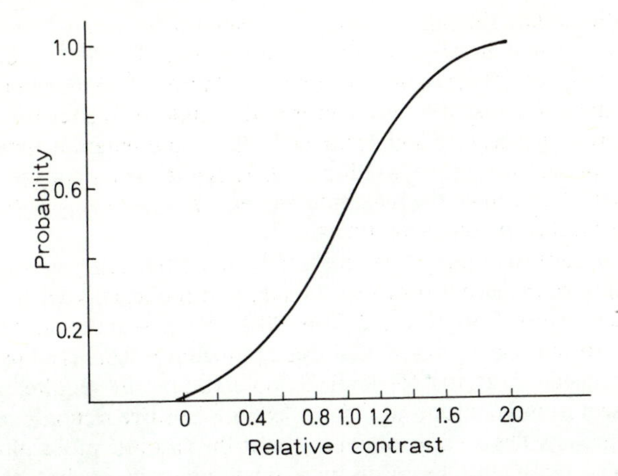

Fig. 3.1. Mean experimental within-observer frequency of seeing curve (Reproduced from Blackwell[1] by courtesy of the Journal of the Optical Society of America).

that the frequency of seeing curves, if plotted on a base of relative contrast, were virtually independent of presentation conditions over the whole gamut of sizes and luminances tested. Further it was claimed that they could be adequately described by a cumulative normal distribution function. The basic form of average observer curve was as shown in Fig. 3.1, where it will be seen that the difference in contrast required to change from 10% probability to 90% probability is of the order of 4:1. The results of many additional studies by Blackwell and coworkers have been collected together in two further papers by Blackwell.[2,3] It is shown that the form of frequency of seeing curve for detection of simple stimuli is remarkably consistent for forced choice viewing over a very wide range of conditions.

For several years we have questioned whether the 'relative contrast' base assumed by Blackwell for fitting most of his data is correct, since it seems inconceivable that there is a finite probability of detection with contrast tending to zero as implied by Fig. 3.1. It should be noted that the sets of stimuli from which Blackwell's frequency of seeing curves were obtained usually had a range of contrasts of 4:1 only – that is, they only, on average, provided data between approximately the 10% and 90% points. With such limitations on range of contrasts it would be difficult to prove that a cumulative normal distribution function with a base of relative contrast was correct with any degree of certainty. It appears from the literature that the only other form of cumulative probability function studied for adequacy of fit was a cumulative normal function with logarithmic contrast as the base. In most cases this was found to yield a worse fit than the function based on contrast ratios.

Several other workers have studied forced choice frequency of seeing and come up with similar findings to those of Blackwell in terms of both shape, spread and constancy irrespective of conditions (e.g. Crozier[4] and Taylor[5]). Blackwell[3] compares a large number of sets of data (mainly obtained by his own co-workers) and finds that the only factor which appears to alter the frequency of seeing curves significantly and predictably is the presentation time. Here he finds that, as presentation time is reduced, there is a tendency for the frequency of seeing curves to steepen, the maximum increase in slope being of the order of 30% with the shortest presentation times.

The lack of confirmation of the absolute form of frequency of seeing curves in the literature must be put down to the fact that the regions which define the detailed shape – from 0 to 10% and from 90 to 100% – are regions where the accumulation of sufficient data to establish a probability score is a huge exercise in itself. For instance, at the 5% level 20 observations are required at a given stimulus presentation condition in order to get one positive response on average. In order to have any degree of confidence about the absolute probability, several hundred observations must be made by a single observer at the one stimulus condition. Not even an experiment the size of the Tiffany one[1] can look at such detail. The alternative, that of examining the required detail at one or two points in the multidimensional threshold space, must leave many questions unanswered about the universality of any findings.

In the light of the above it must be taken as fairly conclusively verified that the frequency of seeing curves for large samples of data under forced choice conditions can be adequately described by a curve of the form shown in Fig. 3.1

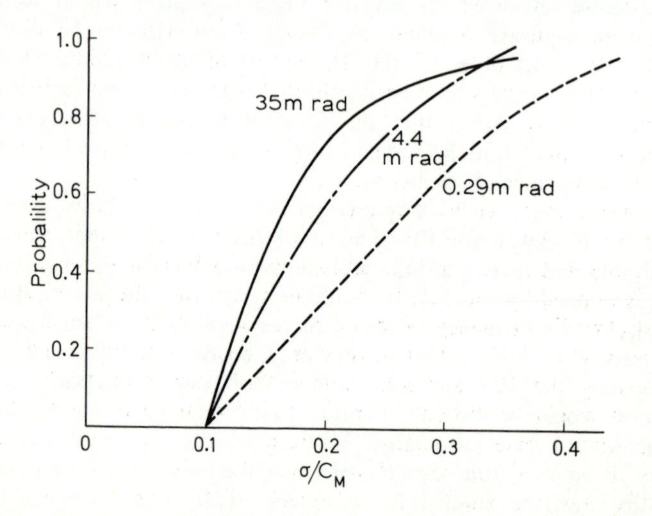

Fig. 3.2. Probability of occurance of individual session frequency of seeing curves having standard deviations less than σ/C_M for various disc stimuli – 1/3 second presentation time, mean of results for 4 observers. (Data from Taylor[5]).

for controlled laboratory experiments involving simple stimuli, although this may not be the *best* descriptor.

If now we look at the form of frequency of seeing curve for individual short experimental periods, instead of Blackwell's overall average, a rather different picture emerges. Silverthorn[6] has chosen to carry out a detailed study of original experimental data obtained by Taylor[5] in terms of the probability of occurance of given standard deviations of frequency of seeing within experimental sessions. The results are summarised in Fig. 3.2. It will be seen that the variation from experimental session to experimental session is very large. However, what is considered most striking, and will be discussed again later, is the constant *positive* deviation independent of size at zero probability – in other words, the implication that the session variability of threshold cannot be less than approximately 0.1 C_M where C_M is the session mean threshold contrast.

3.1.2 Temporal variability of probability

The alternative way to study variability of observer threshold performance is to study the probability of acquisition of a constant form of stimulus presentation within the threshold region as a function of time. Several such studies have been carried out[7–10]. From them it would appear that there are several predominant

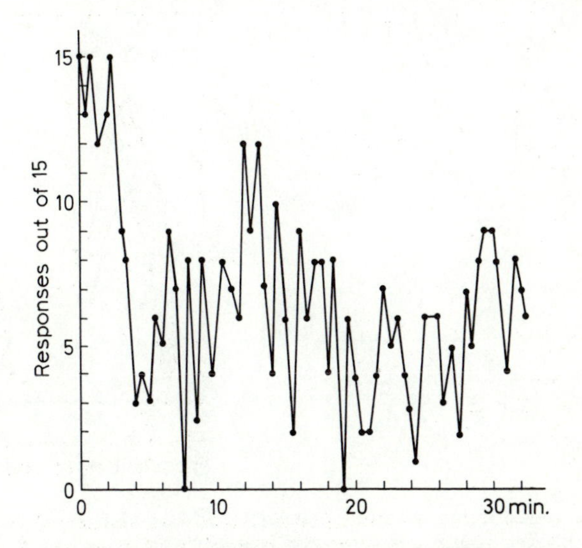

Fig. 3.3. Time course of perception probability for repeated presentations of flash stimuli every 2 s for a period of 30 min. The testing was carried out at 0.175 rad retinal eccentricity, each point representing the number of flashes perceived from a batch of 15. (Reproduced from Ronchi and Brancato[7] by courtesy of S. Karger AG. Basel).

periodic trends in threshold level varying from a few seconds to several hours. For instance, Ronchi and Brancato[7] refer to a periodicity of between 6 and 10 s and a further strong periodicity of around 2–4 min (see Fig. 3.3), whilst Ronchi and Novakova[8] found marked minima in probability at well defined times of the day – particularly mid morning and late afternoon (Fig. 3.4). It is tempting to consider that one may separate out the Blackwell frequency of seeing curve into components on the basis of these periodicities and the probability of various standard deviations discussed in Section 3.1.1. One might then possibly apportion a standard deviation of about 0.1 C_M to the short term variations of a few seconds period, whilst the diurnal variations might be expected to account

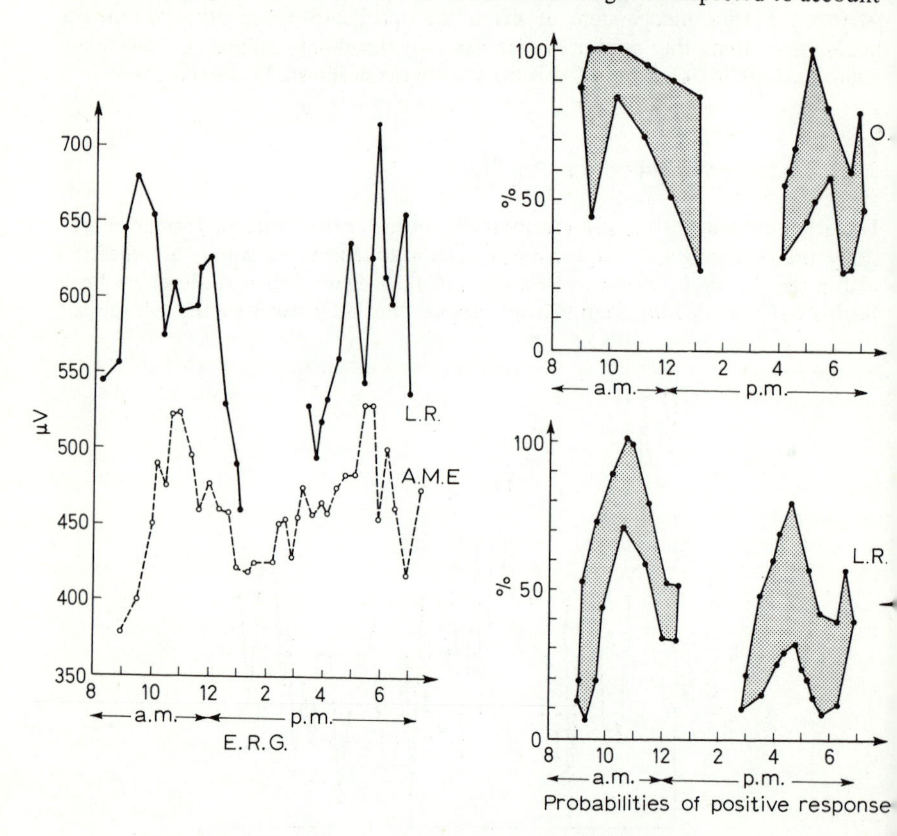

Fig. 3.4. Time course, during the day, of both ERG scotopic response (b-wave amplitude) for observers L.R. and A.M.E. and perception probability for observers L.R. and O.N.. The stimulus was a small, brief test flash at a given point on the dark adapted retina. The dashed areas show intra-session variability. Note the peak of responsiveness mid-morning and mid-afternoon for all observers and sessions. (Reproduced from Ronchi and Novakova[8] by courtesy of the Italian National Institute of Optics, Florence).

for much of the remainder. In addition some part of the total variation might be expected to be on a longer timescale again, according to threshold variations obtained in repeated experiments on visual acuity by Spicer[11] and also according to Sweeney *et al.*[5] although Blackwell[2] refutes such a suggestion.

A possible explanation of part of the short term variation of threshold may be found in considering the actual processes of vision. When an object is being viewed, the vision processes are such that the retinal image is repositioned periodically. The repositioning – known as a saccade – takes place on average about 3 times a second, although it may be as frequent as 10 times a second or as little as possibly once a second on occasion (see for instance Ford *et al*[12]). The shift of image position in a saccade is typically 1.5 mrad. After each saccade a 'fixation' follows during which time the retinal image (of a stationary object) is relatively stable (that is apart from a residual tremor of about ±0.15 mrad and a slow drift – intersaccadic drift – of the order of 6–9 mrad/s). See Lavin,[13] Ditchburn and Foley–Fisher[14] for a comprehensive description of these involuntary eye movements.

Now, as discussed in Chapter 2, the retina contains large numbers of individual receptors – rods and cones – which transmit information about local detail in the retinal image to the brain via the neural networks. As with any detector, each receptor and its associated neural networks must have associated with them some basic sensitivity to light energy. This means that the complete data received by the brain from a stimulus covering several receptors must have some variability about it, dependent on the discrete sensitivities of the set of receptors on which the image falls. Since this group will change from glimpse to glimpse, the effective stimulus, if around threshold, will sometimes appear stronger than others, according to exactly where on the retina it falls. In addition it is unlikely that the decision processes are binary in form – there is most probably some region of indeterminance for a given received stimulus, where the decision as to whether it is present will vary from time to time.

Both the above factors make it imperative that the visual process is probabilistic in nature. Thus there will be a range of contrasts for a given viewing situation where, given a large number of observations, there will be less than unity but greater than zero chance of seeing the stimulus. Thus, at least part of the variance associated with the short time frequency of seeing curve can be directly attributed to signal/noise relationships in the received visual stimulus.

3.2 BETWEEN OBSERVER VARIABILITY

If all observers could be considered identical then, for a given task, a single observer frequency of seeing curve could be expected to predict performance of a group of observers. Unfortunately it is not admissible to consider observers as identical. The 50% probability response may vary from one observer to another due to a number of factors. The primary and most obvious one is visual acuity (see Chapter 5.2), which for uncorrected vision can vary from 6/4 to 6/9 without the observer necessarily being considered to have defective eyesight. With corrected vision it is to be expected that the tolerance is somewhat closer, although tests on many observers at BAC(GW)* suggests that a range from 6/4 to 6/7.5 might still be expected. In addition to this fairly readily determined variability in observer eyesight when looking at high contrast letters there appears to be a considerable interaction between observer ability and task. For instance Spicer and Ensall[15] have found strong interactions between observer abilities when looking at high contrast and low contrast letters. Equally a very considerable lack of agreement in the rank order of performance has been shown for 8 observers at BAC(GW) on a variety of acquisition tasks.[16]

It would seem from the foregoing that one must expect very significant

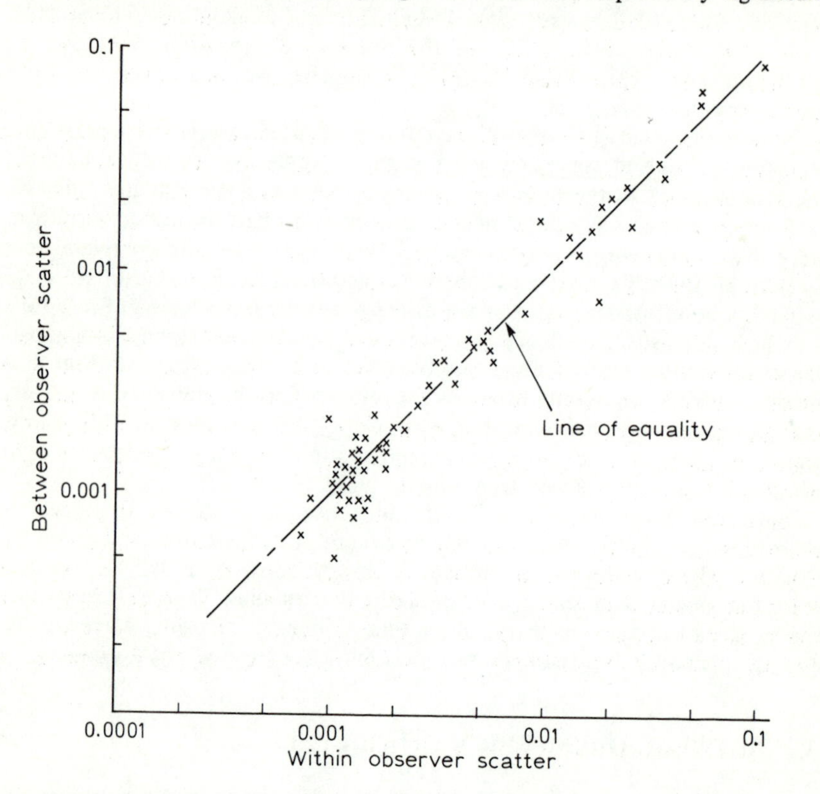

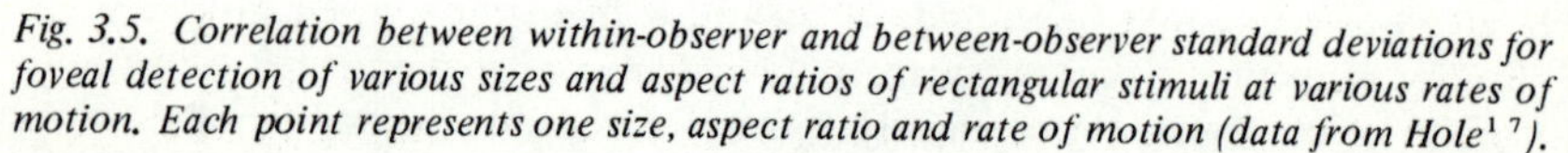
Fig. 3.5. Correlation between within-observer and between-observer standard deviations for foveal detection of various sizes and aspect ratios of rectangular stimuli at various rates of motion. Each point represents one size, aspect ratio and rate of motion (data from Hole[17]).

*British Aircraft Corporation (Guided Weapons Division).

differences in performance between observers for a variety of reasons. Unlike the within observer variability, the extent of this between observer variance is, as far as the author is aware, not well documented. However, it is implied by some authors that the spread is probably of the same order as that shown in Fig. 3.1 for populations of observers with good eyesight. Some limited experimental support for this view is to be found in a recent experiment by Hole[17]. The results of this experiment, basically studying the contrast threshold trends for moving targets as a function of motion, size and shape (see Section 4.10), were analysed to yield standard deviations of frequency of seeing both within and between subjects for each size and shape tested. The individual values of the ratios of within and between observer standard deviation are plotted in Fig. 3.5, where it can be seen that there is a very high correlation.

3.3 THE INFLUENCE OF ENVIRONMENT

In Section 3.1.1, mention was made of the Blackwell threshold being obtained by a *forced choice* presentation method. If we now think for a moment, it will be realised that, in many realistic situations, one is not forced to make a decision. We may thus rightly pose the question 'Is the frequency of seeing curve under *free choice* conditions (when the observer is free to withold a decision) the same as that under 'forced choice' conditions?'

Now it would seem reasonable to assume that, if the decision about a stimulus is 'forced', it will be made at a low confidence level. This is the Blackwell (Tiffany) situation and, indeed, is confirmed by the debriefing of Blackwell's observers, who were said to have no confidence at all in the existence of the stimulus until its contrast was at a level where they were making 90% correct

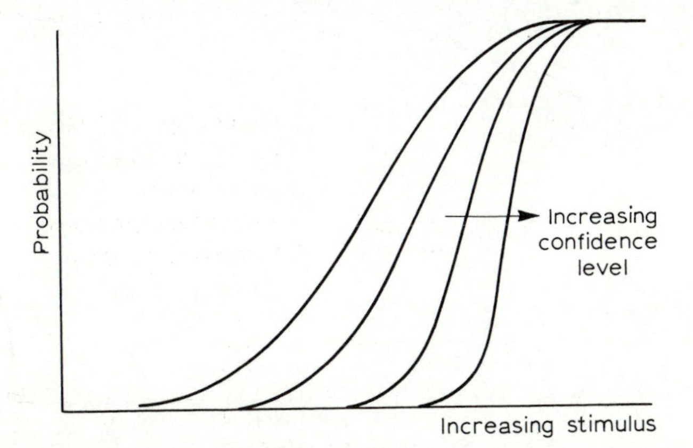

Fig. 3.6. Hypothetical changes of slope of frequency of seeing curves as a function of confidence level.

responses.[1] In the extreme 'free' choice situation, on the other hand, one might expect decisions only to be made when there was nearly 100% confidence. Under such circumstances decisions that a stimulus was present might be expected to be subject to a much closer tolerance in contrast. This might be expected to lead to a set of frequency of seeing curves of the form shown in Fig. 3.6, the value of contrast for 50% probability and the slope of the curve both increasing with increasing freedom of choice. In practice, Blackwell[2] has found larger but *more indefinite* thresholds for free choice than for forced choice situations. However, the greater indeterminacy does not appear to follow a stable and simple cumulative probability form. This same increased randomness in probability functions for free choice viewing has been found by other workers and is not currently understood to the author's knowledge.

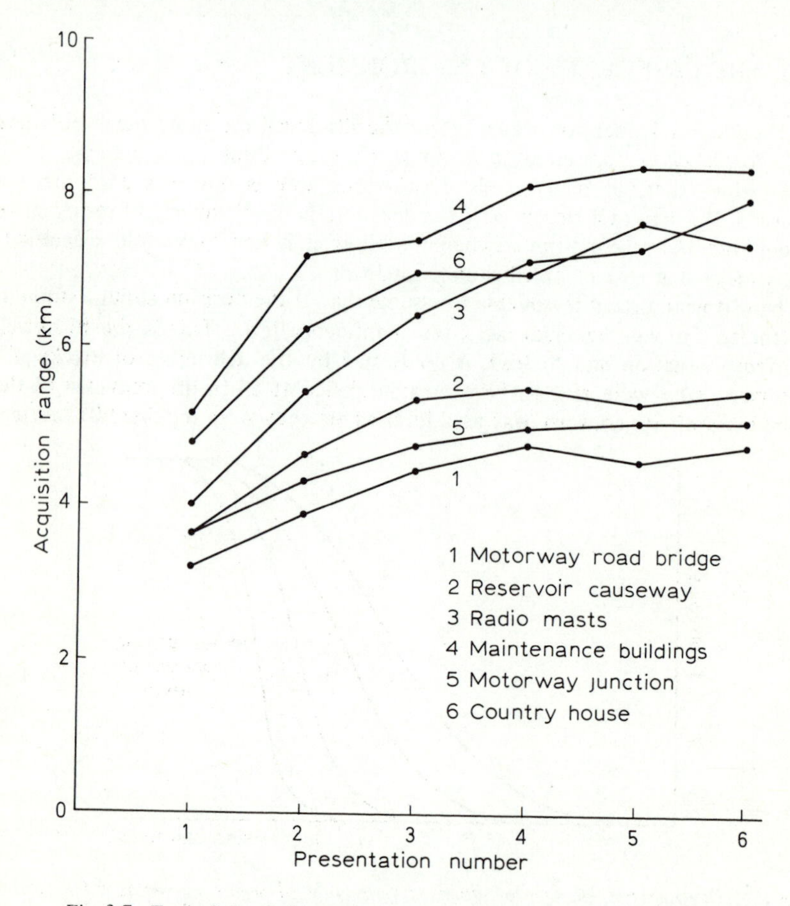

Fig. 3.7. Typical threshold trends due to learning. (after Milnes–Walker[18]).

Other factors which are known to influence the threshold level are the state of learning, motivation and briefing of the task. These three factors may be considered to be loosely related, and to be all a form of state of learning. Studies on the effects of state of learning on laboratory thresholds have been carried out be several workers including Blackwell,[2] Milnes-Walker[18] and Bloomfield.[19] In general it has been found that thresholds improve towards an optimum plateau according to a negative exponential law with increasing number of sets of observations. A typical set of data is shown in Fig. 3.7. Similar trends were found in simulated and real field trials by the present author and colleagues.[20] However, Blackwell found certain interactions between learning and motivation. Other workers, such as Gilinsky[21] and Campbell,[22] have shown an implied form of learning by preadaptation to specific patterns to produce similar enhancements, whilst preadaptation to antagonistic patterns resulted in a depression of thresholds – akin to forgetfulness.

3.4 ATTEMPTED MODELLING OF FREQUENCY OF SEEING

A number of attempts have been made to provide forms of predictive modelling of frequency of seeing curves. Such modelling must, of necessity, attempt to take some account of the reasons for uncertainties of response, and of the basic effects of such factors as the confidence level at which the decision is made. Two of the more major attempts at such modelling will be discussed.

3.4.1 The effect of quantum fluctuations

A basically simple way to consider frequency of seeing is in terms of the number of quanta which need to be absorbed by a retinal receptor in order that a stimulus shall be detected. It will be seen later (Chapters 6, 7, 12 and 13) that this is a great oversimplification of the mechanisms of vision for extended objects, but nevertheless it can serve as a very useful introduction to modelling of vision and is believed by the author to be most probably close to the truth as an explanation of the *input* of information to the neural networks from the retinal image.[32] Pirenne[23] provides a very useful discussion of the concept, this being that a sensation of seeing a flash at absolute threshold will result from the rhodopsin or other receptor photopigments having absorbed a certain critical number of quanta N_p in a given trial. But such a value of N_p must be a variable due to the quantum nature of light, the quantum fluctuations normally being assumed to be Poissonian.[25] Pirenne shows that the forms of probability curves for absorption of N_p or more than N_p quanta in one trial, when the mean quanta absorbed per trial are U_p, are as shown in Fig. 3.8. The present author believes that such a set of functions should also be applicable to differential thresholds if the value N_p is applied to local differences in quanta absorbed. Note the striking similarity between the curves of Fig. 3.8 and those

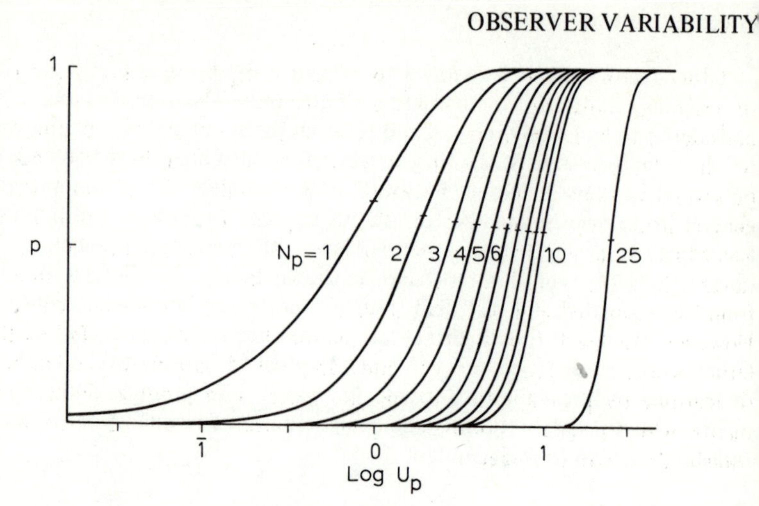

Fig. 3.8. Poissonian cumulative probability functions for various numbers of quanta absorbed when the mean number absorbed is N_p.

suggested by Fig. 3.6 as a function of confidence level. Pirenne shows, as confirmation of the *shape* of his predicted frequency of seeing curves, the results of an experiment carried out by Hecht, Schlaer and Pirenne,[25] showing that good fits to this experiment are obtained by assuming N_p lies between 5 and 7, the exact number varying with observer. It has been shown recently[26] that a Poisson distribution curve with $N_p = 6$ is almost identical in shape, above the 5% point, to a cumulative normal distribution function with $\sigma \approx 0.43\ C_M$ (the average value found to fit a wide range of experimental results by Blackwell[3]). The range of N_p from 5 to 7 as found by Pirenne is thus well matched to the range of σ found by Blackwell. The Poisson function has the merit, however, that it goes to zero for a zero stimulus, thus overcoming a basic objection to the cumulative normal function.

3.4.2 Decision theory

Whilst Pirenne's modelling possibly appears to account for the fluctuations of performance due to fluctuations of *input*, it seems unreasonable to suppose that there are *no* fluctuations and uncertainties within the visual system itself, most particularly at the decision level of the brain. An approach which appears to provide at least some ability to take account of these genuinely 'within observer' variations is decision theory as discussed, for instance, by Tanner and Swets[27] and Swets, Tanner and Birdsall.[28]

Conventionally it is assumed that the frequency of seeing curve is a 'betting' curve, there being normally a finite chance of a positive response when no

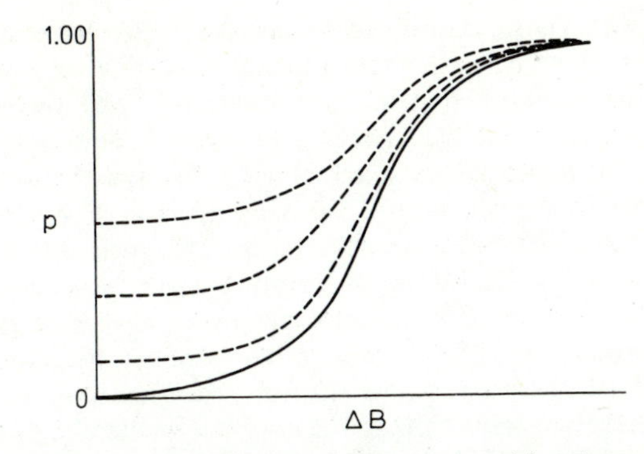

Fig. 3.9. Conventional frequency of seeing curves with various guess rates.

differential stimulus is present (false positive response). The lower the level of confidence at which a decision is made, the larger will be the expected chance of false positives, resulting in a set of curves like those shown in Fig. 3.9. Normally such frequency of seeing curves as Blackwell's are corrected for false positives by applying the formula

$$p = \frac{p' - Q_{fp}}{1 - Q_{fp}} \tag{3.1}$$

where p' is the observed probability of positive responses, p is the corrected probability of positive responses, and Q_{fp} is the false positive response when $\Delta B = 0$.

The justification for such corrections is an assumption that a false positive is a guess which is independent of level of background signal intensity and hence of sensory activity.

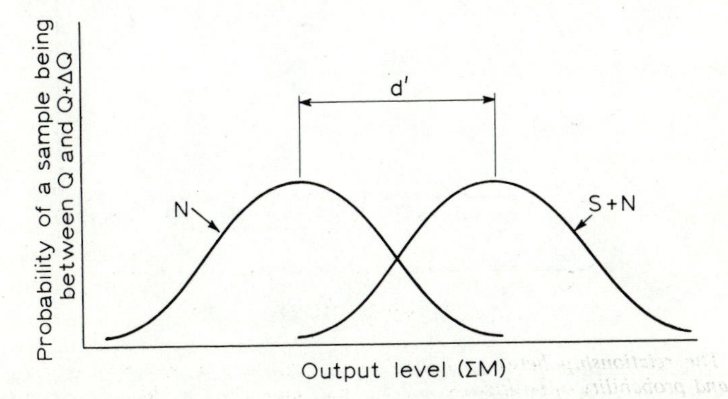

Fig. 3.10. Hypothetical distribution of noise (N) and signal plus noise (S + N). (Reprinted by permission from Tanner and Swets[27]. Copyright (1954) by the American Psychological Association).

Tanner and Swets chose, instead, to assume that false alarm rate and correct detection rate vary together, the level of neural activity being assumed to be a monotonically increasing function of light intensity. Their general concept is then that the sampled noise may have an envelope N as shown in Fig. 3.10, where N includes both internal and external noise. The sampled signal must then have an envelope $(S + N)$ which is of the same form as N, but displaced by a distance d'. The observer makes an observation ΣM, such that the greater the value of ΣM the more likely the observation is to be a signal. Based on experience and the level of ΣM, the observer must make a decision as to the presence or otherwise of a signal. It should be clear that, if a criterion point is chosen such that all observations less than the criterion level are registered as noise and all observations greater than the criterion level are registered as signal, there will be a finite probability of noise being registered as signal, this

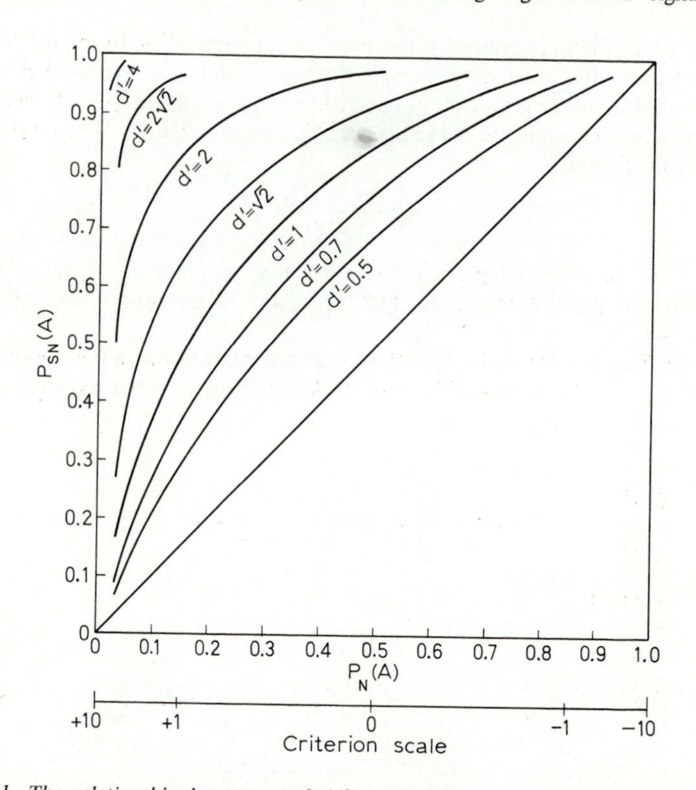

Fig. 3.11. The relationship between probability of positive response to signal plus noise ($p_{SN}(A)$) and probability of positive response to noise alone ($p_N(A)$) for various normalised values of d′. The criterion scale shows the corresponding criteria expressed in terms of R.M.S. noise σ_N from noise mean M_N. (Reprinted by permission from Tanner and Swets[27]. Copyright (1954) by the American Psychological Association).

probability depending on the position of the criterion point. Equally, if the $(S+N)$ envelope is displaced by d', there will be a finite and greater probability of $(S+N)$ being registered as a signal. Tanner and Swets developed diagrams illustrating the joint probabilities of noise being observed as signal ($p_N(A)$) and signal being observed as signal ($p_{SN}(A)$) as a function of criterion level and d'. Such a diagram, assuming the noise distribution to be gaussian (which is a close approximation to Poissonian as already discussed), and with d' being normalised to units of RMS noise, is shown at Fig. 3.11. It will be seen that, as d' tends to zero, the signal and noise probabilities tend to equality. Conversely, as d' becomes greater than 2, the signal detection probability rapidly approaches unity for low noise detection probabilities. It is claimed that such curves can be used to determine maximal behaviour in a given experiment requiring a simple yes/no decision as a point on a particular d' curve having a slope $\beta_{d'}$, given where

$$\beta_{d'} = \frac{1 - p(SN)}{p(SN)} \cdot \frac{(V_{N \cdot CA} + K_{N \cdot A})}{(V_{SN \cdot A} + K_{SN \cdot CA})} \tag{3.2}$$

In the above, $p(SN)$ is the *a Priori* probability that a signal exists, $V_{N \cdot CA}$ is the value of a correct rejection, $K_{N \cdot A}$ is the cost of a false alarm, $V_{SN \cdot A}$ is the value of a correct detection and $K_{SN \cdot CA}$ is the cost of a miss.

Thus, within this theory should exist the basis for determining the forms of frequency of seeing curves *and* the shift of 50% threshold as a function of task motivation and similar psychological factors.

The frequency of seeing curves predicted by this form of modelling are very different from the conventional ones, a set of curves for different criterion levels being as shown in Fig. 3.12. Here it will be seen that, as criterion level reduces

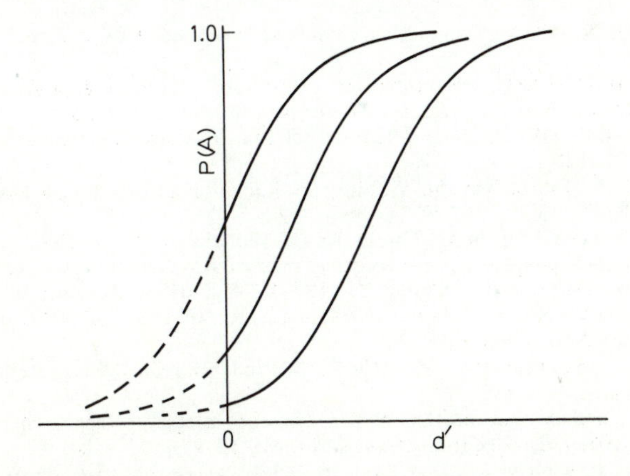

Fig. 3.12. Probabilities of positive response p(A) *as a function of* d′ *for various criterion levels. (Reprinted by permission from Tanner and Swets*[27]*. Copyright (1954) by the American Psychological Association).*

(i.e. a lower confidence, higher chance situation) the probability curve shifts bodily to the left, the dotted portions indicating the incomplete nature of the curve. If frequency of seeing behaviour is really as predicted here then correction of chance false positives according to Equation 3.1 is very definitely wrong!

The reader wishing to pursue this aspect of the subject in more depth should read Tanner and Swets,[27] Swets *et al*,[28] Wald[29] and Peterson *et al.*[30]

3.5 DISCUSSION

It will be seen from the foregoing that the form of frequency of seeing curve for a given data set will be very dependent on a host of factors. In the remainder of this book the majority of acquisition data will be discussed in terms of a 50% probability of seeing. Because of the very considerable dependence of frequency of seeing on so many factors it is inevitable that any set of absolute thresholds will only be relevant to the particular circumstances under which they were obtained and the particular time. Changes of observers, experience, briefing, time of day, season, will all be expected to have a significant influence on experimental results. Thus one should never place too much importance on differences between two sets of threshold data gathered at different times or in different ways, but only on the trends within a balanced set of data. It is essential that the reader remembers this as he proceeds to study the rest of the book.

REFERENCES

1. Blackwell, H. R. (1946). 'Contrast Thresholds of the Human Eye', *J. Opt. Soc. Am.*, **36**, 624
2. Blackwell, H. R. (1953). 'Psychophysical Thresholds: Experimental Studies of Methods of Measurement', *Bull. Eng. Res. Inst.*, University of Michigan, No. 36
3. Blackwell, H. R. (1963). 'Neural Theories of Simple Visual Discriminations', *J. Opt. Soc. Am.*, **53**, 129
4. Crozier, W. J. (1950). 'On the Visibility of Radiation at the Human Fovea', *J. Gen. Physiol.*, **34**, 87
5. Taylor, J. H. (1962). 'Contrast Thresholds as a Function of Retinal Position and Target Size for the Light-adapted Eye', *Proceedings of the NAS–NRL Vision Committee*
6. Silverthorne, D. G. and Garland, N. (1972). 'A Further Account of Progress in Modelling the Visual Acquisition of Ground Targets from the Air', *BAC (GW) Human Factors Study Note Series 7, No. 17*
7. Ronchi, L. and Brancato, R. (1971), 'Periodicities in Visual Responsiveness', *Opthalmologica*, **163**, 189
8. Ronchi, L. and Novakova, O. (1970). 'On the Intensity Range where Vision is Uncertain', *Atti della Fondazione Giorgio Ronchi*, **25**, 97
9. Sweeney, E. J., Kinney, J. S. and Ryan, A. (1960). 'Seasonal Changes in Sensitivity', *J. Ophthal. Soc.*, **50**, 237
10. Wertheimer, M. (1953). 'An Investigation on the 'Randomness' of Threshold Measurements', *J. Exp. Psychol.*, **45**, 294

1. Spicer, P. J. (1972). 'The Measurement of Visual Acuity', App. 3, Section 1 of *Research into Factors affecting the Detection of Aircraft through Optical Sights*, (BAC(GW) Ref. L50/186/1449.)
2. Ford, A., White, C. T. and Lichtenstein, M. (1959). 'Analysis of Eye Movements during Free Search', *J. Opt. Soc. Am.*, **49**, 287
3. Lavin, E. P. (1972). 'A Literature Survey on Retinal Image Motions', App. 3, Section 3 of *Research into Factors affecting the Detection of Aircraft through Optical Sights*, (BAC(GW) Ref. L50/186/1449)
4. Ditchburn, R. W. and Foley-Fisher, J. A. (1967). 'Assembled Data on Eye Movements', *Optica Acta,* **14**, 113
5. Spicer, P. J. and Ensell, F. J. (1973). 'Comparison of Visual Acuity Tests and Viewing Condition Interactions', *Aerospace Medicine*, November, 1290
6. Spicer, P. J. (1972). 'The Significance of Visual Acuity Measurements', App. 3, Section 2 of *Research into Factors affecting the Detection of Aircraft through Optical Sights*, (BAC(GW) Ref. L50/186/1449)
17. Hole, R. J. (1975). 'A Study of the Effects of Target Shape and Target Motion on Visual Threshold Detection Performance', BAC(GW) Ref. ST 13117
18. Milnes-Walker, N. D. (1968). 'The Effect of Repeated Viewing of a Target Run on Recognition Range', *Human Factors Study Note Series 4, No. 22*, BAC(GW) Ref. R47/20/HMF/1703
19. Bloomfield, J. (1970) *'Visual Search'*, PhD. Thesis, University of Nottingham
20. Murphy, M. J., Overington, I. and Williams, D. G. (1965). *Final Report on Visual Studies Contract*, Section 3.5, B.A.C.(GW) Ref. R41S/11/VIS
21. Gilinsky, A. S. (1968). 'Orientation-specific Effects of Patterns of Adapting Light on Visual Acuity', *J. Opt. Soc. Am.,* **58**, 13
22. Campbell, F. W. and Maffei, L. (1970). 'Electrophysiological Evidence for the Existence of Orientation and Size Detectors in the Human Visual System', *J. Physiol.,* **207**, 635
23. Overington, I. (1974). 'An Exploratory Study into the Various Observed Complex Functional Characteristics of Vision and their Compatability with a Unified Simple Modelling' BAC(GW) Ref. ST 12386
24. Pirenne, M. H. (1957). *Vision and the Eye*, Chapter 8, Chapman and Hall, London
25. Hecht, S., Schlaer, S. and Pirenne, M. H. (1942), 'Energy Quanta and Vision', *J. Gen. Physiol.,* **25**, 819
26. Overington, I. (1975). 'The Form of Frequency of Seeing Curves and Sources of Variance', BAC(GW) Ref. ST13105
27. Tanner, W. P. (Jr.) and Swets, J. A., (1954). 'A Decision-making Theory of Visual Detection', *Psych. Rev.,* **61**, 401.
28. Swets, J. A., Tanner, W. P. (Jr.) and Birdsall, T. G. (1961). 'Decision Processes in Perception', *Psych. Rev.* **68**, 301.
29. Wald, A. (1950). *Statistical Decision Functions*, Wiley, New York
30. Peterson, W. W., Birdsall, T. G. and Fox, W. C. (1954). 'The Theory of Signal Detectability', *IRE Trans.,* **PGIT–4**, 171

4 Basic Thresholds for Detection

The most positive point on the frequency of seeing curve discussed in the previous chapter is the 50% point. The 50% probability of carrying out any task is also the most useful measure of performance, especially if the variance is known as it seems to be for many visual processes. It has thus become almost universal practice to quote the results of basic experimentation on visual detection in terms of the 50% probability of correct positive responses. For most of the following chapters we shall follow this line and discuss only the 50% probability points.

The most basic quantity in study of thresholds must be that of differential energy available to different parts of the retina. However the visual processes operate – and this will be discussed in succeeding chapters – some difference of available energy must be present. It is usual to use as the basic practical parameter the threshold of contrast which may be thought of, for objects and backgrounds of the same colour (i.e. possessing no chromaticity difference), as the ratio of the difference brightness or luminance between an object and its surroundings to the luminance of the surroundings,[1] ie

$$C = \frac{(B - B')}{B'} = \frac{\Delta B}{B'}$$

where C is the contrast, B is the luminance of the object of interest, B' is the luminance of the background.

It will be seen in later chapters that such a simple concept leads to problems when dealing with structured fields, but it will serve for the present where we shall be discussing mainly thresholds associated with objects in plain fields. The value of C for 50% probability of detection in a plain field is usually referred to as the liminal contrast and is denoted by the symbol ϵ.

ϵ is in general a function of many parameters associated with the viewing conditions – background luminance, size of illuminated field, object size and shape, edge sharpness, presentation time, position in the visual field, rate of motion, colour, level of confidence at which a decision is made, experience and others. In this chapter we shall be looking at the available experimental data on the variation of ϵ with these various parameters in controlled laboratory environments. Apart from Section 4.11, which is devoted to effects of colour, all other thresholds discussed in this chapter are concerned with neutral shades.

4.1 EFFECT OF OBJECT SIZE

As might be expected, if the size of an object is increased, other conditions being held constant, it becomes easier to see – that is, the value of ϵ reduces. This is almost universally true, except in the case of certain forms of viewing of very large objects. When viewing a simple circular stimulus with very small angular subtense at the eye – dimensions of the order of a millirad or less – the contrast required to see it is universally related to size by the simple law

$$\epsilon\alpha^2 = \text{constant}$$

where α is the angular diameter of the stimulus. This is known as Ricco's Law, and implies that the total differential energy received from the stimulus is the factor defining the threshold.[2] However, as the size is increased above about 3 mrad diameter the deviation from this law becomes more and more marked until, typically, when looking at stimuli with diameters in excess of perhaps 30–60 mrad, further increase in size has little further effect on the value of ϵ. This is particularly true at good light levels approximating to daylight, where the value of ϵ becomes virtually independent of size for $\alpha > 30$ mrad. Sample size/contrast threshold curves are shown in Fig. 4.1, the axes being log α and log ϵ in order that the large range of sizes and contrasts shall be adequately represented. It will be seen that the absolute shapes and positions of the threshold curves are very dependent on values of other test parameters. It will, however, be

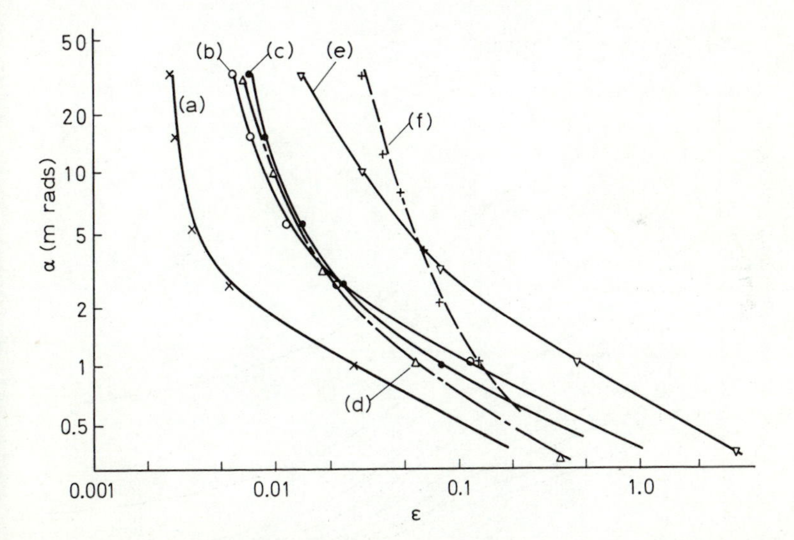

Fig. 4.1. Miscellaneous threshold curves for liminal contrast against size for disc and similar stimuli viewed against plain, extended backgrounds. Curve (a): Infinite viewing time, foveal presentation, 34.3 cd/m², forced choice. (after Blackwell[3]). (b): Infinite viewing time, foveal, 0.343 cd/m², forced choice. (after Blackwell[3]). (c): 8 position search in 6 s, 34.3 cd/m², forced choice. (after Blackwell[3]). (d): 1/3 second presentation time, foveal, 257 cd/m², forced choice. (after Taylor[6]). (e): 1/3 second presentation time, 0.21 rad peripheral, 257 cd/m², forced choice. (after Taylor[6]). (f): 8 position search with slowly growing contrast, 34.3 cd/m², free choice. (after Vos et al.[5]*).*

noted that the Ricco's Law region (a straight line on the log/log plot) is in strong evidence in most cases. For more detailed information on such size threshold trends the reader is referred particularly to Blackwell,[3,4,9] Vos *et al*,[5] Taylor[6,] and Davies.[8]

4.2 EFFECT OF FIELD LUMINANCE

As the luminance of the background field is reduced, so it becomes progressively more difficult to see any particular object. At high field luminances this fall off in visual performance is hardly noticeable, and can only be reliably measured in large and very highly controlled experiments (e.g. Blackwell[3] and Taylor[7]) However, as luminance falls from good daylight to typical interior light levels and on down through twilight, the effect becomes rapidly more marked. In the limit, when the background luminance is taken down to starlight levels the fall off is very rapid. However, an interesting phenomenon occurs at these low light levels. Typical luminance/contrast threshold curves are shown in Fig. 4.2, where it will be seen that there is a discontinuity at around 3×10^{-3} cd/m^2. These

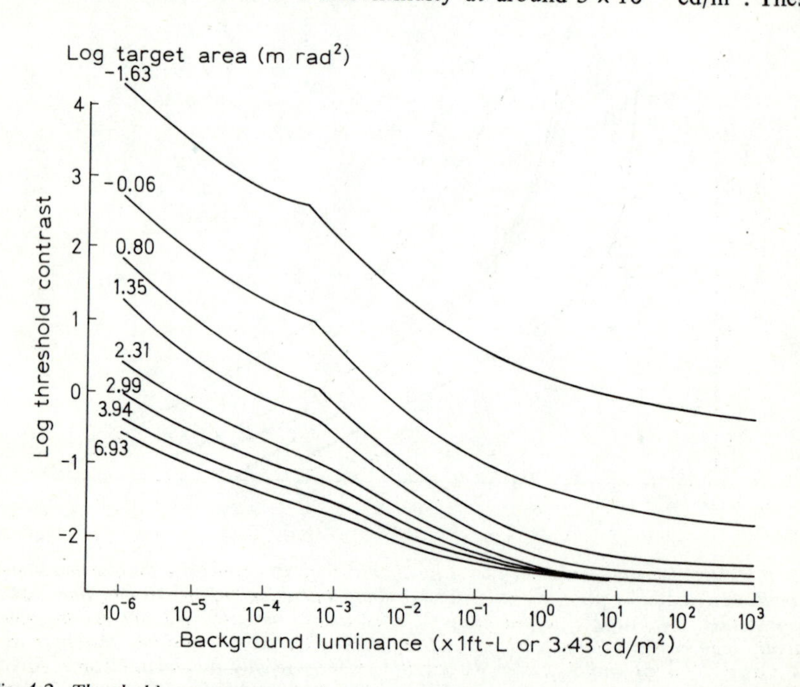

Fig. 4.2. Threshold contrast as a function of background luminance for disc stimuli. Unlimited viewing time. (Reproduced from Blackwell[3] by courtesy of the Journal of the Optical Society of America).

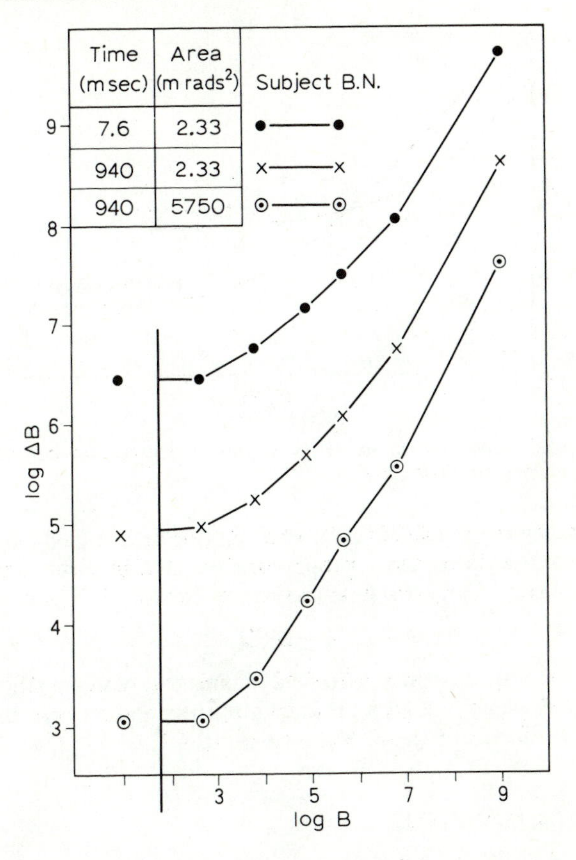

Fig. 4.3. Lot. incremental threshold intensity (ΔB) plotted against log background intensity B for various quanta (507 nm)/s mrad². (Reproduced from Barlow[11] by courtesy of the Journal of Physiology).

discontinuities are found by all experimenters who have carried out unrestricted studies in the luminance domain (e.g. Blackwell[3]) and may be explained as follows. For high scene luminances cone vision is used and a simple detection task usually employs foveal vision (where the cone density is greatest – Section 2.4). As the luminance reduces the visual process becomes progressively less efficient (e.g. Stiles and Crawford[10]). Eventually, at low light levels, the rod vision becomes more efficient than cone vision and there is a fairly well defined shift of fixation to the region of optimal rod vision (about 0.3 rad from the fovea where rod density is greatest – Section 2.4).

The full story may be built up by considering available data on thresholds of rod vision as a function of field luminance. Barlow[11] has shown that the

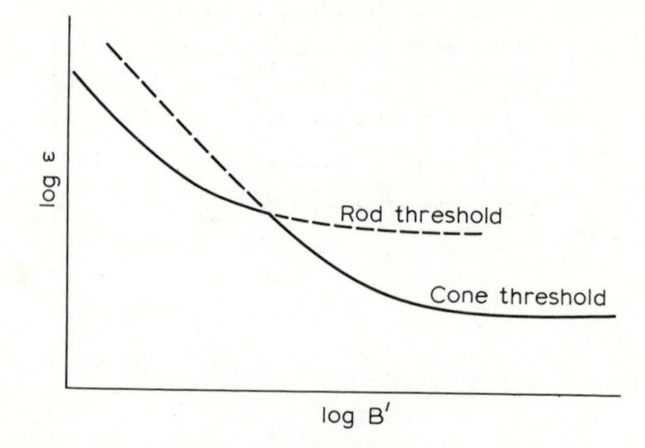

Fig. 4.4. Schematic construction of a composite luminance/contrast threshold from separate rod and cone threshold curves.

luminance increment threshold (ΔB), when plotted against field luminance B' on a log/log plot, tends to show proportionality at high field luminances and absolute constancy at very low field luminances (see Fig. 4.3). But, since

$$\epsilon = \Delta B / B'$$

this means that ϵ is constant at high field luminances. Assuming that cone vision has a similar behaviour, but with the known superior absolute size threshold, one may predict the *form* of Fig. 4.2 as shown in Fig. 4.4.

4.3 ADAPTATION TIME

The threshold curves shown in Fig. 4.2 are only relevant if the eye has become adapted to the background illumination level. As stated in Chapter 2 the process of adaptation can take many minutes, depending on the difference in luminance levels before and after adaptation. Figures 2.8 and 2.9 illustrated typical forms of dark and light adaptation curves. Such adaptation is necessary in order that the eye can detect small changes of luminance whilst having a very wide operating range (some 10^9:1 in luminance levels). If attempts at viewing are made before the processes are complete then the eye will not be working at optimum but will behave rather like an overexposed or underexposed photographic emulsion. This is illustrated in Fig. 4.5 where a series of sensitivity curves are shown for the eye at various levels of adaptation. If the state of adaptation is other than matched to the prevailing field luminance, then the differential output signal for a given differential input will be progressively reduced, the greater the mismatch in field luminance and adaptation state. It is thus imperative that adequate time is given to adapt to the experimental conditions if reliable

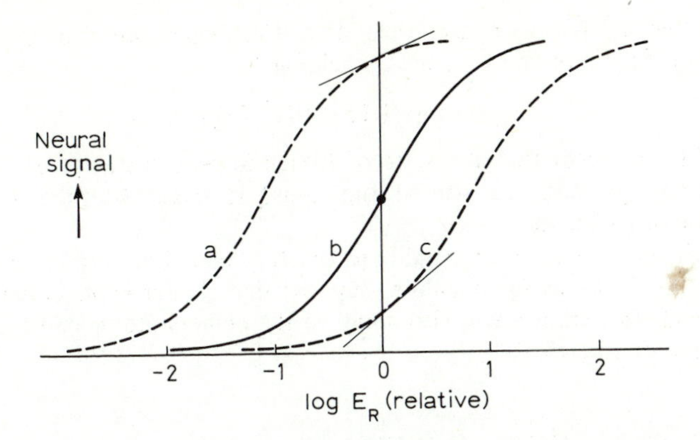

Fig. 4.5. Illustrating the effect of incorrect adaptation on threshold performance. The curves show neural signal level as a function of local retinal illumination level for individual receptors. Log $E_R = 0$ indicates relative background illumination level. Curve (b) shows response function for correctly adapted receptors, Curves (a) and (c) showing response functions for adaptation levels 1.2 log units too low and 0.8 log units too high respectively. Local threshold performance is defined by the tangents to the curves where they intersect the line log $E_R = 0$. (Basic curve after Werblin[12]).

threshold data are to be obtained. A useful discussion on the concept of progressively shifting sensitivity curves is to be found in Werblin.[12]

Rushton and Powell[13] have recently confirmed that the rate of regeneration of rhodopsin after bleaching, which had previously been shown to be closely related to threshold trends for rad vision (Section 2.5 and Fig. 2.12) may be fitted over a wide range of differential adaptation states by the equation

$$(1 - p_t) = (1 - p_o)\, 10^{-t/14} \tag{4.1}$$

where p_t is the fraction of unbleached rhodopsin remaining after t minutes in the dark and p_o is the fraction of unbleached rhodopsin at the start of dark adaptation.

Equation 4.1 may be rewritten as an exponential

$$(1 - p_t) = (1 - p_o) \exp(-t/6.1) \tag{4.2}$$

thus showing the regeneration to be of exponential form with a time constant of approximately 6.1 min.

Rushton and Powell also cite considerable evidence that the state of bleach of rhodopsin after exposure to a flash of energy $(I\, t_o)$ expressed in scotopic trolland seconds can be given by

$$\log (\log 1/p_o) = \log (I\, t_o) - 7.3 \tag{4.3}$$

where p_o is the fraction of photopigment remaining immediately after bleaching, providing that t_o is not greater than 45 s. Here again we may rewrite the equation in the form of an exponential, yielding

$$p_o = \exp(-1.15 \times 10^{-7}\, I\, t_o) \tag{4.4}$$

It would thus seem that the state of bleach in general should be a balance between two opposing exponential functions, as might be predicted for a combination of natural processes.

Although no data of a comparable nature for cone vision are known to the present author it seems reasonable to suppose that similar exponential laws of bleaching and regeneration will also apply to the cone photopigments, but with different time constants.

4.4 SIZE OF BACKGROUND FIELD

In order to be sure that there is no interaction between threshold values and size of background field many experimenters have gone to extreme lengths to provide reasonably uniform illumination over very extended regions of the visual field. Such a practice is difficult to accomplish and is also unrealistic. It can even be uncomfortable if high luminances are used. Several experimenters have thus carried out studies aimed at defining how thresholds vary with size of background field. Lythgoe[14] showed that threshold improved as the size of test field increased up to an angular subtense of the order of 0.25 rad. He also showed that with smaller test fields the provision of a surround field of lower intensity did not much influence the thresholds as long as the surround was not more than 1 to 2 orders of luminance lower than the test area. If, however, the surround was much brighter than the test area then the thresholds quite rapidly deteriorated. Similar studies were carried out by BAC(GW) with similar findings.[15] This is, in general, additional evidence of the high and low key effects in vision which can occur if the eye is not correctly adjusted. Further evidence will be presented in Chapter 13 of the effects of local regions of scene luminance.

4.5 EFFECT OF OBJECT SHAPE

Up to this point, in discussing detailed thresholds, we have been primarily talking about simple circular objects. Many experimenters have realised that simple circles are so different to most real life objects that questions must be asked about effects of shape. For the moment we are only concerned with *detection* and so the concepts of detection of detail are not relevant. These latter will be discussed in Chapter 5. However, what, may we ask, is the effect of aspect ratio on *detection* thresholds? This matter has received thorough investigation by Lamar and co-workers,[16,17] who studied the detection

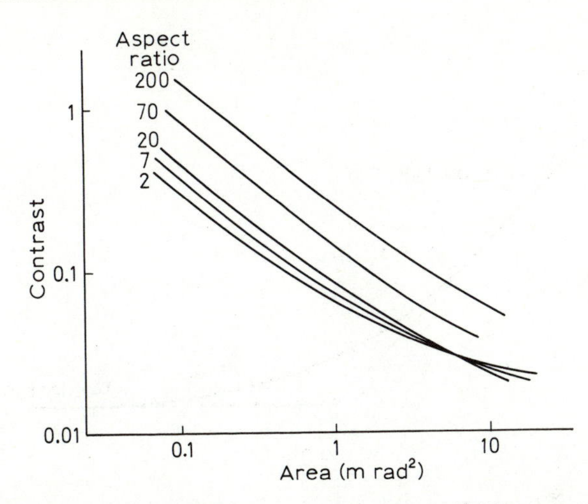

Fig. 4.6. Contrast/size thresholds for rectangular stimuli of various aspect ratios at a background luminance of 60 cd/m². (After Lamar et al.[17]*).*

thresholds of a set of rectangular stimuli with aspect ratio's varying from 2:1 to 200:1. They found that the thresholds were largely independent of aspect ratio up to an aspect of 7:1, after which the thresholds – in terms of area – began to deviate markedly from those of the smallest aspect ratio. A typical variation of threshold as a function of aspect ratio is shown in Fig. 4.6. Lamar suggested that a better agreement of thresholds, relatively independent of aspect ratio, would be obtained if it was assumed that detection took place due to a ribbon of 0.3 mrad width just inside the perimeter. Other studies of the effects of shape on detection thresholds are those of Guth and McNelis[18] who studied threshold functions for various complex targets (parallel bars, Landolt rings, letters, dot patterns, etc.). Their findings were that detection was approximately independent of shape for high luminance and that, although luminance trends were shape dependent, trends for circular stimuli were a good mean.

4.6 EFFECT OF PRESENTATION TIME

No specific mention has been made so far of presentation time in the thresholds discussed. In practice this is another very important parameter. It has long been known that presentation time has an effect on detection thresholds, and that for very short flashes the threshold is inversely proportional to flash duration (i.e. $\Delta B . t_D$ = Constant – Bloch's Law[19]). Equally it appeared that, for long viewing times, thresholds tended to become constant. There was known to be an uncertain region between these two extremes (extending from about 0.01 s to a few seconds) where little was really known about the form of the threshold

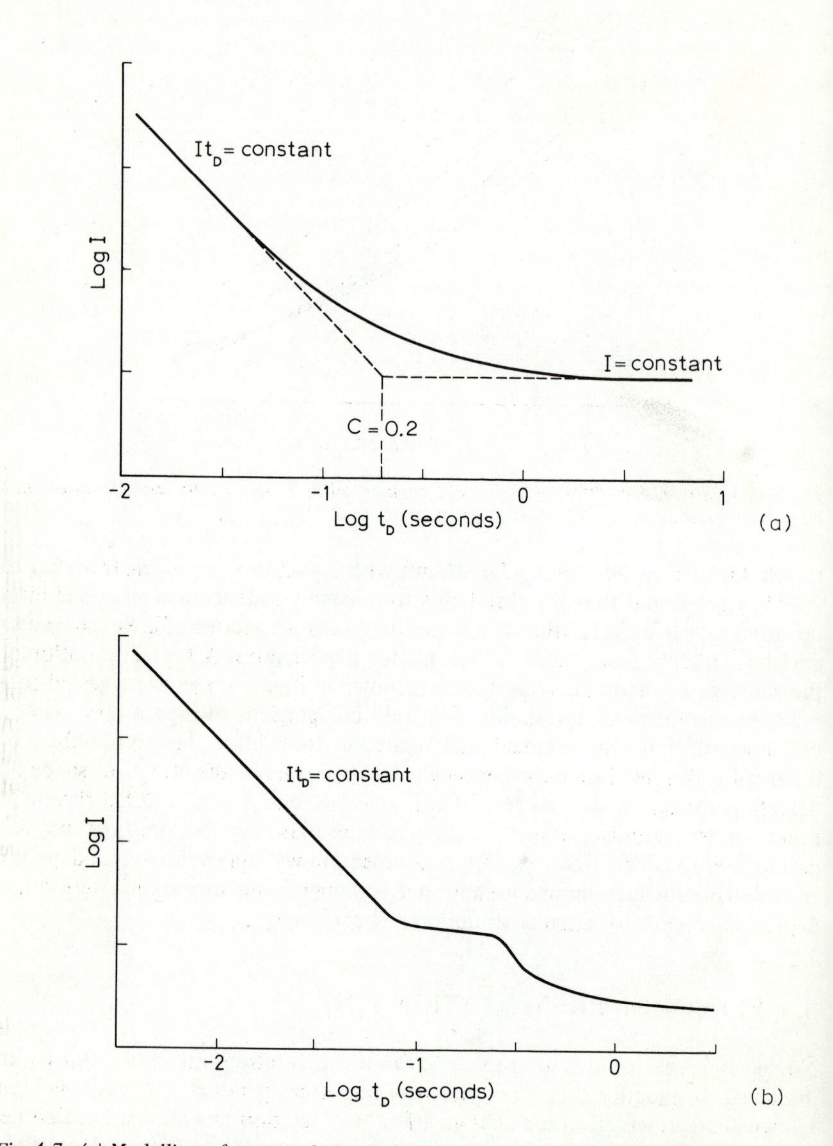

Fig. 4.7. (a) Modelling of temporal thresholds as proposed by Blondel and Rey[24]*. (b) A typical experimental temporal threshold trend showing a major inflection around 0.3–0.4 s. (Reproduced from Taylor*[22] *by courtesy of the Advisory Group for Aerospace Research and Development of NATO).*

characteristics. The first attempt to come to terms with this was by Blondel and Rey[20] who formulated an equation bridging this gap by a gentle curve as shown by the solid line in Fig. 4.7(a). This equation suggests that

$$\Delta B_e = \frac{\Delta B \cdot t_D}{c + t_D} \tag{4.5}$$

where ΔB_e is the effective differential luminance, ΔB is the actual differential luminance, t_D is the presentation time and c is a constant which varies considerably with presentation conditions but is often taken as about 0.2 s. Other workers, invoking the quantum theory, predicted that for long presentation times the relationships should tend to $\Delta B . t_D{}^{0.5}$ = constant.[21] For many years the Blondel-Rey equation held favour, until, in the last two or three decades, a series of carefully controlled studies have repeatedly shown a discontinuity in the curve at some point in the region 0.2 to 1 s (e.g. Blackwell and McCready[4]). A typical curve is shown in Fig. 4.7(b). Possible explanations of the discontinuity are discussed in Sections 6.4 and 7.3.

Taylor[22] has recently reported the historical and current situation in studies on presentation time and is currently engaged in an extended series of experiments to produce more controlled and reliable data. This is necessary since, although there are, according to Taylor, over 600 references on the problem of seeing flashes of light, very few really contribute to a general understanding of the problem due to the great variety of conditions under which experiments have been carried out.

4.7 EFFECT OF EDGE SHARPNESS

Most laboratory experimentation on detection thresholds has been carried out on sharp-edged objects – that is on objects whose luminance cross-sections are of rectangular form. Certain experimenters have realised that in many situations (e.g. astronomy from the earth's surface through a turbulent lower atmosphere, viewing of projected images on screens, viewing through imperfect optical aids) the objects viewed are not sharp-edged. They have therefore set up experiments to study the effect of degraded borders on thresholds. Some of the most extensive of this work has been reported by Ogle[23] where the thresholds for various sizes of out-of-focus aerial images of disc stimuli have been studied. The results of this work (Fig. 4.8) have shown that there is a massive effect of blur on thresholds for very small objects – presumably due to the extensive spreading of energy in the retinal image – but also that there are significant threshold degradations due to blurring for large objects. However, for very large objects it would appear that there is little degradation.

Other work on effects of blur has been carried out by Middleton[24] and Fry,[25] both of whom have studied the degradation in threshold for detection of

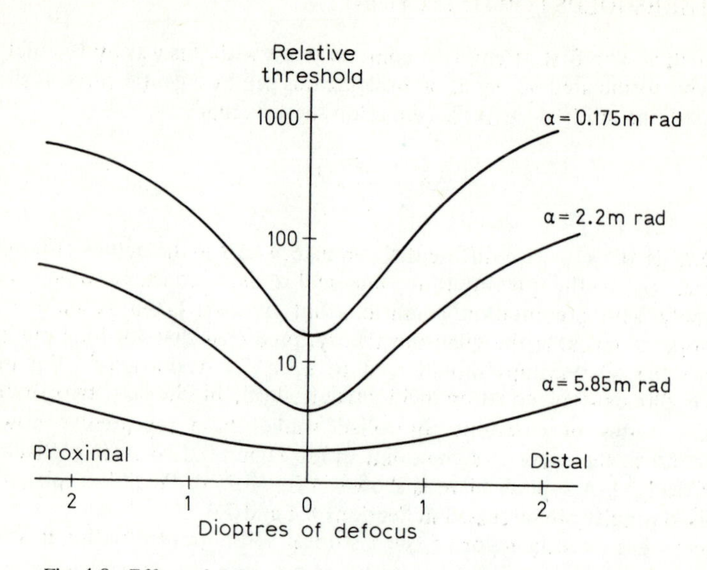

Fig. 4.8. Effect of defocus blur on detection thresholds. (After Ogle[23]).

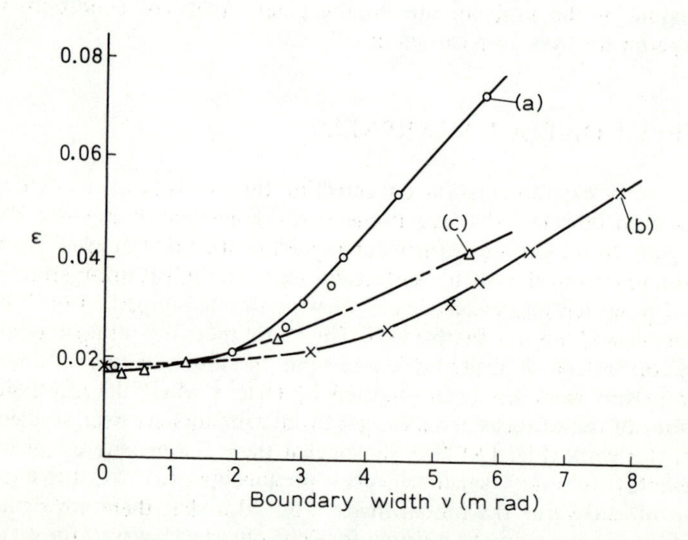

Fig. 4.9. Effect of a diffuse boundary on detection thresholds – boundary has a Gaussian form of rate of change of luminance where v is measured from the 1% to the 99% points on the luminance profile. Curve (a): Middleton[24] – v increasing. Curve (b): Middleton – v decreasing. Curve (c): Fry[25].

a long border between two uniformly illuminated halves of a field when the border is degraded by an effectively gaussian spread function. In both cases they found a very small effect with small border spreads, followed by a very rapid degradation as the effective width of the border (measured between the 5% and 95% points of the differential luminance) increased beyond about 2 mrad. Their relative results are shown in Fig. 4.9 and are in striking contrast to Ogle's findings for large objects. Yet other workers (e.g. Kruithof[26]) have found some degradation of thresholds for moderate blur, subsequently levelling out as blur was further increased, whilst Hood[27] has found the effect of blur on detection thresholds to be a function of exposure time. According to him there is little effect for 1 ms exposures but a considerable effect when exposures are of 1 s or longer.

Recently experiments on the effects of blur on thresholds of disc and diamond-shaped stimuli have been studied at BAC(GW).[28] In our case, in constrast to Ogle's work, we projected the blurred stimuli onto a high quality screen and viewed the *incoherently* blurred image. We were thus able more procisely to define the form of the retinal image than when viewing was of an aerial image (see Section 9.1.5). Here the findings were similar to Ogle's results despite the different image formation, tending to zero degradation for very large stimuli where one might expect some similarity to Fry's and Middleton's blurred borders.

Weighing all the evidence presented it would seem that the effects of blur on detection thresholds are a function of both stimulus size and presentation time, being minimal for single glimpse situations and maximal for long time viewing of a given stimulus. A possible explanation for this behaviour is discussed in Section 7.7.

4.8 EFFECT OF RETINAL POSITION

Most of the threshold data discussed so far have been concerned with optimal viewing – that is, foveal in daylight and optimal peripheral at night. Now, although imagery at other than optimal parts of the retina is not as important as optimal imagery for no search situations, the whole of the visual field can be of considerable importance in many visual tasks. It is thus necessary to know how the threshold performance varies as a function of retinal image position. Major published work in this sphere is by Taylor.[6,7] This shows a rise of thresholds for peripheral viewing in good light conditions with a corresponding decrease out to a few degrees at very low light levels (Fig. 4.10). This is largely in keeping with the concepts of optimal rod and cone vision discussed in Section 4.2. The earlier paper of Taylor's[7] shows a comprehensive set of data for angles up to 0.2 rad from the fovea and high light levels for a variety of stimulus sizes. Other work of Taylor's, known to the author but presently unpublished, provides data out to 0.8 rad from the fovea. Other workers, for instance Kishto,[29] have concentrated

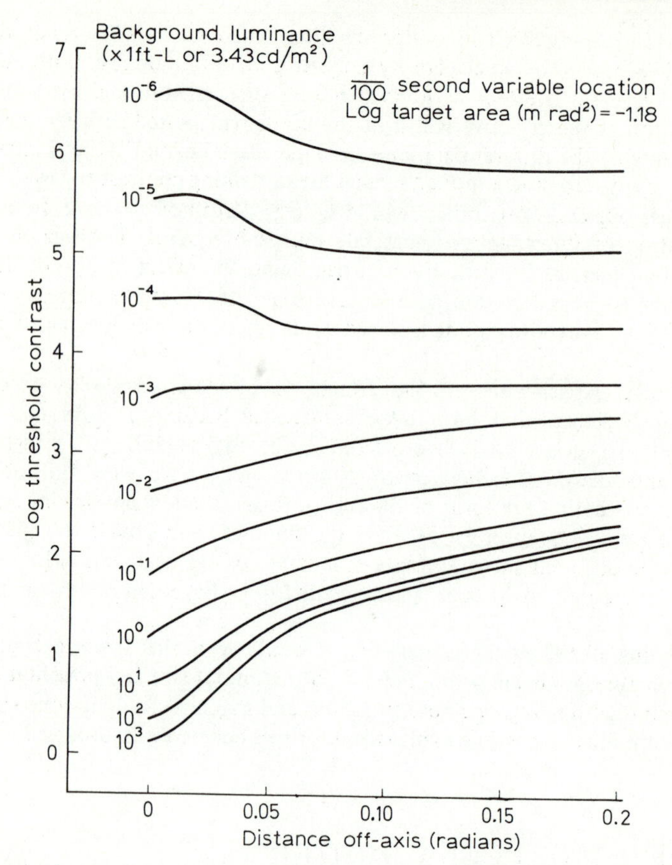

Fig. 4.10. The effect of viewing position at various background luminances. (Reproduced from Taylor[6] by courtesy of the Scripps Institution of Oceanography, San Diego).

on studying the behaviour trends of vision at specific points on the retina and for specific stimuli, whilst Ogle has extended his studies on thresholds for blurred stimuli to peripheral vision[30] and Herrick has recently shown that the effects of presentation time are similar for peripheral vision to those previously observed for foveal vision.[31]

4.9 FLICKER THRESHOLDS

Another form of temporal threshold which is of some relevance in practical situations is the threshold of flicker fusion. If a stimulus or presentation field are presented repetitively, then there will in general be, for a given adaptation level, contrast and stimulus size, a particular frequency of repetition at which the presentation cannot be differentiated from a steady, non-flickering field. This frequency is commonly known as the critical flicker fusion frequency (CFF). Considerable studies of the CFF as a function of various presentation variables have been carried out. From some of the earliest of these the Ferry-Porter law

$$f_c = k_1 \log B_m + k_2 \tag{4.6}$$

was formulated, where f_c is the CFF, k_1 and k_2 are constants and B_m is the mean luminance.

In the above it is usually assumed that the test field is flashed with 100% modulation, that is, it is dark between flashes.

In 1939 Wright[32] reported work showing that the 'constants' k_1 and k_2 were considerably dependent on degree of adaptation, nature of surround illumination and retinal position of the test field. Hylkema[33] subsequently found the area of the test field to influence the CFF for both foveal and peripheral presentation. Hecht and Smith[34] presented CFF versus luminance curves for foveally fixated test fields of various areas. Their summarised results are shown in Fig. 4.11 where it will be seen that, whilst the small test fields (35 and 5 mrad) reasonably obey the Ferry-Porter law other than at very high luminances, as the field size is increased a secondary portion develops. The form of curve for

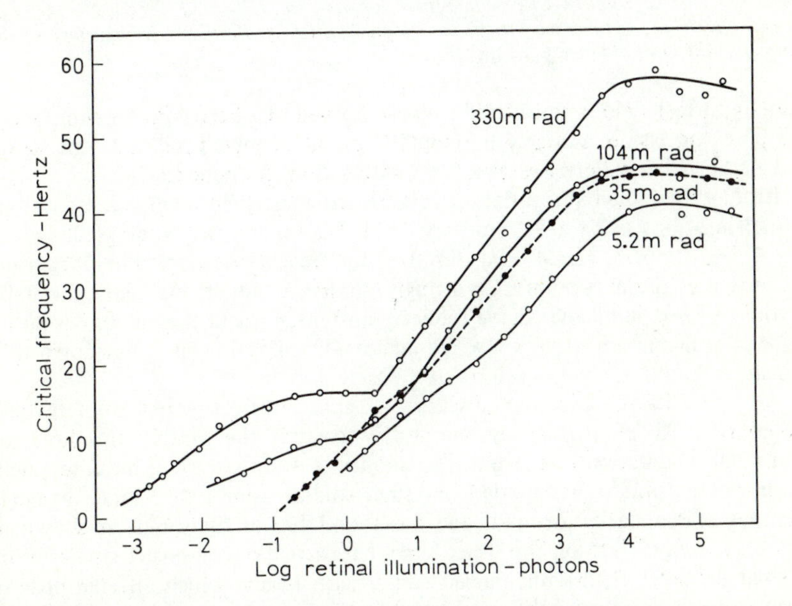

Fig. 4.11. Effect of area of fixation field on critical flicker frequency – diameter of fixation field in mrad marked on curves. (Reproduced from Hecht and Smith[34] by courtesy of the Journal of General Physiology).

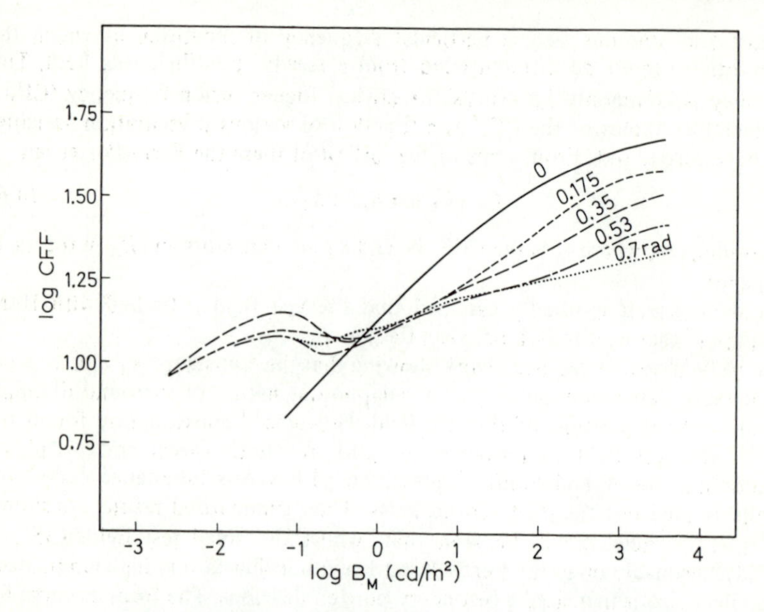

Fig. 4.12. The variation of CFF with retinal location for a 35 mrad test field – angles of eccentricity in radians marked on curves. (Reproduced from Brooke[35] by courtesy of the Journal of the Optical Society of America).

the largest test field is considered to be composed of a Ferrry-Porter component for pure rod vision, a largely horizontal section for mixed rod and cone vision and a second Ferry-Porter section (up to saturation) for cone vision.

Brooke[35] studied the variation of CFF with test field luminance at various retinal locations using a 35 mrad test field. His summarised findings, shown at Fig. 4.12, confirm Hecht and Smith's for foveal viewing. For increasing eccentricities the maximum CFF at high luminance reduces markedly, the slope of the CFF/log luminance curve reduces substantially and the curves develop a secondary maximum at very low luminances. Once again this is put down to a transition from full cone to full rod response.

All the CFF data so far discussed assume 100% square-wave test field modulation. An alternative test method is to study the contrast threshold for sinusoidal modulation of the test field about the mean at various mean test field luminances. Roufs[36] has carried out such studies using a 17.5 mrad diameter stimulus and a dark surround, and has found flicker thresholds as shown in Fig. 4.13. Kelly,[37,38] on the other hand, has carried out extensive studies using a 1 rad diameter field with blurred edges, with results which differed little in *form* from those shown in Fig. 4.13 although the absolute values were, as might be expected, different. This sinusoidal modulation approach is particularly useful for studying the behaviour of the visual system to general temporal

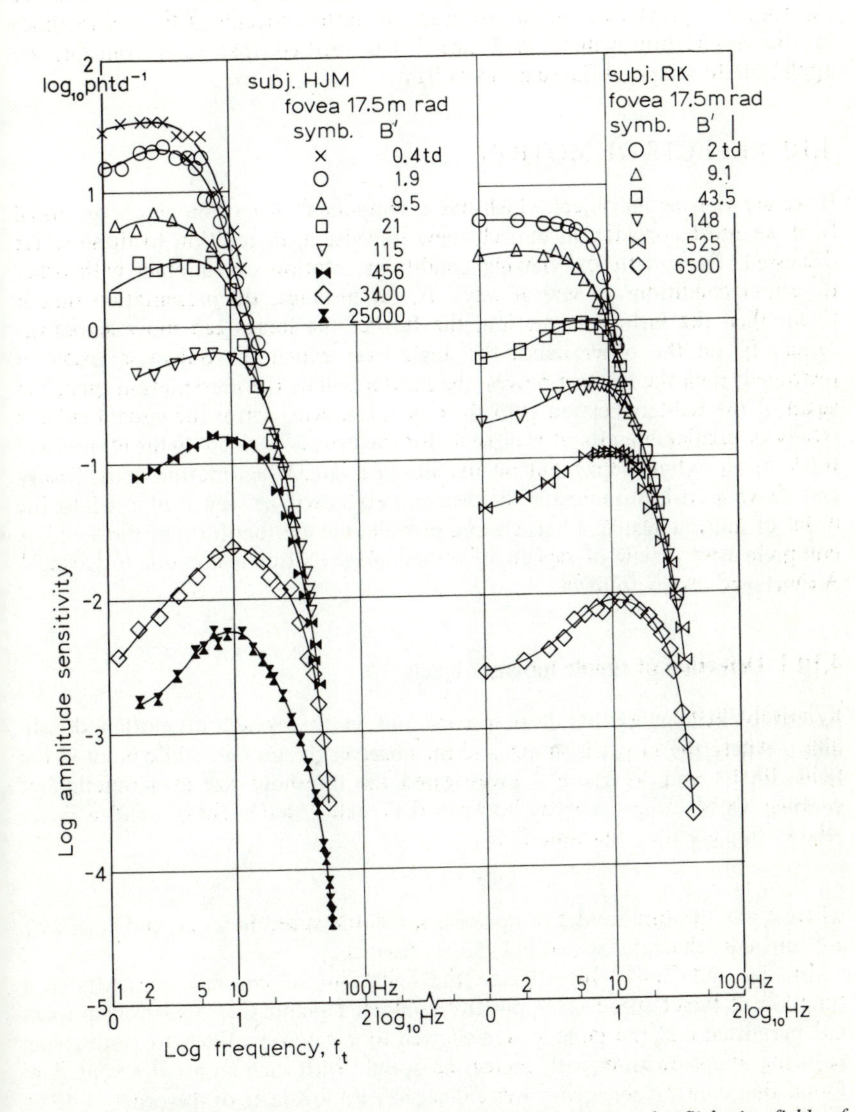

Fig. 4.13. Typical amplitude sensitivity as a function of frequency for flickering fields of 17.5 mrad diameter viewed against various backgrounds B'. *(Reproduced from Roufs*[36] *by courtesy of Pergamon Press).*

variations of test field luminance by the application of Fourier methods. Cornsweet[39] provides a useful introduction to the concepts of Fourier methods in this application whilst de Lange[40] has verified that such concepts are applicable to temporal fluctuations of light.

4.10 EFFECTS OF MOTION

If we are viewing an object which has a component of motion across the visual field we must consider an entirely new dimension, in addition to those so far discussed. Dependent on viewing conditions, motion can interact with other threshold conditions in several ways. If, for instance, the presentation time is fixed, then the faster the motion, the further the image can move across the retina. If, on the other hand, the angle over which the object is visible is restricted, then the faster it moves, the shorter will be the presentation time. Yet again, if the field associated with the object is moving across the visual field at a steady or predictable rate, it is possible for the eye to 'latch on' to the motion and track to an extent dependent on the absolute rate. When motion is oscillatory and above a critical frequency the resultant effect will be seen as blur, whilst the onset of motion against a background provides yet another form of threshold. A comprehensive review of aspects of retinal image motion is that due to Lavin.[41] A shortened review follows.

4.10.1 Detection of simple moving objects

Relatively little work has been carried out in this sphere of motion thresholds – where the target is simple and the observer fixates on a static point in the field. Bhatia and Verghese[42] investigated the threshold size as a function of crossing speed using 7 speeds between 0.17 and 2 rad/s. They found a linear relationship given by the equation

$$\alpha_T = c_1 + c_2 v$$

where α_T is the threshold size in radians, v is the speed in rad/s, and c_1 and c_2 are constants characteristic of individual observers.

Ercoles and Zoli[43] investigated the behaviour of contrast sensitivity as a function of target speed using positive contrast Landolt C's. Ocular pursuit was not permitted and the targets were allowed to move over a fixed arc, thus giving reducing exposure time with increasing speed. With such an arrangement they found that contrast sensitivity was enhanced with low rates of the order of 40 or 50 mrads/s, before beginning to fall rapidly. Typical results for two observers and 4 stimulus sizes are shown in Fig. 4.14. Ercoles and Zoli deduced from their results a theory of suppression of visual sensation during saccades. Hole[44] has studied the effects of motion on rectangular stimuli of various aspect ratios presented for 1/3 s exposure. He finds that there is a complex interrelationship between size, aspect ratio and rate of motion. Fig. 4.15 shows his results for

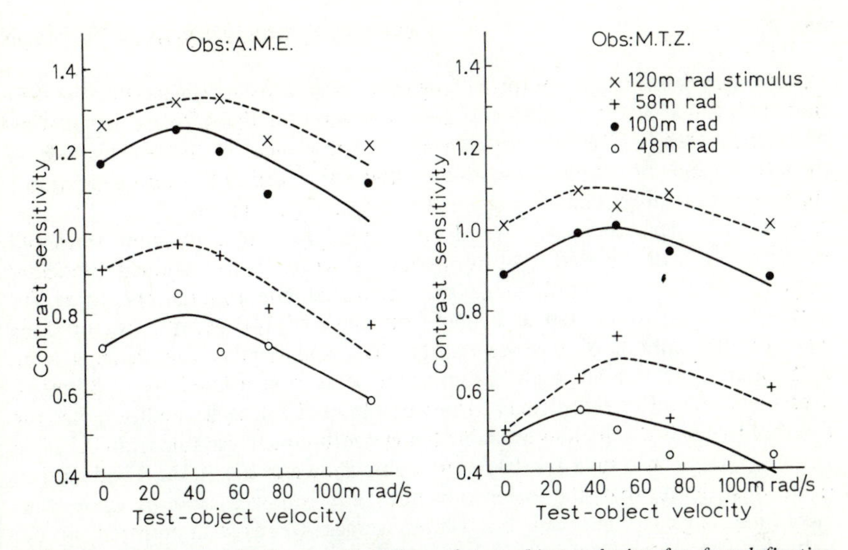

Fig. 4.14. Contrast sensitivity as a function of test object velocity for foveal fixation. Exposure time decreases with increasing speed. (Reproduced from Ercoles and Zoli[43] by courtesy of the Italian National Institute of Optics, Florence).

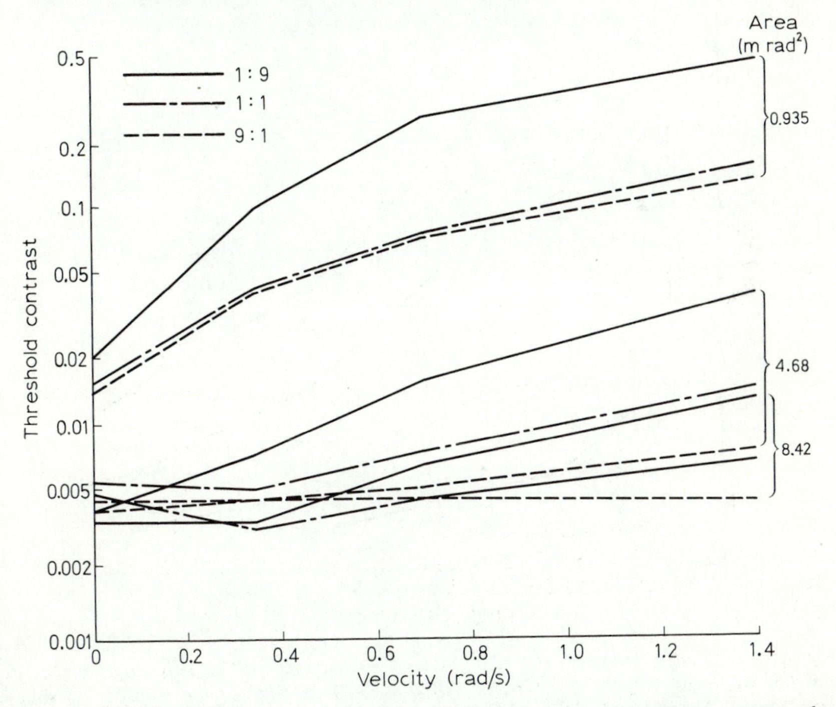

Fig. 4.15. Sample effects of aspect ratio on dynamic contrast thresholds for rectangular targets. Aspect ratios give dimension in direction of motion first. (after Hole[44]).

aspect ratios from 9:1 to 1:9 for rectangles of three constant areas. It can be seen that the results for the smallest area appear to show an almost linear degradation of contrast threshold with increasing speed. The results for the larger areas, on the other hand, not only show very complex interaction but also a signficant enhancement of threshold at low rates for certain aspect ratios.

Lavin and Spicer[45] have recently attempted to extend dynamic threshold performance with fixated eyes to peripheral vision. They studied binocular response to moving stimuli along the equatorial line through the fovea for eccentricities up to 1.5 rad at a field luminance of 140 cd/m^2. Presentations were of disc stimuli of positive contrast 0.3 and of five sizes ranging from 2.75 mrad to 18.6 mrad, each presentation being over a fixed arc of 87 mrad, irrespective of mean eccentricity. The variable used for each condition was the threshold velocity, velocities available being continuously variable from 0.175 to 7 rad/s. The mean results for 10 observers are shown in Fig. 4.16. It will be seen from this figure that the behaviour does not follow a smooth curve with increasing angle of eccentricity. These average trends are mirrored in the individual performance curves and it is suggested by Lavin and Spicer that the

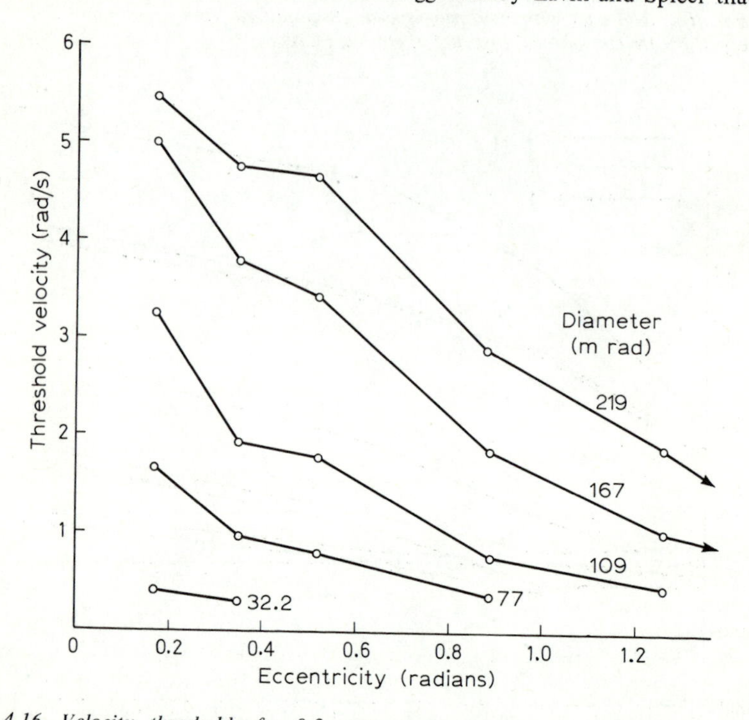

Fig. 4.16. Velocity thresholds for 0.3 contrast disc targets of various diameters as a function of mean viewing eccentricity. All presentations were over a constant 87 mrad viewing arc. (After Lavin and Spicer[45]).

total behaviour consists of three components, a fully binocular regime covering the majority of the field from 0 to 0.9 rad eccentricity, a monocular regime above about 0.9 rad eccentricity caused by the obstruction of vision from one eye by the bridge of the nose and an anomolous region around 0.3 rad eccentricity due to partially monocular vision occasioned by the blind spot (see Fig. 2.4).

4.10.2 Ocular pursuit

If the eye is permitted to attempt to follow a moving stimulus during presentation, and if that stimulus has such structure that the whole is readily seen, as with a Landolt C, then a detection exercise becomes one of inspection of the stimulus whilst attempting to follow its motion. One might expect that such a task is very different from the fixation task since, at least at low rates, it should be possible to 'lock-on' to the prominent whole of the stimulus to some degree in order to carry out detail inspection. As expected the findings in practice are that under these conditions performance remains substantially constant for low rates of motion, only beginning to fall when the rates are greater than 0.4 or 0.5 rad/s. Since such investigations of necessity require complex stimuli in order to provide the cues to motion it is hardly relevant that they should be discussed under the heading of the present chapter. Instead they will be discussed at length in Chapter 5.

4.10.3 Motion sensitivity

Another important form of motion threshold is the ability to detect that an object is moving. In many situations such an ability is possibly of more value than the basic ability to see an object, since it may be only by the fact that it moves that an observer may be able to detect that it is not part of the background. Whilst the main use of awareness of motion must be in structured fields, some work has been carried out by Miller and Ludvigh on detection of cessation of motion of simple stimuli in homogenous fields.[46] The results of this work led the experimenters to propose a concept of 'contrast of velocity'. This postulates that a stimulus travelling at high speed and for a relatively long time produces a more definite perception of velocity than one travelling slowly and observed for a shorter period of time. Therefore when the former stimulus is stopped suddenly the resulting contrast of velocity is great and the response to it is rapid. Conversely, when the latter stimulus is similarly stopped, the resulting contrast of velocity is slight and its perception, therefore, slower. Using this hypothesis leads to the suggestion that the time delay for detection of motion cessation will be the reaction time t_r plus some inverse function of the velocity v and the duration of the stimulus. A function suggested by Miller and Ludvigh, and one which gives moderate agreement with their results, is

$$t_p = t_r + \frac{1}{cv^2} \tag{4.7}$$

where c is dependent on stimulus duration, increasing with increasing duration.

Other work on velocity thresholds in empty fields is that of Aubert[47] and Boyce[48] who found that the velocity threshold for sustained motion in the absence of a frame of reference is of the order of 2–3 mrad/s. At lower velocities the detection of motion is uncertain, depending largely on the time of display of motion. Boyce explains his results by suggesting a simple sampled data model which assumes that the visual system samples the position of a stimulus at regular intervals of time, each sample being an estimate of the average position during a sampling interval. Movement is then presumed only to be perceived

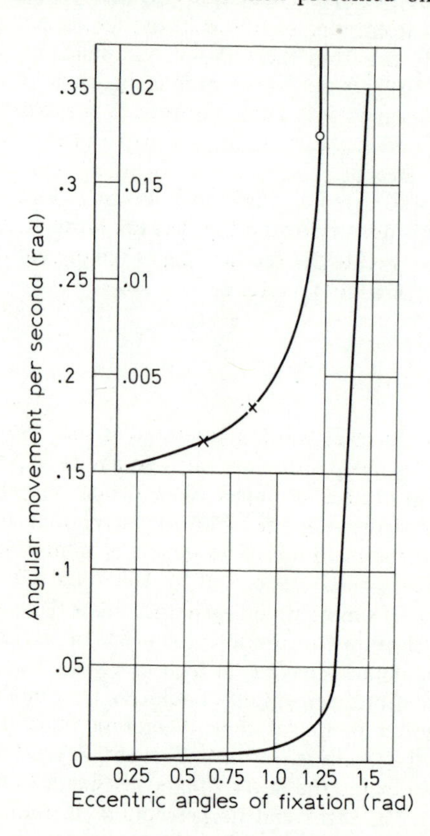

Fig. 4.17. The threshold for detection of motion in the periphery of the retina. The inset shows the toe of the main curve on a scale ten times as large. (Reproduced from Salaman[52] by courtesy of H.M.S.O.).

when the difference between two successive samples is significant because each sample eliminates the memory of earlier samples. This gives a simple explanation of failure of perception at very low velocities, the ultimate experimental limit appearing to be between 0.3 and 0.6 mrad/s.

As with other visual functions the ability to detect motion is dependent on retinal position and field luminance. Stratton,[49] Basler,[50] Laurens[51] and Salaman[52] have made studies of the variation of motion thresholds with retinal position at photopic luminances. The general finding is that for motion to be perceived in peripheral vision the velocity has to be increased as a function of eccentricity. A typical curve is shown in Fig. 4.17. This decrease in performance is, however, considerably less than the decrease in static visual performance functions for a given eccentricity. The capacity is thus *relatively* great in the periphery of the retina. The specialisation of the peripheral retina in the appreciation of motion is well seen, for example, in studies of the extreme periphery.[53] In this region a stationary object is invisible but is readily seen when moved. Studies of motion acuity at scotopic levels of luminance have been carried out for both foveal and peripheral presentation by Warden, Brown and Ross.[54] The results, in general, approximated to a linear relationship between log velocity and eccentricity, but with very large subject to subject differences.

4.11 EFFECTS OF COLOUR

Most experimentation on visual thresholds is carried out on neutral objects viewed against essentially neutral backgrounds. This practice is good as far as it goes, in that it allows reasonable comparisons between many different sets of data without the need to understand the intricacies of colour perception by the visual system. However, we must face the fact that most of the population have well developed colour vision, and even amongst those 8% or so who suffer from some form of colour blindness only a very small percentage have no colour sense at all.[55] Colour at low light levels, where rod vision is employed, may be handled simply, of course, by determining the effective relative luminosity on the scotopic luminosity curve. This section thus only mentions studies carried out at photopic luminance levels. At such levels of luminance colour can have one of three effects. Firstly, if both object and background are of the same colour, but significantly non-neutral, there can be an effect on threshold as a result of this shift of spectral balance. However, in practice this effect appears to be small.[56] Secondly, if the object is of a different hue to the background there will be a threshold associated with this difference of hue when the photopic luminosities of object and background are equal. The variation of chromaticity threshold with mean wavelength has been reported in particular by Regan and Tyler[56] and by Luria.[57] The magnitude and wavelength dependence appear to be somewhat dependent on the form of stimulus presentation, in the main tending to show a maximum sensitivity in the yellow/orange and blue/green with

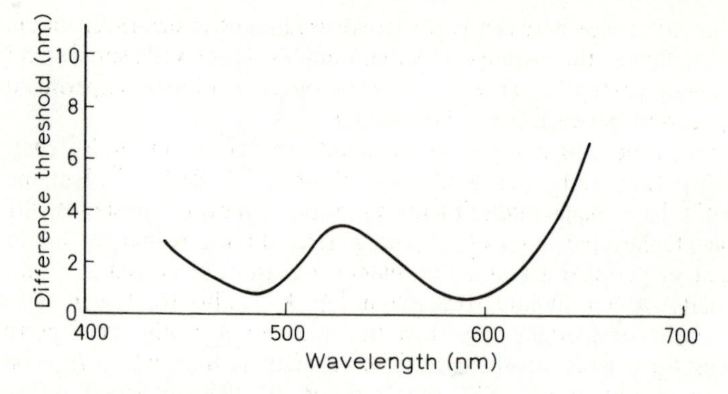

Fig. 4.18. The variation of differential wavelength threshold as a function of mean wavelength. (After Luria[57]).

reduced sensitivity in the green and poor sensitivity at extreme blue and red (see Fig. 4.18). Guth and Lodge,[58] amongst others, discuss the extent of disagreement of wavelength sensitivities between various threshold conditions.

Finally there is the situation where both hue and luminosity are different. Here it has been suggested, on the basis of limited experimental work, that it is adequate to describe contrast as the root mean square of luminosity and chromaticity contrast.[59] That is

$$C = \sqrt{C_L{}^2 + C_c{}^2}$$

where C_L is the luminosity contrast and C_c is the chromaticity contrast.

Owing to the difficulty of interpreting basic colour threshold data for a specific practical situation, and since one very important sphere of BAC(GW)'s work is concerned with detection of aircraft, a specific experiment was mounted to check whether there was any difference in detecting a neutral object against grey or sky blue backgrounds.[60] This experiment was purposely limited to the study of objects of areas similar to those which were likely to be presented by approaching aircraft at realistic acquisition ranges. The results showed conclusively that, for objects in this size range, there was a highly significant difference in detection against grey and sky blue backgrounds at equal luminosity.

4.12 EFFECT OF STATE OF EYE FOCUS

To this point it has been assumed that, whatever the quality of the information presented to the eye at threshold, the eye has been able to view the information in optimal focus. However, in practice the eye is not always in sharp focus even if it is well corrected. The naturally accommodated eye has been shown to

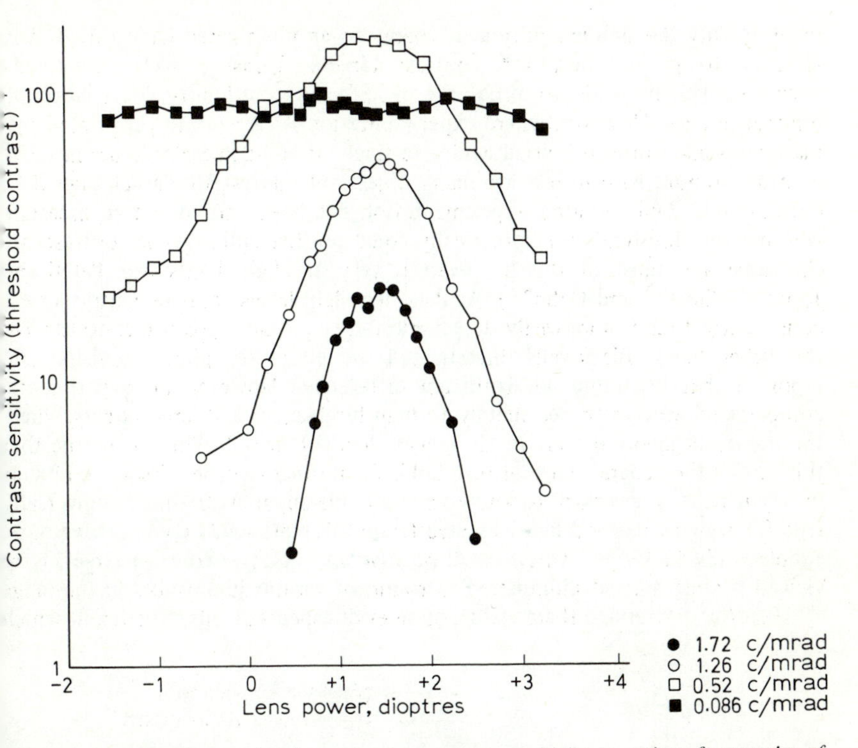

Fig. 4.19. Typical effects of defocus on the contrast sensitivity to various frequencies of spatial sinusoidal bar patterns. (Reproduced from Campbell and Green[72] by courtesy of the Journal of Physiology).

exhibit some 0.75 dioptres of defocus (or equivalent aberration) at the extremes of accommodation,[61–63] and under such circumstances it must be expected that performance will be restricted. In the case of persons with imperfect refractive correction there may well be additional defocus present in normal viewing, without recourse being taken to the wearing of spectacles. The extent to which defocus affects performance has been studied extensively by Campbell and co-workers.[64,65] A typical curve showing fall off of response to a grating pattern is shown in Fig. 4.19.

4.13 POSITIVE VERSUS NEGATIVE CONTRAST

The majority of simple projection stimuli studied in laboratory experimentation are of positive* contrast against their background – primarily because this

*Positive contrast is when the object of interest is brighter than the background.

involves only the simple addition of energy to an illuminated background. Any attempt to project negative* contrast targets against a wide illuminated background field leads to problems of either resolution or local surround luminance cues. However, where experimentation is carried out on printed test material – e.g. opticians' visual acuity testing – it is much easier to use negative contrast stimuli. Also in real life many objects of interest are darker than their backgrounds. Only limited experimentation has been carried out to ascertain whether the thresholds for numerically equal positive and negative contrasts are the same and most of this has been at very low light levels (e.g. Patell and Jones,[66] Short[67] and Cohn[68]). At these low light levels negative contrasts were consistently found more easily detectable than equivalent positive contrasts. On the other hand, Blackwell[3] investigated the effect for photopic vision and reported that he found no significant differences. However, his experimental comparisons were restricted mainly to high luminances and large targets, where the contrasts involved were rarely higher than 0.1 or 0.2. Whilst it is true that this covers the general range of thresholds for normal viewing of isolated objects in daylight, it is necessary to know whether this apparent finding is universally true for viewing in good light. To investigate this matter BAC(GW) carried out a simple study in 1963[69] where small positive and negative contrast targets were viewed against a local illuminated surround of various luminances in the range 7–70 cd/m^2. Although there were a number of aspects of this experiment which

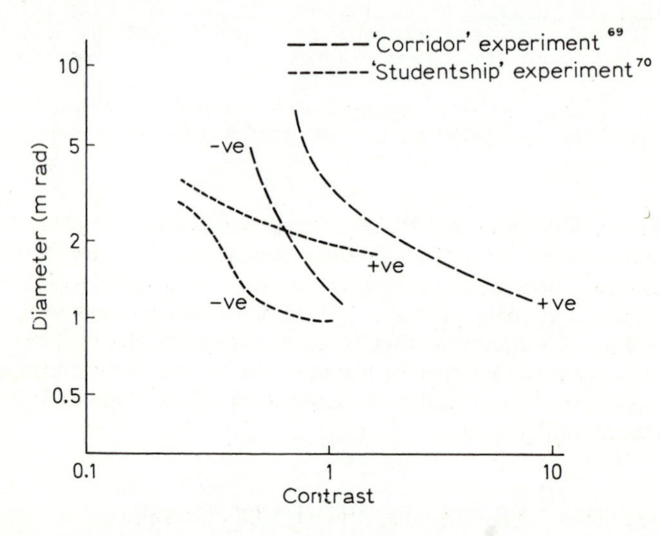

Fig. 4.20. Collected results of the two B.A.C. (GW) experiments studying the relationship between thresholds for positive and negative contrast stimuli.

*Negative contrast when the object is darker than the background.

could be faulted on psychometric grounds, it appeared that there was a very strong bias towards negative contrast stimuli of numerical contrast greater than about 0.2 being more readily visible than equivalent positive contrast stimuli. A second experiment was thus set up which was carefully designed to overcome the main criticisms of the first.[70] Although the absolute thresholds in this case were markedly superior, once again there was a very definite bias in favour of negative contrasts when the numerical contrast was in excess of about 0.2. The results of the two experiments are summarised in Fig. 4.20. To substantiate rigorously the effects at all contrasts would have required an extensive experimental program. It was thus decided to carry out a well controlled experiment to determine the positive contrast required to yield the same size threshold as one high negative contrast. The highest negative contrast readily achievable in a form of projection presentation was –0.75. A statistical threshold size was determined for each of 4 observers for fixated viewing of a circular stimulus of this contrast. A set of positive contrast stimuli were then prepared, matched for individual observers, and presented at various positive contrasts. It was found[71] that a positive contrast of between 1.2 and 1.5 was required to match the negative contrast of –0.75. This is considered to give some indication of the magnitude of possible differences between positive and negative contrast stimuli when viewed in good light.

4.14 CONTRAST SENSITIVITY

So far all thresholds studied have been associated with isolated objects. One simple form of scene complexity which should yield valuable information on the way the visual process works is the response to one-dimensional periodic patterns. Such a measurement must have a close relationship to the MTF described in Chapter 2. To distinguish it from instrumental MTF, and at the same time to retain an absolute measure of threshold, the inverse of the threshold contrast for such an experiment as a function of spatial frequency is known as the Contrast Sensitivity Function (e.g. Campbell and Green[72]). This function usually has a very characteristic shape, having a maximum value (i.e. minimum contrast threshold) for a spatial frequency of around 0.15–0.6 c/mrad and falling off at both higher and lower frequencies for sinusoidally modulated patterns, although the exact form and absolute value of the function depends considerably on such factors as mean luminance level and presentation time (e.g. Nachmias,[73] Nes and Bouman,[74] Valois *et al*[75]). A typical curve is shown at Fig. 4.21. The fall-off at higher frequencies is to be expected – it is due to the progressive attenuation of higher frequencies by optical defects and limitations, as is characteristic of all MTF curves. The fall-off at low frequencies is at first unexpected. However, a little thought allows one to realise that, if one is fixating a local part of a periodic pattern which has a frequency tending to zero, there will be only a very slow rate of change of luminance due solely to the period of the pattern. In fact

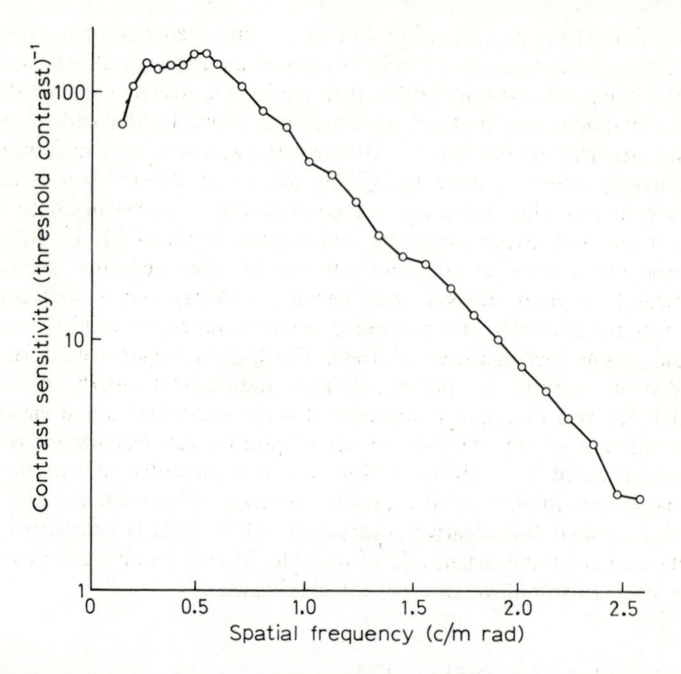

Fig. 4.21. Typical contrast sensitivity function for an in-focus eye viewing sinusoidally modulated spatial bar patterns. (Reproduced from Campbell and Green[72] by courtesy of the Journal of Physiology).

the local rate of change will tend to be inversely proportional to frequency for very small spatial frequencies. Thus one way of explaining the low frequency fall-off of performance would be to consider that what the observer 'sees' is the maximum slope. This will be discussed further in later chapters.

As a further aid to understanding how the visual system handles spatially periodic data, Campbell and Robson have studied the contrast sensitivity function for other than sinusoidal modulation.[76] In addition to studying the thresholds as a function of frequency for *detection* of square waves, sawtooth waves and rectangular waves of 10:1 duty cycle (pulse occupying 0.1 of a period), they have also studied the thresholds for discrimination of the waveform not being sinusoidal.[77] The main findings, characterised by the comparative threshold for sinusoidal and square wave modulation shown in Fig. 4.22, are that for medium to high spatial frequencies the detection threshold is controlled by the strength of the fundamental alone, whilst the discrimination threshold is controlled only by the strength of the lowest harmonic present (i.e. the 3rd harmonic for the square wave, the 2nd harmonic for other waves). At low frequencies all waves tested other than the sine waves tend to a constant detection threshold level. A possible explanation will be

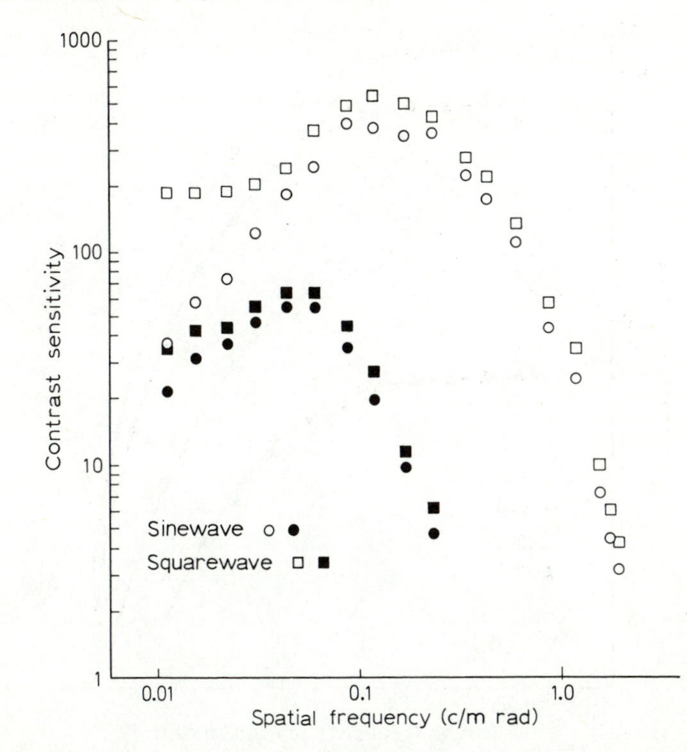

Fig. 4.22. Contrast sensitivity for sine wave and square wave gratings. open symbols for luminances of 500 cd/m², filled symbols for 0.05 cd/m². (Reproduced from Campbell and Robson[76] by courtesy of the Journal of Physiology).

discussed in Chapter 12. It has been found by other workers (e.g. Kelly[37,38]) that there is an analogous behaviour to that observed for spatially modulated bar patterns if these are presented with temporal modulation. Most striking, however, is the effect of both temporal and spatial patterns together (e.g. Robson[78]), where it is found that there is a pronounced interaction between the two at low frequencies as shown in Fig. 4.23. It has been recently shown by Kelly that this spatio-temporal behaviour is equally characteristic of the separate colour mechanisms.[79]

4.15 ORIENTATION EFFECTS

Since the advent of contrast sensitivity testing it has been possible to investigate the variability of visual threshold performance as a function of pattern orientation in an objective manner. It has been found that performance is

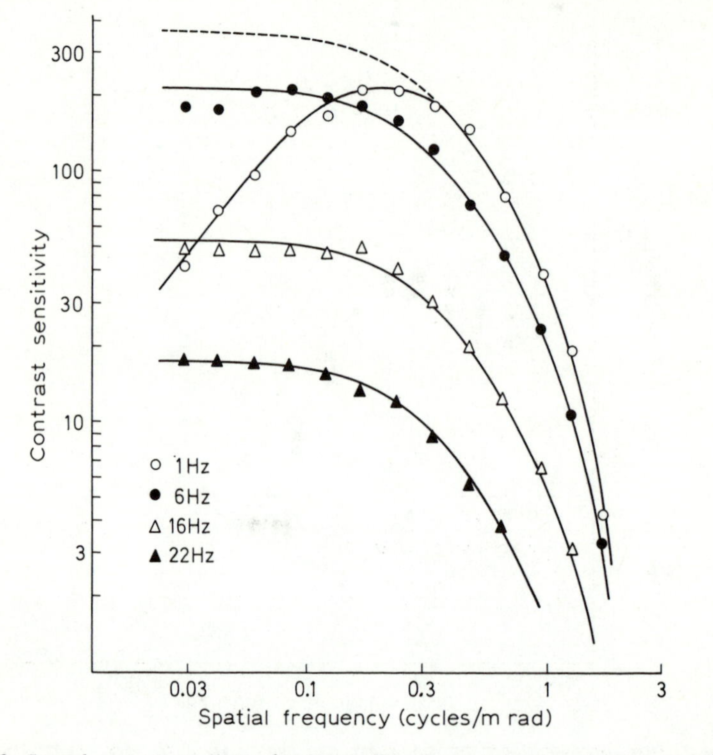

Fig. 4.23. Spatial contrast sensitivity functions where the spatial pattern is being modulated temporally. (Reproduced from Robson[78] by courtesy of the Journal of the Optical Society of America).

maximal to vertical edges with a secondary maximum to horizontal edges and pronounced minima for patterns inclined at approximately $\pi/4$ radians to the vertical (e.g. Campbell *et al*[80–82]). Ginsburg[83] has recently interpreted this variation of threshold in terms of a complex two-dimensional contrast sensitivity function and has developed computer programs for studying the two-dimensional filtering effects of such a function.

4.16 OTHER FACTORS

When attempts are made to compare one set of threshold data with another obtained under what appear to be similar conditions, one is often faced with the problem that there appear to be marked discrepancies. It is easy to assume that this is all due to observer differences, as indeed some of it inevitably must be (see Section 3.2). However, when full account is taken of observer differences many discrepancies still remain. Many of these remaining differences can be explained subjectively in terms of such factors as confidence level at which the decision is made, task motivation, experience (i.e. state of learning), state of eye accommodation, etc. For instance a very marked effect is produced according to whether the experiment is conducted on a forced or free choice basis (see Chapter 3), whilst Blackwell[84] has shown major effects due to task motivation. Equally, state of learning can play a very important part. Blackwell, in reference 4.3, reported training of some thousands of observations before any experimental results were analysed. Particular studies of learning have been carried out by BAC(GW)[85,86] and Miller and Ludvigh.[87] Typically these resulted in thresholds which improved towards a plateau according to an exponential law with increasing number of presentations (see Section 3.3). That there is an effect of viewing distance might be predicted from the data on eye quality as a function of accommodation, as presented in Section 2.3, although such an effect has not been demonstrated conclusively in a laboratory experiment to the knowledge of the author. Finally there is the effect of what may be termed vigilance. Work reported by Mackworth[88] has shown that detection thresholds, in common with many other human behavioural capabilities, degrade logarithmically with time of prolonged vigilance, increasing by a factor of 2 in a period of some 30 min. On the other hand Blackwell[84] has shown constancy of performance during an experimental session, suggesting that any effect due to vigilance is more than a simple fatigue effect.

It is important that the foregoing effects should be borne in mind when designing threshold experiments and one must always be prepared to expect significant differences in mean results, dependent on the experimental procedures used.

REFERENCES

1. Middleton, W. E. K. (1958). *Vision through the Atmosphere*, p. 60, University of Toronto Press
2. Ronchi, L. (1971). 'Ricco's Law', *Atti della Foundazione Giorgio Ronchi*, **26**, 751
3. Blackwell, H. R. (1946). 'Contrast Thresholds of the Human Eye', *J. Opt. Soc. Am.*, **36**, 624
4. Blackwell H. R. and McCready, D. W., (1952). 'Foveal Detection Thresholds for Various Durations of Target Presentation', *Minutes and Proceedings of NAS–NRC Vision Committee*, AGSIL/53/4405, 249
5. Vos, J. J., Lazet, A. and Bouman, M. A. (1956). 'Visual Contrast Thresholds in Practical Problems', *J. Opt. Soc. Am.*, **46**, 1065
6. Taylor, J. H. (1969). 'Factors underlying Visual Search Performance', Scripps Institution of Oceanography Report No. 69–22, November
7. Taylor, J. H. (1962). 'Contrast Thresholds as a Function of Retinal Position and Target Size for the Light-adapted Eye', *Proceedings of the NAS–NRC Vision Committee*
8. Davies, E. B. (1969). 'Visual Theory in Target Acquisition', RAE Technical Memo, WE 1301
9. Blackwell, H. R. (1963). 'Neural Theories of Simple Visual Discriminations', *J. Opt. Soc. Am.*, **53**, 129

10. Stiles, W. S. and Crawford, B. H. (1934). 'The Liminal Brightness Increment for Whi Light for Different Conditions of the Foveal and Parafoveal Retina', *Proc. R. Soc. B* **116**, 55
11. Barlow, H. B. (1957). 'Increment Thresholds at Low Intensities considered a Signal-noise Discriminations', *J. Physiol.*, **136**, 469
12. Werblin, F. S. (1973). 'The Control of Sensitivity in the Retina', *Scientific American* 71
13. Rushton, W. A. H. and Powell, D. S. (1972). 'The Rhodopsin Content and the Visua Threshold of Human Rods', *Vision Research*, **12**, 1073
14. Lythgoe, R. J. (1932), 'The Measurement of Visual Acuity', MRC. Special Repor No. 173, HMSO, London
15. Murphy, M. J., Overington, I. and Williams, D. G. (1965). 'Final Report on Visua Studies Contract', BAC(GW) Ref. R41S/11/VIS, App. 3.6.I
16. Lamar, E. S., Hecht, S., Schlaer, S. and Hendley, C. D. (1947). 'Size, Shape an Contrast in the Detection of Targets by Daylight Vision. I. Data and Analytica Description', *J. Opt. Soc. Am.*, **37**, 531
17. Lamer, E. S., Hecht, S., Hendley, C. D. and Schlaer, S. (1948). 'Size, Shape an Contrast in the Detection of Targets by Daylight Vision. II. Frequency of Seeing an The Quantum Theory of Cone Vision', *J. Opt. Soc. Am.*, **38**, 741
18. Guth, S. K. and McNelis, J. F. (1969). 'Threshold Contrast as a Function of Targe Complexity', *Am. J. Optom.*, **46**, 98
19. Velden, H. A. van der (1944). 'About the Minimum Light Quanta necessary for sensin by the Human Eye', *Physica*, **11** 179
20. Blondel, A. and Rey, J. (1912). 'The Perception of Lights at their Range Limits', *Trans I.E.S.*, London. 7, 625
21. Bouman, M. A. and van der Velden, H. A. (1948). 'The Quanta Explanation of Visio and the Brightness Impession for Various Times of Observation and Visual Angles', *J. Opt. Soc. Am.*, **38**, 231
22. Taylor, J. H. (1972). 'Air to Ground Visibility of Lights at Low Background Levels' *AGARD Conference Proceedings No. 100*, (Ed. H. F. Huddleston), London, November p. B8–1
23. Ogle, K. N. (1961). 'Foveal Contrast Thresholds with Blurring of the Retinal Image and Increasing Size of Test Stimulus', *J. Opt. Soc. Am.*, **51**, 862
24. Middleton, W. E. K. (1937). 'Photometric Discrimination with a Diffuse Boundary', *J. Opt. Soc. Am.*, **27**, 112
25. Fry, G. A. (1965). 'The Eye and Vision', in *Applied Optics and Optical Engineering Vol. 2.* p. 46, (Ed. R. Kingslake), Academic Press
26. Kruithof, A. M. (1950). 'Perception of Contrasts when the Contours of Details ar Blurred', *Phillips Technical Review*, **11**, 333
27. Hood, D. C. (1973), 'The Effects of Edge Sharpness and Exposure Duration or Detection Threshold', *Vision Research*, **13**, 759
28. Crowther, A. G. and Overington, I. (1975). 'Experiments on the Detection of Blurred Targets', BAC(GW) Report No. ST10840
29. Kishto, B. N. (1970). 'Variation of the Visual Threshold with Retinal Location', *Vision Research*, **10**, 745
30. Ogle, K. N. (1961). 'Peripheral Contrast Thresholds and Blurring of the Retinal Image for a Point Light Source', *J. Opt. Soc. Am.*, **51** 1265
31. Herrick, R. M. (1973). 'Increment Thresholds for Multiple Identical Flashes in the Peripheral Retina', *J. Opt. Soc. Am.*, **63**, 1261
32. Wright, W. D. (1939). *The Perception of Light*, Chemical Publishing Co., New York
33. Hylkema, B. S. (1942). 'The Fusion Frequency of Intermittent Light', (in Dutch), Thesis, University of Amsterdam, Van Gorcum & Co., Arsen
34. Hecht, S. and Smith, E. L. (1936). 'Intermittent Stimulation by Light. VI. Area and the Relation between Critical Frequency and Seeing', *J. Gen. Physiol.*, **19**, 979

35. Brooke, R. C. (1951). 'The Variation of Critical Fusion Frequency with Brightness at Various Retinal Locations', *J. Opt. Soc. Am.*, **41**, 1010
36. Roufs, J. A. J. (1972). 'Dynamic Properties of Vision', *Vision Research*, **12**, 261 and 279
37. Kelly, D. H. (1961). 'Flicker Fusion and Harmonic Analysis', *J. Opt. Soc. Am.*, **51**, 917
38. Kelly, D. H. (1969). 'Flickering Patterns and Lateral Inhibition', *J. Opt. Soc. Am.*, **59**, 1361
39. Cornsweet, T. N. (1970). *Visual Perception*, Chap. 14, Academic Press
40. Lange, H. de (1954). 'Relationship between Critical Flicker-frequency and a Set of Low-frequency Characteristics of the Eye', *J. Opt. Soc. Am.*, **44**, 380
41. Lavin, E. P. (1972). 'A Literature Survey on Retinal Image Motions', App. 3, Sect. 3 of *Research into Factors affecting Detection of Aircraft through Optical Sights*, (BAC(GW) Ref. L50/186/1449)
42. Bhatia, B. and Verghese, C. A. (1964). 'Threshold Size of a Moving Object as a Function of its Speed', *J. Opt. Soc. Am.*, **54**, 948
43. Ercoles, A. M. and Zoli, M. T. (1968). 'Contrast Threshold for Moving Landolt Rings', *Atti della Fondazione Giorgio Ronchi*, **23**, 515
44. Hole, R. J. (1975). 'A Study of the Effects of Target Shape and Target Motion on Visual Threshold Detection Performance, BAC(GW) Ref. ST13117
45. Lavin, E. P. and Spicer, P. J. (1972). 'The Measurement of Extra-foveal Dynamic Thresholds', App. 3, Sect. 5 of *'Research into Factors affecting the Detection of Aircraft through Optical Sights'*, (BAC(GW) Ref. L50/186/1449)
46. Miller, J. W. and Ludvigh, E. (1961). 'The Perception of Movement Persistence in the Ganzfeld', *J. Opt. Soc. Am.*, **51**, 57
47. Aubert, H. (1886). 'The Threshold of Visual Sensation of Movement', (in German), *Pfug. Arch. ges. Physiol.*, **39**, 347.
48. Boyce, P. R. (1965). 'The Visual Perception of Movement in the absence of an External Frame of Reference', *Optica Acta*, **12**, 47
49. Stratton, G. M. (1900). 'A New Determination of the Minimum Visible and its bearing on Localisation and Binocular Depth', *Psych. Review.* **7**, 429
50. Basler, A. (1917). *A. g. Phys.* 'Brightness Threshold for Moving Panels' (in German), **167**, 98
51. Laurens (1914). 'Local Differentiation Capability by Mesopic Vision' (in German), *Z. Sinnesorgane*, **48**, 233
52. Salaman, M. (1929). 'Some Experiments on Peripheral Vision', MRC Special Report 136, HMSO, London
53. Duke-Elder, W. S. (1944). *Textbook of Ophthalmology*, Vol. 1, C. V. Mosby Co., St. Louis
54. Warden, C. J., Brown, H. C. and Ross, S. (1945). 'A Study of Individual DIfferences in Motion Acuity at Scotopic Levels of Illumination', *J. Exptl. Psych.*, **35**, 57
55. Wald, G. and Brown, P. K. (1965). 'Human Colour Vision and Colour Blindness', *Symposia on Quantitative Biology*, Cold Spr. Harb., New York. **30**, 345
56. Regan, D. and Tyler, C. W. (1971). 'Some Dynamic Features of Colour Vision', *Vision Research*, **11**, 1307
57. Luria, S. M. (1966). 'Colour Vision', *Physics Today*, 34
58. Guth, S. L. and Lodge, H. R. (1973). 'Heterochromatic Additivity, Foveal Spectral Sensitivity and a New Colour Model', *J. Opt. Soc. Am.*, **63**, 450
59. Anon (1944), 'Influence of Colour Contrast on Visual Acuity', Nat. Def. Res. Com., Off. Sci. Res. Dev. Rep. No. 4541, Eastman Kodak Co., Rochester, New York
60. Spicer, P. J. (1972). 'The Colour Experiment', Study Note No. 2 of MOD(PE) Contract K64A/77/CB64A. BAC(GW) Ref. L50/186/1570
61. Westheimer, G. (1955). 'Spherical Aberration of the Eye', *Optica Acta*, **2**, 151
62. Ivanoff, A. (1956). 'About the Spherical Aberration of the Eye', *J. Opt. Soc. Am.*, **46**, 901.

63. Overington, I. (1975). 'On the Interaction between State of Accommodation and Retinal Image Quality', *Atti della Fondazione Giorgio Ronchi,* **29**, 909
64. Green, D. G. and Campbell, F. W. (1965). 'Effect of Focus on the Visual Response to a Sinusoidally Modulated Spatial Stimulus', *J. Opt. Soc. Am.,* **55**, 1154
65. Campbell, F. W. and Gubisch, R. W. (1967). 'The Effect of Chromatic Aberration on Visual Acuity', *J. Physiol.* **192**, 345
66. Patel, A. S. and Jones, R. W. (1968). 'Increment and Decrement Visual Thresholds', *J. Opt. Soc. Am.,* **58**, 696
67. Short, A. D. (1966). 'Decremental and Incremental Visual Thresholds', *J. Physiol.,* **185** 646
68. Cohn, T. E. (1974). 'A New Hypothesis to explain why the Increment Threshold exceeds the Decrement Threshold', *Vision Research,* **14**, 1277
69. Achurch, I. C. (1963). 'Initial Analysis of Data from the Contrast Recognition Experiments' Assessment Group Memo No. 154, BAC(GW), Bristol
70. Clark, M. I., Herdan, B. L., Lloyd-Bostock, P. M. R. and Tyrwhitt-Drake, B. G., (1960 'Visual Studies – Studentship Experiment', Optics Group, BAC(GW), Bristol
71. Hawkins, K. and Church, N. T. (1969). 'Contrast Sign Dependence', Study Note No. 2 of *Research into Factors affecting the Detection of Aircraft through Optical Sights* BAC(GW) Ref. L50/20/PHY/186/1059
72. Campbell, F. W. and Green, D. G. (1965). 'Optical and Retinal Factors affecting Visual Resolution', *J. Physiol.,* **181**, 576
73. Nachmias, J. (1967). 'Effect of Exposure Duration on Visual Contrast Sensitivity with Square Wave Gratings', *J. Opt. Soc. Am.,* **57**, 421
74. Nes, F. L. van and Bouman, M. A. (1967). 'Spatial Modulation Transfer in the Human Eye', *J. Opt. Soc. Am.,* **57**, 401
75. Valois, R. L. De, Morgan, H. and Snodderly, D. M. (1974). 'Psychophysical Studies of Monkey Vision – III Spatial Luminance Contrast Sensitivity Tests of Macaque and Human Observers', *Vision Research,* **14**, 75
76. Campbell, F. W. and Robson, J. G. (1968). 'Application of Fourier Analysis to the Visibility of Gratings', *J. Physiol.,* **197**, 551
77. Campbell, F. W., Howell, E. R. and Robson, J. G. (1971). 'The Appearance of Grating with and without the Fundamental Fourier Component', *J. Physiol.,* **217**, 17
78. Robson, J. G. (1966). 'Spatial and Temporal Contrast Sensitivity Functions of the Visual System', *J. Opt. Soc. Am.,* **56**, 1141
79. Kelly, D. H. (1974). 'Spatio-temporal Frequency Characteristics of Color-vision Mechanisms', *J. Opt. Soc. Am.,* **64**, 983
80. Campbell, F. W. and Kulikowski, J. J. (1966). 'Orientational Selectivity of the Human Visual System', *J. Physiol.* **187**, 437
81. Campbell, F. W., Kulikowski, J. J. and Levinson J. (1966). 'The Effect of Orientation on the Visual Resolution of Gratings', *J. Physiol.* **187**, 427
82. Campbell, F. W., Cleland, B. G., Cooper, G. F. and Enroth-Cugell, C. (1968). 'The Angular Selectivity of Visual Cortical Cells to Moving Gratings', *J. Physiol.* **198**, 237
83. Ginsburg, A. P. (1973). 'Pattern Recognition Techniques suggested from Psychological Correlates of a Model of the Human Visual System', *NAE CON'73 Record*, 309
84. Blackwell, H. R. (1953). 'Psychophysical Thresholds: Experimental Studies of Methods of Measurement', *Bull. Eng. Res. Inst. University of Michigan,* No. 36
85. Milnes-Walker, N. D. (1968). 'The Effect of Repeated Viewing of a Target Run on Recognition Range', Human Factors Study Note Series 4, No. 22, BAC(GW) Ref R47/20/HMF/1703
86. Murphy, M. J., Overington, I. and Williams, D. G. (1965). 'Final Report on Visual Studies Contract', Sect. 3.5, BAC(GW) Ref. R41S/11/VIS
87. Miller, J. W. and Ludvigh, E. (1957). Kresge Eye Institute Project No. NM170199 Subtask 2, Rep. 13
88. Mackworth, J. F. (1970). *Vigilance and Attention*, Penguin Science and Behaviour Series

5 Recognition Thresholds

In the preceding chapter we have been concerned with laboratory performance at the most basic level of acquisition – that of detecting the *presence* of an object. By definition such detection must always imply no ability to extract detail information about an object such that it can be recognised as of a particular type. In real life it is rare indeed that the visual task involves nothing more than simple detection. Much more frequently it is necessary for an observer to recognise that an object is of the correct class – a square rather than a circle, a Landolt 'C' rather than an annulus, a given letter out of a possible set of letters, a vehicle rather than a bush. For any of these tasks it is not enough to be aware of the presence of the object – one must be able to see some of the structure. What is by no means obvious is exactly what structure it is necessary to see in order to effect recognition in a given situation. Nor is it obvious how to relate the detectability of certain local structure to the detection thresholds of isolated simple shapes. In this chapter we shall consider some of the various laboratory experiments which have been carried out to attempt to determine the thresholds associated with a variety of recognition tasks. The coverage can be by no means comprehensive owing to the enormous field covered by the term recognition. Rather we shall only summarise general statements and present some of the more clear cut threshold behavioural trends. The reader wishing to pursue this particular facet of acquisition in depth is recommended to start with Yves le Grand's book 'Form and Space Vision'[1] for general reading or Zusne's book 'Perception of Form'[2] for a study in depth. The latter reference itself contains some 2 500 references to information related to recognition – an indication of the scope of the subject.

Before proceeding to a study of various forms of recognition it is necessary to discuss a few general points about the subject of recognition. Firstly, whilst detection may be considered as a decision that a local difference in energy exists, recognition can by no means be treated so simply. Ability to recognise *must* depend, at least to some extent, on such factors as the number of possible stimuli, complexity of form, previous experience of particular forms, orientation of retinal image and association with the particular field of view in addition to the factors found to influence detection. Since the many psychological facets of recognition are beyond the scope of this book we shall in general restrict ourselves to a study of the physical aspects of recognition. However, it is impossible to separate the physical and psychological factors into watertight compartments, so some background of interactions between physical and psychological is important for the reader.

The site of the main recognition processes is variously assumed to be at the cortex – the global view of the Gestalt theorists (e.g. Marshall and Talbot[3]) – or in the vicinity of the retina (e.g. Byram[4] and Gibson[5]). The Gestalt school consider that the ease of recognition should be associated with the 'goodness' c 'simplicity' of a form, this being effectively a measure of the complexity of th shape[2]. On such a scale the circle might be considered to be the simplest forn and hence it would be expected to have the lowest recognition threshold. Th has been shown to be false by a number of experimenters, several of whom hav found rectangles and triangles to be more easily recognised than circles (e.g Helson and Fehrer[6]). In contrast to the Gestalt school, Gibson has proposed tha most basic processing towards recognition should be associated with periphera rather than central processes, the main cues being difference information o various forms at the retina. He cites such factors as convergence of line (perspective) and differential sharpness as providing cues to perception of dept and suggests that such other factors as progressive change of hue saturation an scale of texture also provide the basis for perception in 3 dimensions. Equall one is led to the concepts of differences in luminance, differences in contou direction and similar difference functions as mainly controlling recognition in dimensions.

With this preamble let us now look at some of the recognition threshold dat available.

5.1 SIMPLE SHAPES

The simplest form of recognition which it is possible to think of is that o recognising that a given object is one of a selected small number of simpl shapes. The task might be, for instance, the recognition of a square from a set o squares, discs, rectangles and triangles. Alternatively it might be the recognitio of a star shape from a set of stars and various polygons. As already stated i might be expected that in such experiments those forms with the simplest an most symmetrical outlines would be easiest to recognise. In practice the finding vary from experimenter to experimenter. Helson and Fehrer[6] and several othe groups have found that, when the choice lies between rectangles, discs an triangles, amongst other forms, the rectangles are easiest to recognise with th triangles generally being next easiest. An experiment at BAC(GW)[7] using 4: rectangles, equilateral triangles, discs and squares of equal area and variou contrasts produced similar results. In this experiment the four classes of stimul were presented in random order at various contrasts to approximately 10(observers who were required to walk forward slowly, pausing when necessary until they could recognise which type of object was being presented. Th findings were that, on average, there was little to chose between the recognitio of triangles, squares and discs of equal area and contrast, but that the 4:1 rectangles could be recognised at considerably greater ranges than the other 3 types of stimuli. An interesting subjective observation was made by severa observers concerning the triangles and squares, i.e. in both cases, prior to positive recognition, they had been often aware of certain recognition cues, in that both

FORMS

FIGURES		Ellipse	Rectangle	Triangle	Diamond	Cross	Star
1	MD	0.202	0.254	0.272	0.254	0.240	0.235
	P	0.662	0.716	0.817	0.716	0.882	0.720
2	MD	0.269	0.287	0.311	0.311	0.258	0.272
	P	0.683	0.766	0.828	0.745	0.985	0.944
3	MD	0.310	0.327	0.373	0.358	0.318	0.304
	P	0.748	0.828	0.890	0.803	1.242	1.112
4	MD	0.414	0.373	0.411	0.414	0.411	0.334
	P	0.949	0.897	0.995	1.035	1.635	1.391
5	MD	0.518	0.421	0.455	0.518	0.534	0.433
	P	1.157	0.983	0.992	1.064	2.132	2.099

ig. 5.1. The experimental forms and figures studied by Casperson. Maximum dimensions MD) and perimeter (P) are given in centimetres for the smallest figures used. Larger figures ith the linear dimensions scaled by factors of 1.5, 2.0, 2.5, 3.0, 4.0 and 5.0 were also udied. (Reprinted from Casperson[12] by permission. Copyright (1950) by the American sychological Association).

riangles and squares tended to scintillate, growing points which varied between & 4 and not always in the correct locations. For instance it was a common xperience to see the triangle upside down before resolving it the correct way p. Such subjective comment is important to an understanding and modelling of he visual process and is presumed to be due to the combination of effects due o the retinal mosaic and the involuntary eye movements. Other very useful ubjective comments on appearance of objects at recognition threshold are to be ound in Salaman's studies[8].

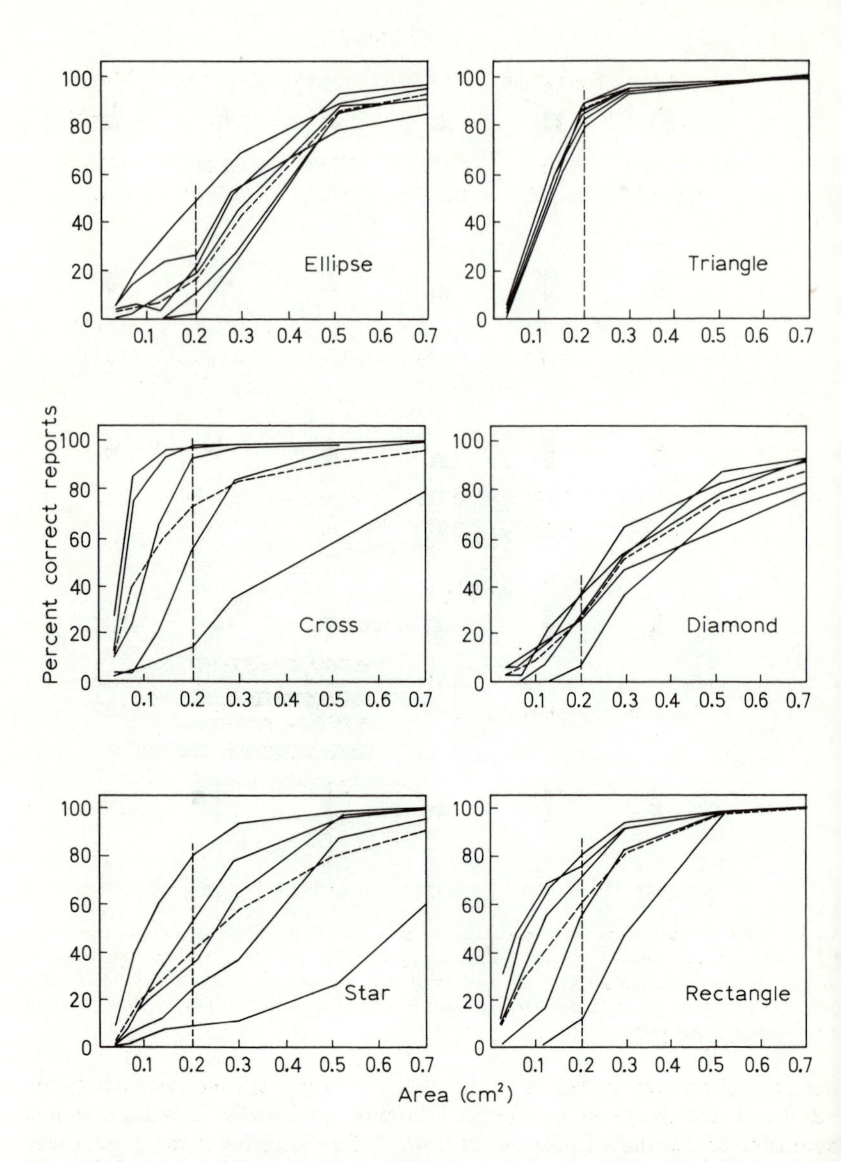

Fig. 5.2. The relationship between area and percent correct reports found by Casperso (each point represents 480 judgements). Solid lines represent the functions for the variou figures within a particular form. The dashed lines represent the mean curve for each forn (Reprinted from Casperson[12] by permission. Copyright (1950) by the American Psych logical Association).

Results of several studies show recognition threshold varying directly with compactness'* (e.g. Bitterman, Krauskopf & Hochberg[9] and Engstrand and Moeller[10]). Others find that, whilst form does affect recognition threshold, neither perimeter, area nor P/A are good predictors (e.g. Fox[11]). Casperson[12], n an attempt to resolve some of the conflict, set up an experiment utilising 30 hapes, five each of various forms of rectangles, ellipses, triangles, diamonds, rosses and stars. His stimuli, all presented at high negative contrast, are eproduced in Fig. 5.1. It will be seen that these stimuli include a disc, a square nd an equilateral triangle as three of the most compact forms. He presented hese stimuli in random order at various sizes to 20 observers, each figure being een by each observer at each presented size 24 times. He then computed the ercentage correct responses for each stimulus as functions of area, perimeter nd maximum dimension. He found no common behaviour for the different orms. For instance, whilst the area was a good predictor of threshold for all riangles it was a very poor predictor of threshold for stars and crosses as shown n Fig. 5.2. On the other hand, the maximum dimension or perimeter were ound to be good predictors for stars but very poor predictors for ellipses and iamonds. There thus seems to be no one simple measure of form which can be sed as a universal predictor. Casperson realised that his results may possibly be bserver dependent, so he carried out an analysis of variance on the results. This howed that, whilst there was a significant observer difference, this was ompletely swamped by the difference between forms.

In an attempt to approximate real life more nearly, whilst still using ssentially geometrical forms, BAC(GW) carried out a recognition experiment sing 4 stylised shapes[13]. These shapes, shown in Fig. 5.3, were all formed from

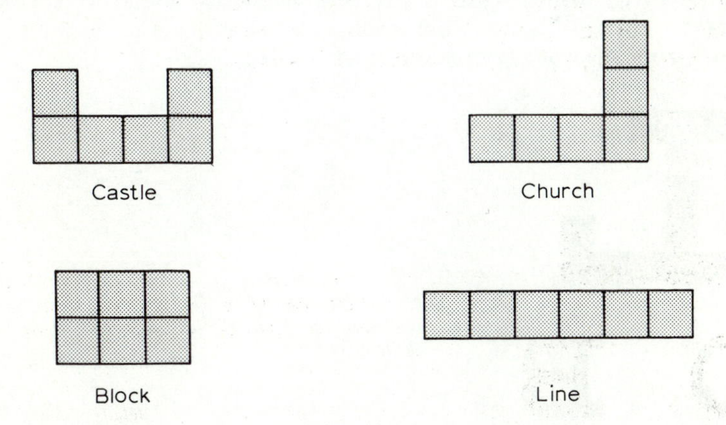

Fig. 5.3. The 4 stylised shapes used in the B.A.C. (GW) recognition experiment.

*Compactness is a term used by Zusne[2] and others to describe forms of low aspect ratio and imple contour – characterised by a low perimeter to area ratio (P/A ratio).

6 unit blocks, and were presented at high negative contrast in random order after observers had been thoroughly briefed. It was found that the recognition thresholds for these forms were such that the unit blocks subtended around 0.3 mrad at the observer's eye. If the subtense of the unit blocks is compared with the detection threshold for a square or circular target of similar contrast it is found to compare closely. Hence such a finding has been taken by some as a tentative confirmation that recognition may be equated to detection of detail for approximate predictive purposes.

5.2 SNELLEN LETTER TESTS

A recognition task familiar to most people because of its widespread use by opticians as an eye test is the Snellen letter test. This, in its normal usage, consists of the reading of a series of high contrast black letters on a white background in rows of decreasing size until an error is made. Hence it is really a legibility test. The letters used obey standard laws of construction[14] (see Fig. 5.4) and it is from them that a person's eyesight is defined in terms of a ratio such as those commonly quoted (6/x or 20/x). In these ratios the numerator refers to the viewing distance in metres (6) or feet (20) respectively and the denominator refers to the distance at which the *detail* of the letters subtends 0.292 mrad. Normal vision is usually taken as 6/6 or 20/20, although some eyes have acuity as high as 6/3. This form of eyesight testing is useful in modern western civilisation, since it effectively defines a person's ability to read standard print. However, as a measure of recognition threshold it leaves something to be desired since it has been shown by several workers (e.g. Lythgoe[15], Ludvigh[16] and Mandlebaum and Sloan[17]) that contrast and luminance both affect ability to interpret letter shapes.

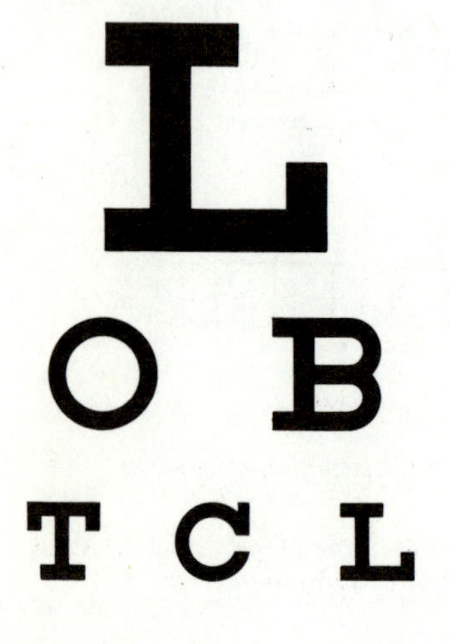

Fig. 5.4. Illustrating the construction of Snellen test letters and charts. The lines and serifs have a thickness one fifth of the height or width of a whole letter.

In an endeavour to make Snellen letter testing more universally applicable to recognition tasks, some workers have attempted experimentation using Snellen forms of letters of various contrasts and at various background luminance levels. Some have also attempted to identify this form of testing more closely with other threshold experimentation by presenting random letters from a limited set in a statistical fashion and by recording mean recognition threshold as that for 50% correct recognition rather than approaching 100% as used by opticians. Typical of work on these lines is that carried out in our own laboratory by Spicer[18,19], who studied the effect of both contrast, field luminance and binocular/monocular viewing on mean recognition thresholds for limited sets of Snellen letters, together with observer interactions. In these experiments Spicer used typical opticians' Snellen displays reproduced at various contrasts. On each line representing visual acuities of 6/9 and better one letter was cut out and a system arranged so that any one of a limited set of letters of the same size could be presented at random. Luminances of 41 and 6 900 cd/m^2 were used in combination with contrasts of – 0.09 and – 0.85. In order to reduce variance, and to limit the set of letters to be presented in these experiments, the results of a study by Coates[20] were invoked. This study implied that some 11 letters of the alphabet were a sufficient set to use for this form of experimentation.

The findings from Spicer's experiments were as follows:

(1) That the contrast of the letter has a major effect on recognition threshold.

(2) That there is an interaction of contrast and scene luminance – at high contrast the lower the scene luminance the lower the recognition threshold whilst at low contrast the lower the scene luminance the higher the recognition threshold.

(3) That the recognition threshold for binocular viewing is only marginally lower than for monocular viewing (of the order of 5 to 10% in linear size). This is in marked contrast to the difference in detection threshold for simple shapes between binocular and monocular viewing which in general is a factor of 40% (i.e. $\sqrt{2}$) in contrast or 20% in size (see Section 2.10).

(4) That there are significant observer differences in relative performance as a function of contrast and luminance.

Having found the foregoing significant effect of contrast and luminance, Spicer proceeded to investigate, for a group of 8 of the observers used in the Snellen tests, the correlation of the various visual acuity scores with performance in various detection tasks[21]. The detection tasks chosen were the detection, from photographs, of high and low contrast aircraft seen against structured (cloud) and unstructured backgrounds, and of 4 miscellaneous ground features (bridges and isolated buildings) as seen from the air. These together were considered to be typical of the more simple forms of realistic detection exercise. Non-parametric rank statistics were used to compare various measures. The only significant correlations obtained for the ground features were for one of two isolated buildings, whilst the only aircraft situation providing high correlation was that of a high contrast aircraft against a structured background, this latter providing high correlation with *all* the visual acuity scores. Some of the other comparisons showed very low correlations and serve to imply that Snellen testing is not a good measure of relative visual performance for many real-life detection tasks (although it is of course possible that it may yet be found to be good measure for real-life recognition performance).

5.3 OTHER LEGIBILITY TESTS

Several other forms of what may be described as legibility tests have been used over the years (e.g. Le Grand[1]). By far the most common of these – and one which has become a standard experimental stimulus for many purposes – is the Landolt 'C'. This stimulus, shown in Fig. 5.5, is really an annulus with thickness to outside diameter ratio of 1:5 from which a slice, of width equal to the annulus, has been removed. As such it provides a very versatile stimulus for form of recognition task, since below recognition threshold it is indistinguishable from an annulus. Thus gap orientation can be used as the recognition task, simplifying what can otherwise be a problem of preparation and presentation of varied stimuli. Alternatively the Landolt 'C' may be included with several annuli providing a well defined search task.

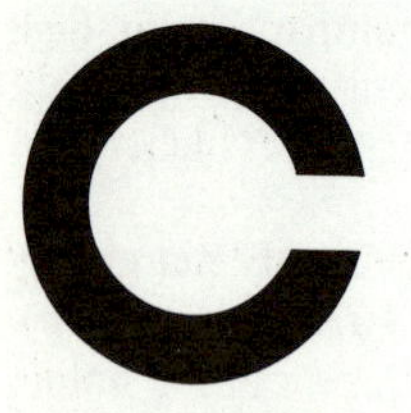

Fig. 5.5. Illustrating the construction of the Landolt C. The line and gap widths are each one fifth of the outer diameter.

As with Snellen letter testing, the majority of work with Landolt C's has been carried out at high negative contrast. Results of such experiments are normally considered to yield a performance in terms of visual acuity, defined as the reciprocal of gap width for recognition in this case. A typical trend of this form of visual acuity with luminance is shown in Fig. 5.6 (from Pirenne and Denton[22]).

Other forms of legibility test are various forms of bar pattern. Of these, some of the more common are the Foucalt chart, the 2-bar Cobb element and the American 3-bar pattern (see Fig. 5.7). When such patterns are used for threshold studies rather than the Landolt C, slightly different trends of performance with variation of luminance are found (e.g. Le Grand[1]). Such patterns are also

*Often called detection, but the term recognition is preferred by the author since the annulus defines a specific local search task.

sensitive to orientation, yielding 7–20% inferior performance if oblique than if horizontal or vertical, and are highly susceptible to astigmatism (e.g. Leibowitz[23]).

Limited studies have been carried out using bar patterns of low contrast for short presentation times and with inverse contrast. As reported for simple

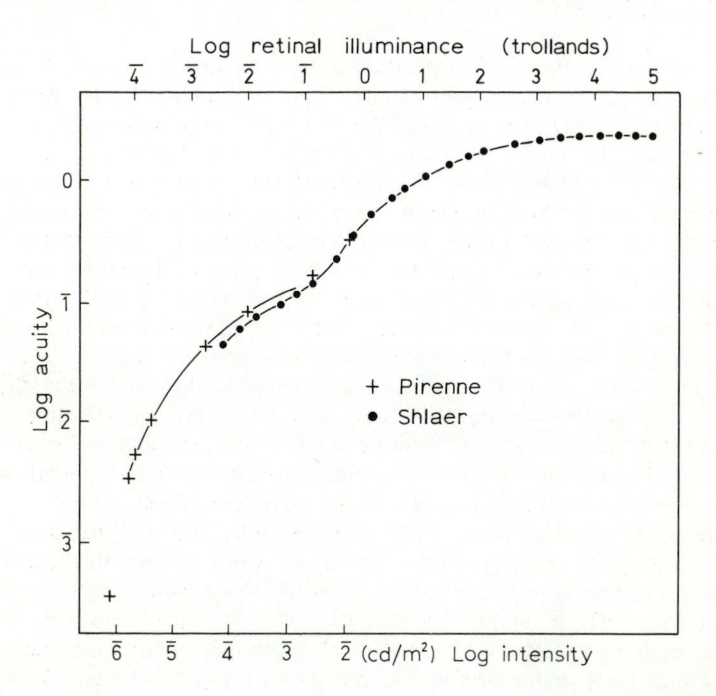

Fig. 5.6. Variation of visual acuity measured with Landolt C's as a function of light intensity. (Reproduced from Pirenne and Denton[22] by courtesy of Nature*).*

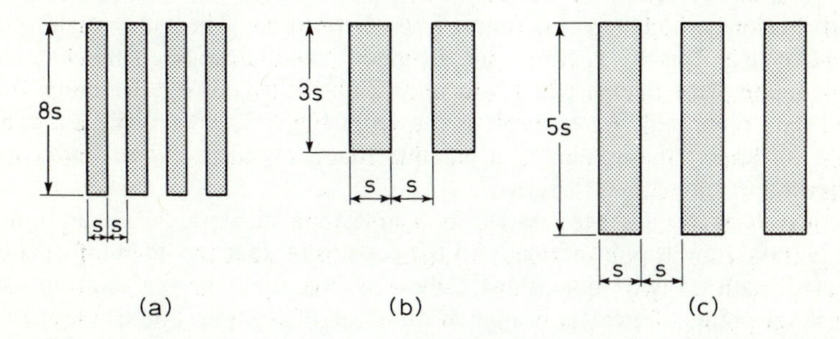

Fig. 5.7. Illustrating the construction of elements of various test patterns. (a) Foucalt, (b) Cobb, (c) American 3-bar.

detection thresholds in Chapter 4, it appears from these studies that gap detection is also considerably dependent on contrast, presentation time and, under certain conditions, sign of contrast[24,25].

5.4 COMPLEX SHAPES

So far our considerations of recognition have been largely confined to choice between a variety of simple geometrical shapes or standard letters. In normal visual activities other than reading, the form of recognition which must be employed may be vastly different to these very specific situations. It thus becomes necessary to consider what controls the recognition of complex shapes. Such complex shapes may be one of a very few or one of a very many, and may have luminance structure within them or may have a luminance interaction with their immediate surround. In order to be able to make predictions about visual recognition performance in such situations it is necessary to study what it is about a stimulus that permits its recognition in a given situation. Although a considerable amount has been written about this problem in recent years (see Zusne[2]) the full understanding still remains somewhat obscure. It is the author's opinion, in common with that of Zusne, that a very considerable amount of work is yet necessary before we shall be in a position to predict performance in other than the simplest cases. In the meantime a number of items of work which have been reported are recommended for the reader's attention.

Attneave[26-28] chose to approach the problem by assuming that possibly it was the local parts of the profile of an object which allowed its recognition. Following this line he produced a series of outline shapes with irregular contours and, allowing each of several observers 10 dots with which to do their best at defining each figure, produced a statistical impression of the 'importance' of various local parts of the profile. Fig. 5.8 shows a typical data set. As can be seen, there appears to be a strong concentration of 'importance' in the regions of maximum rate of change of contour direction, with virtually no importance given to straight regions of the contour. This finding leads on to a range of optical illusions which tend to support the hypothesis. The range of illusions referred to (e.g. Zusne[2], Postman and Brunner[29] and Gregory[30]) are characterised by incomplete figures which are universally 'filled in' by the brain from limited data presented. An example is shown in Fig. 5.9, where only a minimal amount of data at the corners of a possible triangle is sufficient for immediate visualisation of the whole triangle.

Attneave's work has been hailed as a milestone in work on perception of form. Not only did it point strongly to the conclusion that the main information associated with a two dimensional shape is contained in the contour, and primarily at points of greatest change of direction of contour, but also it enabled methods to be proposed for generation of random 2-dimensional shapes of better equivalence than used previously[28].

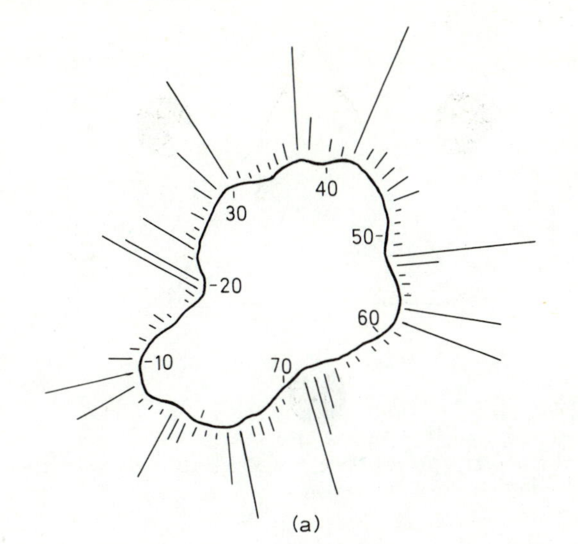

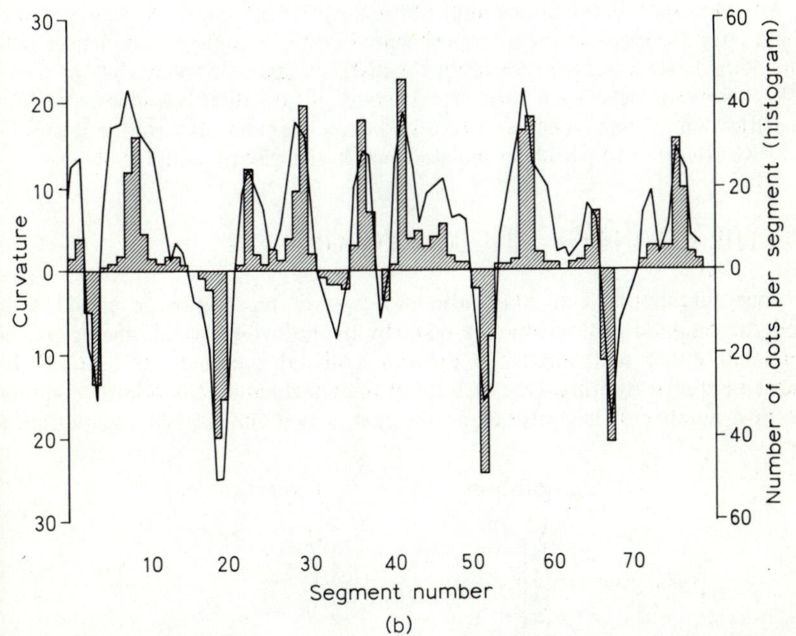

Fig. 5.8. Subjective importance of parts of a contour. (a) Typical contour studied – radial bar lengths are proportional to number of selections of that part of the contour as being important. (b) The number of selections of local importance compared with the local curvature of the contour, showing high correlation. (Reproduced from Attneave F. (1951) Research Note P & MS: 51–8 by courtesy of the Human Resources Center, San Antonio).

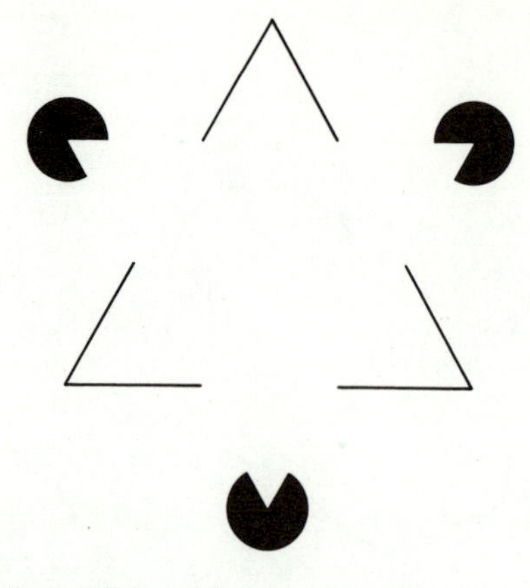

Fig. 5.9. A typical illusionary figure. There is a strong subjective impression of a solid triangular figure although no such figure exists.

An important point concerning the recognition of complex shapes is that found by Goodnow[31]. He experimented with aeroplane silhouettes and schemetised faces, in each case varying 3 characteristics. He found that, in such a task, observers only ever used one or two of the three variables to effect identification. Zusne[2] claims that other experimenters also find evidence of selective attention to particular physical cues in the case of complex objects.

5.5 THE RETINAL IMAGE AT RECOGNITION

A powerful approach to the basic processes of recognition is provided by modern computational techniques, whereby the retinal images of objects may be generated either mathematically through a digital computer or optically by means of spatial filtering. One or both of these techniques are in fairly common use in a number of laboratories. In the first, a two dimensional convolution is

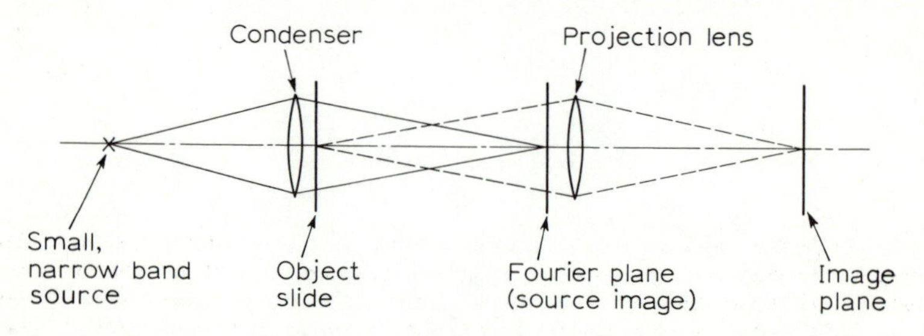

Fig. 5.10. The basic optics necessary for Fourier plane filtering.

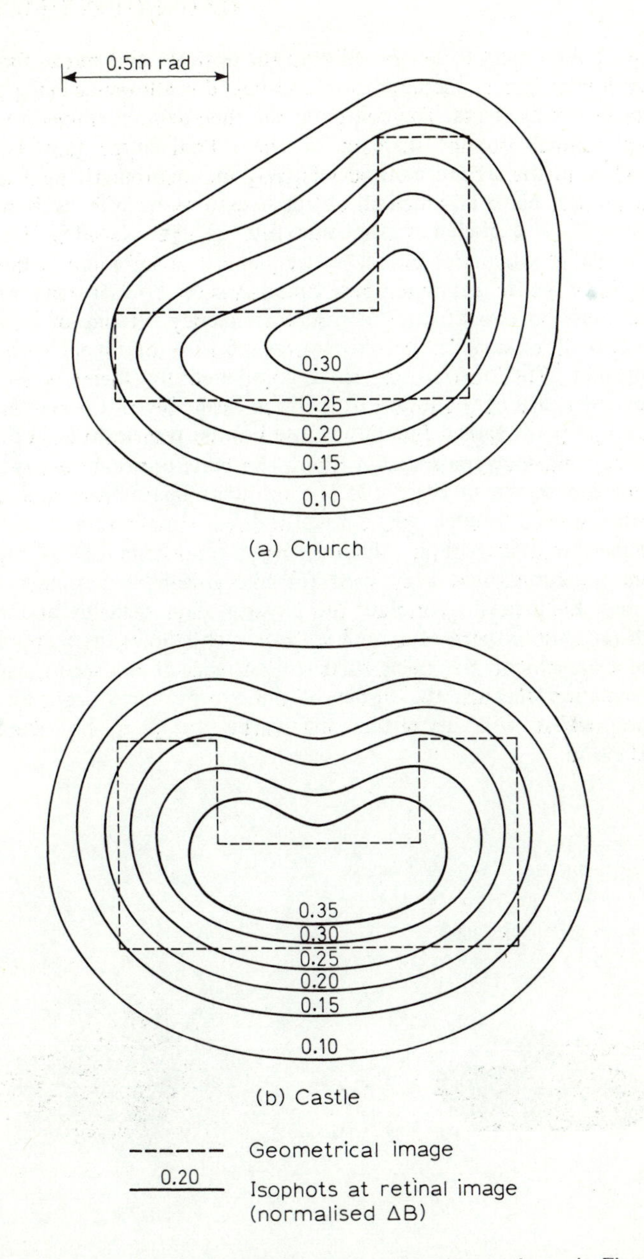

Fig. 5.11. Retinal image isophots for two of the stylised shapes shown in Figure 5.3 when viewed at an angular subtense resulting in a 50% probability of recognition (free choice) from the set of four.

carried out of the object to be viewed with the optical point spread function of the average human eye at the appropriate scaling, due allowance being made for involuntary eye movements. The computer can then be programmed to produce a two dimensional isophot diagram of the retinal image (e.g. Lavin and Overington[32]). In the second method relatively monochromatic light is used to generate a Fourier plane in which all object structures are effectively presented as frequencies[33]. If a graded transmission filter is then placed in this Fourier plane, such that it attenuates the higher frequencies, a softening of the picture will result. Since the frequency response function is the Fourier transform of the line spread function (see Chapter 10), such frequency filtering of a controlled nature in two dimensions is akin to convolution of the object with a point spread function. The optical system is completed by fitting a lens which reconstructs an image of the object as if it had been viewed through an optical system having a point spread function whose Fourier transform was of the form built into the frequency transmission filter. The basic optics necessary for such an operation are shown in Fig. 5.10. The resultant image from such a system may be either viewed through an eyepiece for direct visualisation or recorded on film for subsequent inspection. Much enlarged representations of the retinal image thus presented have been used for 'recognition' experiments so that attempts may be made to correlate the physical data available at the retinal image with recognition performance[34,35]. For such studies there appears to be an optimal enlargement, this being sufficient to mean that a second passage of image information through the optics of the eye produces negligible further degradation, whilst still permitting the entire image to be observed by near-foveal vision.

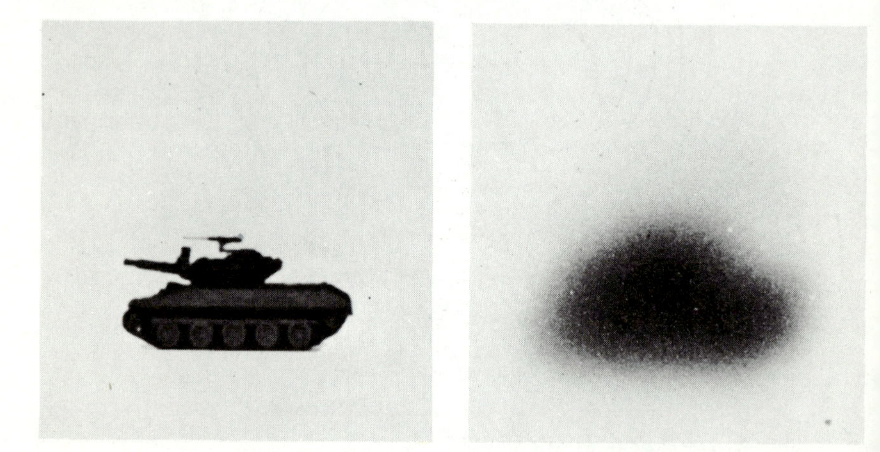

Fig. 5.12. A side view of a tank and an optically-generated representation of the retinal image when viewed to subtend an angle permitting 50% probability of recognition from a set of 30 military vehicles and confusible objects by trained personnel.

The resulting pseudo-retinal images of objects at recognition range by either computer convolution or optical degradation are striking for their lack of sharpness and detail. Figure 5.11 shows isophot plots of the retinal images of the stylised stimuli labelled *castle* and *church*, as shown in Fig. 5.3 and described in Section 5.1, when at 50% probability of correct recognition from a set of six similar shapes. They will be seen to have retained only minimal shape. Figure 5.12, on the other hand, shows the side view of a tank and the typical 'retinal image' of the tank at 50% recognition threshold for military observers, as produced by optical spatial filtering[33]. Again it will be seen to have retained only minimal shape and structure. The implications of the foregoing on attempted modelling of the visual processes will be discussed in Chapter 12.

5.6 EFFECTS OF ORIENTATION AND LEARNING

Experience, expectancy and familiarity are very important factors in recognition. Forms normally seen only in one position (e.g. an outline of a country on a map) are difficult to recognise when rotated, even though one may be very familiar with them. Such mono-orientated objects are hardest to recognise when inverted. Everyday objects which we are accustomed to seeing in several positions (e.g. pencils, keys, books) are much less sensitive to orientation[36].

'Normal' orientation may be defined in either of two ways – in the environment or on the retina. This raises a question of how recognition is affected when the form is tilted but the head remains upright, when the head is tilted and the form remains upright or when both are tilted together. The consequences of the two definitions of orientation were first studied by Oetjen[37], who found recognition performance was only good when the orientation of a form on the retina was maintained. Thouless[38] obtained results using ambiguous figures which tended to confirm these findings. On the other hand Rock[39] obtained results which suggested that the direction in the environment is most important and that retinal orientation is of little importance. Yet again, Braine[40] found that establishing a frame of reference on the retina by stating exactly how forms would appear did not help when used in addition to simple instructions that disorientated forms would appear.

If a form was learnt in one part of a visual field, Dees & Grindley[41] found that this was an assistance when the form was presented in a different part of the visual field. Conversely Lordahl *et al*[42] appeared to find that such learning had an adverse effect.

5.7 FORM FIELDS

So far in this chapter, where thresholds have been considered, it has been mostly assumed that foveal fixation was employed for interrogation. Obviously there are occasions when it is desirable to know the extent of the visual field within which particular recognition can be accomplished and, as with detection, performance *must* be expected to fall off with increasing angle of eccentricity. According to Anstis[43], the reciprocal of visual acuity for Snellen letters increases linearly with distance from the fovea out to an angle of eccentricity of about 0.5 rad. Similar findings for other acuity measures are reported by Le Grand[1]. Day[40] proposed a theory for foveal acquisition which postulated that at area 17 of the striate cortex, usually considered to be the first stage of central processing[45], there will be an activity about the boundary of figure and background which must reach one level for detection, a higher level (some 15 times that for detection) for indefinite form perception and a higher level still for definite form perception. He then applied this concept to peripheral form perception[46] and made various predictions as to what would be seen as various objects were brought progressively towards the fovea from the extreme periphery (e.g. white spaces enclosed in a figure would be seen first as filled in, L and T shapes would be seen first as roughly triangular, cross shapes would be seen first as a hazy blur and then like a diamond). All these predictions were then confirmed in an unpublished experiment. Graefe[47] agrees in general with Day's theory but considers it is incomplete. He suggests for instance, that, whilst some corners may be seen blurred, others may be seen sharpened. Equally, unjoined lines may be seen joined up and *vice-versa.*

The extent of form fields is, in general, dependent on a variety of factors including stimulus size, stimulus contrast, field luminance, contrast/area interaction, etc.. The only consistent finding regarding the size of form fields is that, for geometric shapes, the triangle is consistently seen best at the greatest eccentricities. Equally there seems good agreement that, within geometric shapes, the hexagon and octagon have the smallest form fields[48]. Of particular practical importance is the recent work of Bouma[49], who has studied the legibility of strings of letters presented in the parafovea. He has studied both random letters and words and has found that, in this complex recognition task, scores for end letters farthest from the fixation point are superior to scores for end letters closest to the fixation point. Thus in a complex field there appears to be a strong interaction between local visual acuity and data complexity.

5.8 PROFILE RECOGNITION

Another aspect of recognition which may form part of the total human visual capability is the ability to recognise the form of the profile – that is, whether it is sharp or of a given blur form. Campbell[50] has provided a useful insight into this in the extensions of his Contrast Sensitivity detection threshold work reported in Chapter 4. He has studied the threshold levels for recognition that a one dimensional bar pattern has other than a sinusoidal spatial luminance modulation. Work carried out using square, rectangular and sawtooth modulations has produced strong evidence that the threshold for recognition of profile form of periodic patterns by an experienced observer is governed by the threshold of the harmonic whose presence is necessary to define the waveform. For instance, for recognition of square from square or sinusoidal it is necessary to reach the detection threshold for the *3rd* harmonic. Conversely, for recognition of a sawtooth from sawtooth or sinusoidal it is necessary to reach the detection threshold for the *2nd* harmonic. Campbell has interpreted these findings as evidence of the existence, in the neural system, of effectively tuned circuits responding to each of several narrow bands of spatial frequency. Other studies related to this idea of neural tuned circuits are discussed in Chapter 12.

5.9 DYNAMIC EFFECTS

If one is attempting to recognise moving objects, then it may be that the object in question is large and appears unexpectedly for a short time, or it may be that the object is moving rapidly but in a relatively known fashion. In the former of these cases the recognition decision must be made on interpretation of a smeared retinal image. Such is, for instance, the situation when a threshold experiment is carried out with moving Landolt C's. Since the recognition (or gap detection) in this case consists of the recognition of a smeared gap it should be fairly clear to the reader that ambiguities can creep in due to the orientation of the gap. For horizontal motion, if the gap is at North or South one would get a different answer than if the gap is at East or West. In addition it is found[51] that there is an ambiguity between gap at East and West positions. This latter effect is believed to be due to after-imagery.

In the case of study of objects moving predictably, or in a known part of the visual field, because the basic stimulus for recognition is always well above detection threshold, it is usually possible for the eyes to 'lock on' to the moving stimulus, thus converting the task to one approximating to a static recognition task. This form of viewing – known as ocular pursuit – will result in performance degraded from the static to a lesser or greater degree according to the predictability of the stimulus motion and its rate of angular motion. Some statistical studies of eye motions in attempting ocular pursuit tasks have recently been published by Cheng & Outerbridge[52]. For relatively low frequency sinusoidal motions it has been shown possible to achieve perfect ocular pursuit with training. Under such conditions the performance would be expected to be the same as for static viewing. As the frequency or amplitude of sinusoidal motion is increased, there comes a time when the eyes fail to follow adequately. Under such conditions the performance may be expected to approximate to the normal dynamic visual acuity for the residual motion. Finally, at very high angular rates the eye fails to track at all and the performance falls to that associated with the unpredictable motion situation.

The most systematic studies of degradation of foveal visual acuity during ocular pursuit appear to have been carried out by Ludvigh and Miller. In his early work Ludvigh found, using Snellen letters as test objects, that when motion is in a horizontal plane, visual acuity falls markedly as the angular velocity is increased from 0 to 2.2 rad/s[53]. Subsequently[54] he extended his studies to include velocities up to 3.5 rad/s. During his studies he noted that high intensities of illumination particularly improve visual acuity when the test object is moving. In later work[55-57] Ludvigh and Miller found that their experimental results could be fitted by an equation of the form

$$1/u = 1/u_s + \omega^3/u_d \tag{5.1}$$

where u is the visual acuity and ω is the angular velocity.

The quantity u_s in Equation 5.1 is thus an estimated value of the static acuity whilst u_d is an effective measure of dynamic acuity, being large when acuity does not deteriorate rapidly as anglular velocity is increased. The value of u_d is found to be very dependent on particular test conditions and observers but a very high correlation for given conditions and observers was found by Miller and Ludvigh[58] between the u_d values obtained for horizontal and vertical motions. Typical forms of the degradation of performance with increasing angular velocity are shown in Fig. 5.13. In this figure are shown the cumulative data of Ludvigh and Miller and the cumulative data of Rose[59] fitted by functions of the form of Equation 5.1. The reason for the striking difference in rate of deterioration is that Ludvigh and Miller's data represent fall-off of performance

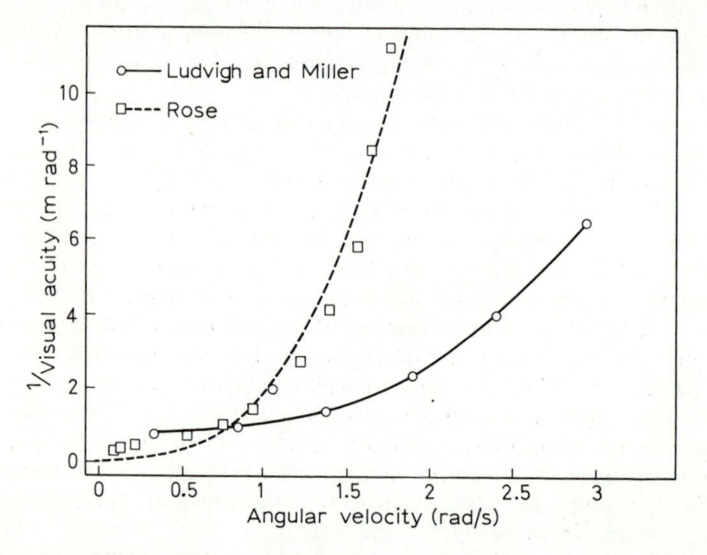

Fig. 5.13. Typical forms of degradation of visual acuity with increasing velocity. (Reprinted from Miller and Ludvigh[57] by permission. Copyright (1962) by the Williams & Wilkins Co., Baltimore).

with constant exposure time whilst Rose's data represent fall-off with a constant angular presentation, the exposure time thus being reciprocally related to the velocity. A useful survey of the above and other work on dynamic visual acuity is to be found in Lavin[60].

Recent work by Lavin[51] has suggested that the cubic fit to data obtained with a constant angular presentation is only adequate over a limited range of velocities, it being contended that at high velocities there is a need to consider a limiting velocity above which there is no chance of seeing anything. As an approximation Lavin has found it possible to obtain a very good fit to his own collected data for a group of observers on a log/linear plot, despite very considerable observer differences, by an equation of the form

$$1/u = y_0 \exp(1.44v) + y_1 \exp(18.3v) \tag{5.2}$$

where u is the visual acuity, v is the limiting velocity in rad/s and y_0 and y_1 are constants peculiar to individual observers.

This collected fit is shown in Fig. 5.14. In this figure the data have been grouped by taking the data for one observer and then overlaying data for all

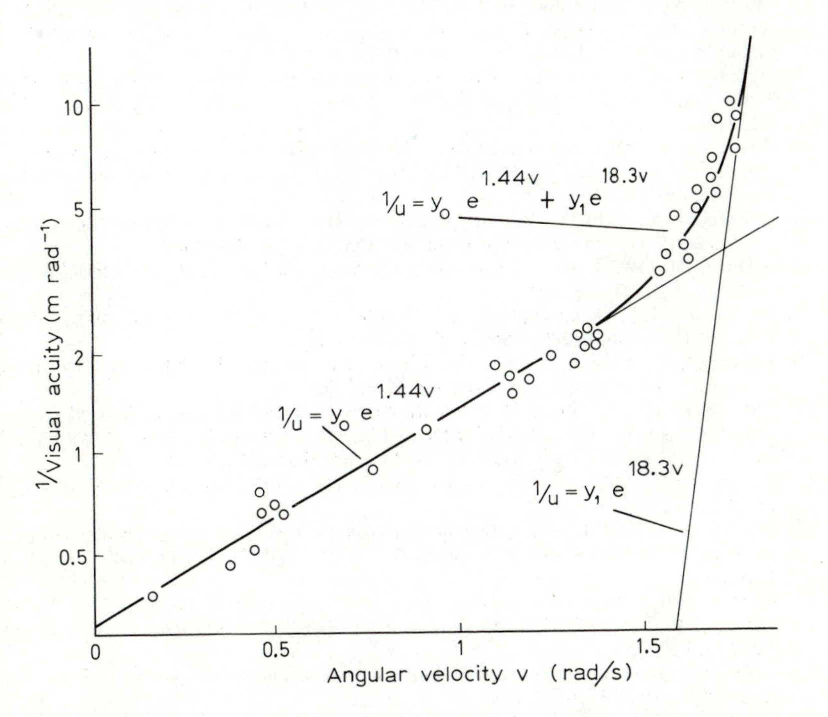

Fig. 5.14. The collected results of the foveal dynamic threshold study of Lavin (from Lavin[51]).

other observers by means of simple orthogonal shifting of the size and velocity axes.

5.10 DISCUSSION

In this chapter many facets of the problem of recognition have been referred to. However, it will be clear to the reader that the present situation is not a very tidy one. As Zusne says 'It would appear that in thirty years of studying form identification we have learnt little that can be stated with any certainty'. Nevertheless, it is hoped that the information cited will at least lead the reader to a fuller appreciation of the problems of this facet of acquisition.

REFERENCES

1. Le Grand, Y. (1967). *Form and Space Vision.* (Chapter 5), (Translation by Millodot, M. and Heath, G. G.) Indiana University Press
2. Zusne, L. (1970). *Visual Perception of Form*, Academic Press
3. Marshall, W. H. and Talbot, S. A. (1942). 'Recent Evidence for Neural Mechanisms in Vision leading to a General Theory of Sensory Acuity' in *Biological Symposia, Vol. 7: Visual Mechanisms*, 117, (Ed. H. Kluever), Ronald Press, New York
4. Byram, G. M. (1944). 'The Physical and Photochemical Basis of Visual Resolving Power: Pt. II, Visual Acuity and the Photochemistry of the Retina', *J. Opt. Soc. Am.*, **34**, 718
5. Gibson, J. J. (1950). *The Perception of the Visual World*, Houghton Mifflin: Boston
6. Helson, H. and Fehrer, E. V. (1932). 'The Role of Form in Perception', *Amer. J. Psychol.*, **44**, 79
7. Achurch, I. C. (1963), 'Initial Analysis of Data from the Contrast Recognition Experiment', Assessment Group Memo. No. 154, B.A.C.(GW) Bristol
8. Salaman, M. (1929). 'Some Experiments on Peripheral Vision', *M.R.C. Special Report No. 136.* H.M.S.O., London
9. Bitterman, M. E., Krauskopf, J. and Hochberg, J. E. (1954). 'Threshold for Visual Form: A Diffusion Model', *Amer. J. Psychol.*, **67**, 205
10. Engstrand, R. D. and Moeller, G. (1962). 'The Relative Legibility of Ten Simple Geometric Figures', *American Psychologist*, **17**, 386
11. Fox, W. R. (1957). 'Visual Discrimination as a Function of Stimulus Size, Shape and Edge Gradient' in *Form Discrimination as Related to Military Problems*, (J. W. Wulfech and J. H. Taylor, Eds.) *Nat. Acad. Sci. N.R.C.*, Washington D.C., 168
12. Casperson, R. C. (1950). 'The Visual Discrimination of Geometric Forms', *J. Exptl. Psychol.*, **40**, 668
13. Brown, M. B. (1972). 'The Effect of Complex Backgrounds on Acquisition Performance' in *AGARD Conference Proceedings No. 100*, (Ed. H. F. Huddleston) p. B5–1, London
14. Emsley, H. H. (1963). *Visual Optics*, Vol. 1, p. 63, Hatton Press, London
15. Lythgoe, R. J. (1932). 'The Measurement of Visual Acuity', *MRC. Special Report No. 173*, H.M.S.O., London
16. Ludvigh, E. (1941). 'Extra-foveal Visual Acuity as measured with Snellen Test Letters', *Am. J. Ophthal.*, **24**, 303
17. Mandlebaum, J. and Sloan, L. L. (1947). 'Peripheral Visual Acuity with Special Reference to Scotopic Illumination' *Am. J. Ophthalmol.*, **30**, 581

18. Spicer, P. J. (1972). 'The Measurement of Visual Acuity', App. 3, Sect. 1 of *Research into Factors affecting the Detection of Aircraft through Optical Sights*, (B.A.C.(GW) Ref. L50/186/1449)
19. Spicer, P. J. and Ensell, F. J. (1973). 'Comparison of Visual Acuity Tests and Viewing Condition Interactions', *Aerospace Medicine*, November, 1290
20. Coates, W. R. (1935). 'Visual Acuity and Test Letters', *Trans. Inst. Ophthal. Opt.*, September
21. Spicer, P. J. (1972). 'The Significance of Visual Acuity Measurements', App. 3, Sect. 2 of *Research into Factors affecting the Detection of Aircraft through Optical Sights*, (B.A.C.(GW) Ref. L50/186/1449)
22. Pirenne, M. H. and Denton, E. J. (1952). 'Accuracy and Sensitivity of the Human Eye', *Nature,* **170,** 1039
23. Leibowitz, H. (1953). 'Some Observations and Theory on the Variation of Visual Acuity with the Orientation of the Test Object', *J. Opt. Soc. Am.,* **43,** 902
24. Cobb, P. W. and Moss, F. K. (1928). 'The Four Variables of Visual Threshold', *J. Frank. Inst.,* **205,** 831
25. Wilcox, W. W. (1932). 'The Basis of Dependence of Visual Acuity on Illumination', *Proc. Nat. Acad. Sci.,* **18,** 47
26. Attneave, F. (1954). 'Some Information Aspects of Visual Perception', *Psychol. Review,* **61,** 183
27. Attneave, F. (1955). 'Perception of Place in a Circular Field', *Am. J. Psychol.,* **68,** 69
28. Attneave, F. and Arnoult, M. D. (1956). 'The Quantitative Study of Shape and Pattern Perception', *Psychol. Bull.,* **53,** 452
29. Postman, L. and Bruner, J. S. (1952). 'Hypothesis and the Principle of Closure: the Effect of Frequency and Recency', *J. Psychol.,* **33,** 113
30. Gregory, R. L. (1974). *Concepts and Mechanisms of Perception*, Duckworth, London
31. Goodnow, R. E. (1954). 'The Utilization of Partially Valid Cues in Perceptual Identification', Upublished Doctoral Dissertation, Harvard University
32. Lavin, E. P. and Overington, I. (1972). 'Visual Modelling', Annex E of *Final Report on the Third Visual Studies Contract*, (B.A.C.(GW) Ref. L50/196/1535). Sect. 5.4
33. Hobson, R. D. (1973). 'The Recognition of Military Vehicles. The Preparation of Blurred Material', B.A.C.(GW) Study Note ST9219
34. Clare, J. N. (1973). 'Recognition of Military Vehicles. Report of First Study', B.A.C.(GW) Study Note ST8611
35. Seale, S. J. (1973). 'Recognition of Military Vehicles. Report of Second Study', B.A.C.(GW) Study Note ST9325. June
36. Gibson, J. J. and Robinson, D. (1935). 'Orientation in Visual Perception: the Recognition of Familiar Plane Forms in Differing Orientations', *Psychol. Monog.,* **46,** 210
37. Oetjen, F. (1915). 'The Importance of the Orientation of Reading Materials to the Reader and of the Orientation of Random Shapes for them to be Recognised as the Same', (in German), *Z. Psychol.,* **71,** 321
38. Thouless, R. (1947). 'The Experience of 'Upright' and 'Upside-down' in looking at Pictures', *Miscell. Psychol. Albert Michotte,* 130
39. Rock, I. (1956). 'The Orientation of Forms on the Retina and in the Environment', *American J. Psychol.,* **69,** 513
40. Braine, L. G. (1965). 'Disorientation of Forms: and Examination of Rock's Theory', *Psychon. Sci.,* **3,** 541
41. Dees, V. and Grindley, G. (1947). 'The Transposition of Visual Patterns', *Brit. J. Psychol.,* **37,** 152
42. Lordahl, D. S., Kleinman, K. M., Levy, B., Massoth, M. A., Pessin, M. S., Storandt, M., Tucker, R. and Plas, J. M. van der, (1965). 'Deficits in Recognition of Random Shapes with Changed Visual Fields', *Psychon. Sci.,* **3,** 245

43. Anstis, S. M. (1974). 'A Chart demonstrating variations in Acuity with Retina Position', *Vision Research,* **14**, 589
44. Day, R. H. (1956). 'Application of the Statistical Theory to Form Perception', *Psychol Review,* **63**, 139
45. Zusne, L. (1970). *Visual Perception of Form,* Chap. 2, Academic Press
46. Day, R. H. (1957). 'The Physiological Basis of Form Perception in the Periphera Retina', *Psychol. Review,* **64**, 38
47. Graefe, O. (1964). 'Qualitative Studies about Contour and Flatness in Optica Perception', (in German), *Psychol. Forsch.,* **27**, 260
48. Collier, R. M. (1931). 'An Experimental Study of Form Perception in Indirect Vision' *J. Comp. Psychol.,* **11**, 281
49. Bouma, H. (1973). 'Visual Interference in the Parafoveal Recognition of Initial and Final Letters of Words', *Vision Research,* **13**, 767
50. Campbell, F. W. and Robson, J. G. (1968). 'Application of Fourier Analysis to the Visibility of Gratings', *J. Physiol.,* **197**, 551
51. Lavin, E. P. (1972). 'The Measurement of Dynamic Visual Acuity', App. 3, Sect. 4 of *Research into Factors affecting Detection of Aircraft through Optical Sights,* (BAC(GW) Ref. L50/186/1449)
52. Cheng, M. and Outerbridge, J. S. (1974). 'Inter-saccadic Interval Analysis of Optokinetic Nystagmus', *Vision Research,* **14**, 1053
53. Ludvigh, E. (1948). 'Visual Acuity while viewing a Moving Object', *Arch. Ophthal.,* **42**, 14
54. Ludvigh, E. (1952). 'Control of Ocular Movements & Visual Interpretation of Environment', *Arch. Ophthal.,* **48**, 442
55. Ludvigh, E. and Miller, J. W. (1958). 'Study of Visual Acuity during the Ocular Pursuit of Moving Test Objects. I. Introduction', *J. Opt. Soc. Am.,* **48**, 799
56. Miller, J. W. (1958). 'Study of the Visual Acuity during the Ocular Pursuit of Moving Test Objects. II. Effects of Direction of Movement, Relative Movement and Illumination. *J. Opt. Soc. Am.,* **48**, 803
57. Miller, J. W. and Ludvigh, E. (1962). 'The Effect of Relative Motion on Visual Acuity', *Survey Ophthal.,* **7**, 83
58. Miller, J. W. and Ludvigh, E. (1953). Kresge Eye Institute Project No. NM001 – 110 – 501, Rep. No. 2
59. Rose, A., (1952). *Proc. Armed Forces N.R.C. Vision Committee,* Washington D.C., 77
60. Lavin, E. P. (1972). 'A Literature Survey on Retinal Image Motions', App. 3. Sect. 3 of *Research into Factors affecting Detection of Aircraft through Optical Sights,* (B.A.C.(GW)Ref. L50/186/1449).

6 Modelling of Vision, 1: A Historical Survey

Having discussed the experimentally evaluated threshold performance of human vision in considerable depth it is time to consider how to predict human visual performance. Now, as was seen in Chapter 2, the human visual system is very complex. Hence at first sight it seems unlikely that any simple modelling will allow predictions of performance over other than a very restricted range of conditions. This complexity has not deterred many researchers from attempting to provide reliable modelling of at least some facets of vision. Such modelling has, until recently, inevitably had to be to a large extent empirical, since it is only recently that some of the physical parameters associated with the eye (such as eye movements and the optical spread function of the refraction optics) have been successfully determined. This chapter is devoted to a historical survey of the progressive development of empirical and semi-empirical models of parts of the spectrum of visual performance and a short description of the more complete of these models which are in common use today.

6.1 SIMPLE LAWS GOVERNING LIMITED REGIONS OF VISUAL PERFORMANCE

6.1.1 Ricco's Law

One of the first laws postulated to describe limited aspects of visual performance was that attributed to Ricco[1], which describes the relationship between threshold contrast and object size for small objects. This law simply states that, for small objects,

$$\epsilon \times \alpha^2 = \text{Constant} \tag{6.1}$$

where ϵ is the psychometric contrast and α is the angular subtense of an assumed circular or square stimulus.

For very small (unresolved) objects such a law must obviously be true, since under such conditions the retinal image size is constant (limited by the point spread function of the refraction optics, eye movements and retinal diffusion). Also, for a given flux received, the retinal illuminance distribution is constant and hence the retinal stimulus is constant. If one inspects any typical threshold curve for isolated circular objects viewed at constant luminance – plotted conventionally as log area against log contrast – the Ricco's law region should be represented by a straight line of unit negative slope. That there is such a region can be seen by inspection of any of the size/contrast threshold curves in Blackwell[2], Blackwell and McCready[3] or Taylor[4], for example, or of those

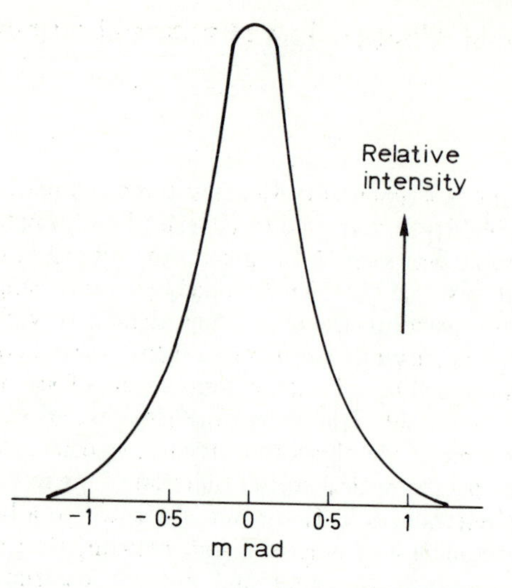

Fig. 6.1. Typical spread function of the naturally accommodated eye (including tremor, drift and retinal diffusion).

collected in Fig. 4.1 of this book. However, considering the typical magnitude of the composite spread function of the eye as shown in Fig. 6.1, it is clear that this region of stimulus constancy is only *guaranteed* over stimulus sizes up to perhaps 1 mrad diameter. For sizes larger than this the continued applicability of Ricco's law must depend on the area integration properties of the retina, and it will be seen from Fig. 4.1 that in most cases the law ceases to hold for stimulus diameters greater than 2 or 3 mrad.

6.1.2 Piper's Law

Looking again at any of the size/contrast threshold curves referred to in Section 6.1.1 it will be noticed that, for sizes above those covered by Ricco's law, the curves progressively steepen, many of them eventually becoming asymptotic to a line of constant log contrast. In attempting to describe mathematically the region between the constant slope covered by Ricco's law and the constant log contrast region, Piper[5] found that an acceptable approximation was to assume

$$\epsilon \times \alpha = \text{constant} \tag{6.2}$$

This has become known as Piper's law.

It is thus possible to produce an adequate approximate fit to any size/contrast threshold curve for many purposes by splitting the curve into three components

$$\epsilon = \text{constant for large sizes}$$

$$\epsilon \times \alpha = \text{constant for intermediate sizes}$$

$$\epsilon \times \alpha^2 = \text{constant for small sizes}$$

6.1.3 **Bloch's Law** (otherwise known as the Bunsen-Roscoe law)

If we transfer our attentions from the domain of size and contrast to that of presentation time and contrast of flash stimuli we find that, for very short presentation times, the threshold is predicted by the law

$$\Delta B \times t_D = \text{constant} \tag{6.3}$$

where ΔB is the difference luminance between the flash stimulus and the background field and t_D is the presentation time.

This is known as Bloch's law, or the Bunsen-Roscoe law, (e.g. Velden[6]) and is the same as saying

$$\epsilon \times t_D = \text{constant} \tag{6.4}$$

since

$$\frac{\Delta B}{B'} = \epsilon$$

where B' is the background luminance.

Once again, as with Ricco's law, this is a law which is to be expected for very short presentation times, since again it is a statement that, under such conditions, total energy received is what matters. The law starts to break down with presentation times much longer than 10 ms (e.g. Clark and Blackwell[7]), a fact which should provide some idea of the effective integration time of the visual system.

6.1.4 The Blondel-Rey Theory

In addition to the Bloch's law region it was realised a long time ago that, for very long presentation times (many seconds), thresholds appeared to become independent of time of presentation. What was required was a functional relationship for the region covering presentation times between 10 ms and several seconds. In 1911 Blondel and Rey proposed that this region should be best fitted by a smooth curve obeying the law

$$I_e = \frac{I \times t_D}{c + t_D} \tag{6.5}$$

where I is the intensity of the flash, I_e is the threshold intensity for long presentation times and c is a constant which varies considerably with experimental conditions but is often taken to be about 0.2 s[8,9].

Such a formula only strictly applies to square pulses, and Blondel and Rey subsequently produced an integral form of Equation 6.5 to apply to any pulse shape – viz.

$$I_e = \frac{\int_{t_2}^{t_1} I \cdot \mathrm{d}t}{0.21 + (t_2 - t_1)} \tag{6.6}$$

where t_1 and t_2 are the start and finish of the pulse respectively.

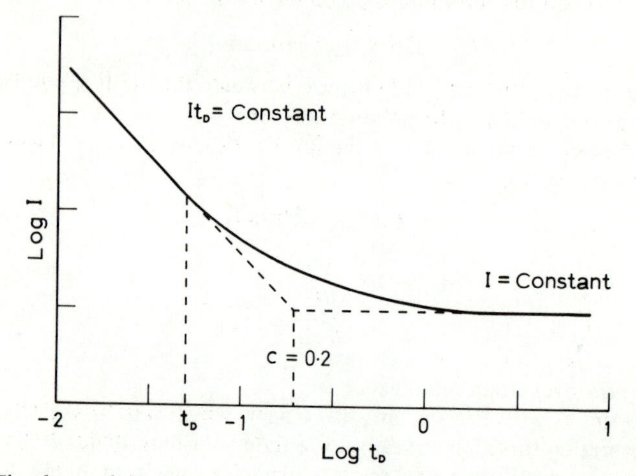

Fig. 6.2. The shape of the exposure time versus intensity function as predicted by Blondel and Rey. t_D indicates the point at which the curve departs from Bloch's law – called the critical duration.

The complete temporal domain for various presentation times of a given size of stimulus as predicted by Blondel and Rey is as shown in Fig. 6.2. Although a good approximation, that this simple form of temporal law is not the full story is amply illustrated by the work of Clark and Blackwell[7] and in a valuable survey paper by Taylor[10]. Alternative explanations more in keeping with Clark and Blackwell's findings, are discussed in Section 6.5 and in Chapter 7.

6.2 THE QUANTUM APPROACH

So far in the discussion of empirical modelling of human vision we have only considered the prediction of very specific sets of threshold data. In order to go further it is necessary to set up some hypothesis of the visual processes operating in a detection situation. The first major progress in this direction started with uggestions made in 1933 by Barnes and Czerny[11] that the statistical luctuations in the arrival of photons may limit the contrast perception of the ye. Several more years were to elapse before other workers expanded on this opic[12-15]. Subsequently Rose developed the topic into perhaps the best known ull statistical approach[16]. In this approach Rose claims that photographic film, elevision pick-up tubes and the human eye are all known to require pproximately the same number of incident quanta for generation of the 'visual ct' – that is, rendering of grain developable in the case of film, release of hoto-electrons in the case of pick-up tubes and production of visual sensation n the case of the eye. The approximate number he puts in the neighbourhood of 00. His subsequent aim in the paper is to determine the limits imposed by luctuation theory, it being then claimed that any other mechanisms operating an only serve to reduce performance.

An ideal pick-up device is defined as one where performance is limited by andom fluctuations in absorption of light quanta in the primary photo process of the device. Then, if the average absorption under given conditions is N_q, this vill be associated with a root mean square deviation of $N_q^{1/2}$. This relationship rises from an assumed Poissonian distribution for arrival and absorption of quanta (e.g. Hecht *et al*[12]).

Accepting the above distribution law as a starting point, Rose defined the smallest perceptible change as

$$\Delta N_q = \kappa N_q^{1/2} \tag{6.7}$$

where κ is an experimentally determined threshold signal/noise ratio. Rose observed that κ had often previously been assumed to be unity, but his own experimental findings suggested that its value should be taken as approximately 5.*

Now the scene luminance B' may be stated as

$$B' \propto \frac{N_q}{X^2} \tag{6.8}$$

where N is the number of quanta received from a scene element and X is the length of the side of the element. Also

$$\frac{\Delta N_q}{N_q} = \frac{\Delta B}{B} = \epsilon,$$

which, from Equation 6.7, yields

$$N_q \propto \frac{1}{\epsilon^2}.$$

*Other workers have assumed other threshold values between unity and 5.

Thus

$$B' \propto \frac{1}{\epsilon^2} X^2 .$$

But $X \propto \alpha$, where α is the angle subtended by element X at the observer. Hence

$$B' = \text{constant} \quad \frac{1}{\epsilon^2 \alpha^2} \qquad (6.9$$

where the constant includes factors such as effective storage time (or integratio time) of the eye, quantum efficiency and various parameters. It is pointed ou that, whilst there is no limit to the applicability of Equation 6.9 for small size and low contrasts in terms of fluctuation theory, in practice there must b limitations associated with the smallest angle resolvable by the eye. Thus i practice, performance might be expected to approach limiting values c subtended size and contrast asymptotically.

A complete equation for vision, according to Rose, is

$$B' = 0.95 \left(\frac{\kappa^2}{D_p{}^2 t\eta_q} \right) \left(\frac{1}{\epsilon^2 \alpha^2} \right) \frac{cd}{m^2} \qquad (6.10$$

where D_p is the diameter of the pupil in mm, t is the quantum yield (η_q = means 100% quantum efficiency), α is the angular size of the test object in mra and ϵ is the percentage contrast (= $\Delta B/B' \times 100$).

In arriving at this equation Rose took one lumen of white light to equa 1.3×10^{16} quanta/s.

Experimental evidence is reported which subjectively confirms the value of κ as approximately 5 and also that the storage time of the eye is around 0.2 s Rose then uses his model to predict the threshold data of Connor an Ganoung[17] and Cobb and Moss[18] and shows good agreement. He also draw attention to the fact that, since Connor and Ganoung's data (using an exposur of 1 s) and those of Cobb and Moss (exposure 0.18 s) are very similar in absolut terms, this is further justification for taking the storage time of the eye to be around 0.2s. Considering next the ability of his model to predict Blackwell's (1946) data[2], Rose finds the model only adequate over a limited range of intermediate sizes, deviating considerably for both large and small sizes. The degree of agreement for these acknowledged comprehensive data for circular disc stimuli against plain backgrounds is shown in Fig. 6.3.

By plotting Connor and Ganoung's data, Cobb and Moss's data and Blackwell's data in terms of ϵ against $1/(B^{1/2}\alpha D_p)$ Rose illustrates that, at leas over a range of moderate sizes of stimulus, the results fall between two lines defined by

$$\frac{\kappa^2}{t\eta_q} = 2800$$

and

$$\frac{\kappa^2}{t\eta_q} = 28\ 000$$

Assuming the values of κ and t to be as previously stated this determines that the quantum efficiency of the eye varies from around 0.5% at high adaptation luminance to around 5% at luminances of the order of 10^{-4} cd/m^2. Reference is made to confirmatory evidence for the quantum efficiency of around 5% at low light levels as found by Hecht[19] and Brumberg *et al*[20]. No other evidence is cited for the reduction to 0.5% at high light levels. It is pointed out that, although this variation of quantum efficiency may appear large it is nevertheless far too small a variation to account for dark adaptation which, according to Rose, covers a range of sensitivities of 1 000:1. In this treatment Rose does not appear to differentiate between rod and cone sensitivity (see Chapter 2).

In considering certain practical applications of this quantum approach Rose compares data of Jones and Higgins[21] for the signal/noise ratio associated with

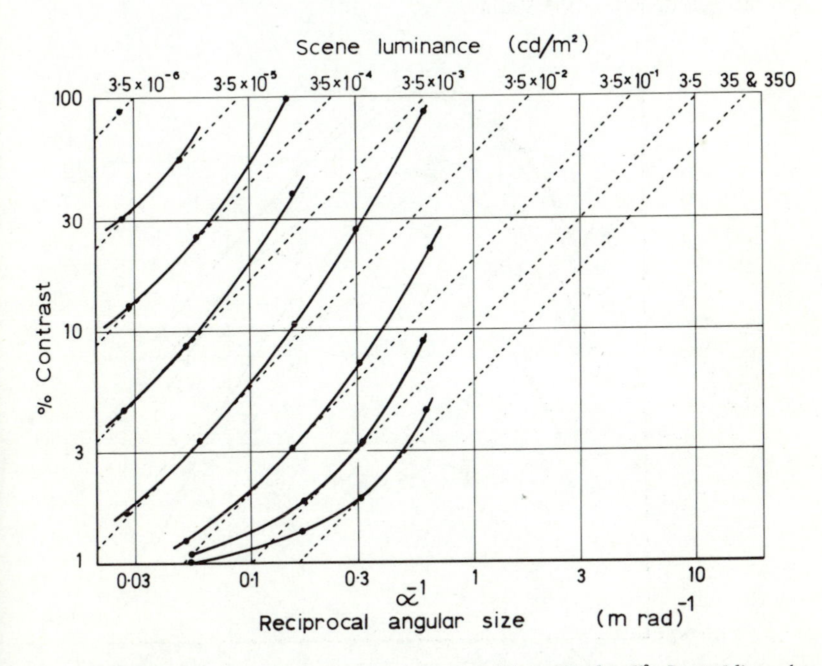

Fig. 6.3. Performance data of the human eye computed from Blackwell[2]. Dotted lines show the performance curves to be expected from an ideal quantum detector (note that each dotted curve is drawn tangent to the best observed performance at a given luminance, not necessarily implying constant quantum efficiency between lines). (Reproduced from Rose[16] by courtesy of the Journal of the Optical Society of America).

TABLE 6.1

Film	*Measured Signal/Noise ratio*	*Computed Signal/Noise ratio*
Tri X	11	13
Super XX	23	22
Pan X	36	39
Fine Grain	77	58

viewing of a variety of films with predictions from his model and shows goo agreement as seen in Table 6.1.

In this table the measured signal/noise ratios were for 40 μm diameter disc on the film at a density of around 0.4. The computed values assumed a storag time of 0.2 s and an 0.5% quantum efficiency. The ability to model th graininess of film must be taken as an important virtue of this model, eve though it has been shown by Rose himself to be limited in coverage o conventional detection thresholds.

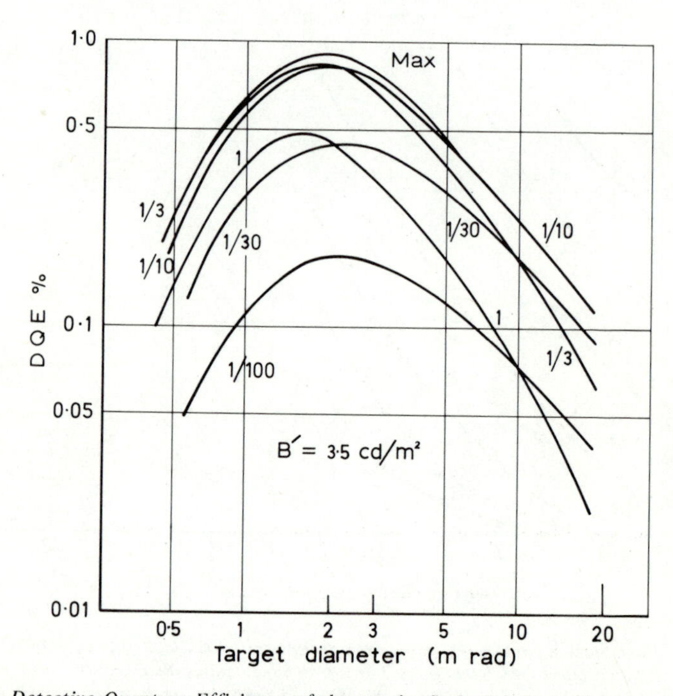

Fig. 6.4. Detective Quantum Efficiency of the eye for flash stimuli as a function of angular size for various flash durations (Flash durations in seconds). (Reproduced from Jones[22] by courtesy of the Journal of the Optical Society of America).

Whilst Rose was developing his quantum model of vision Bruscaglioni[15] was developing a parallel, but significantly different, model in Italy. This parallel approach led Bruscaglioni to formulate an equation of the form

$$\frac{B' = K_{\mathrm{min}}}{\alpha^2 \epsilon D_p{}^2 t \eta_r} \tag{6.11}$$

where η_r is the 'rendimento' defined as 'the way by which the energy is utilized by the detector' and K_{min} includes the signal/noise constant and other parameters.

The main point of variance – and a very significant one – is thus that Bruscaglioni finds threshold contrast inversely proportional to luminance, as opposed to the square root of luminance as proposed by Rose! A possible explanation of the discrepancy is to be found in Section 4.2.

Some years after Rose's work, Jones[22] developed Rose's theory a stage further. He introduced the term 'detective quantum efficiency' (DQE), including now in the performance index not only the concept of an ideal image-detecting device but also an ideal decision making device. This eliminates the

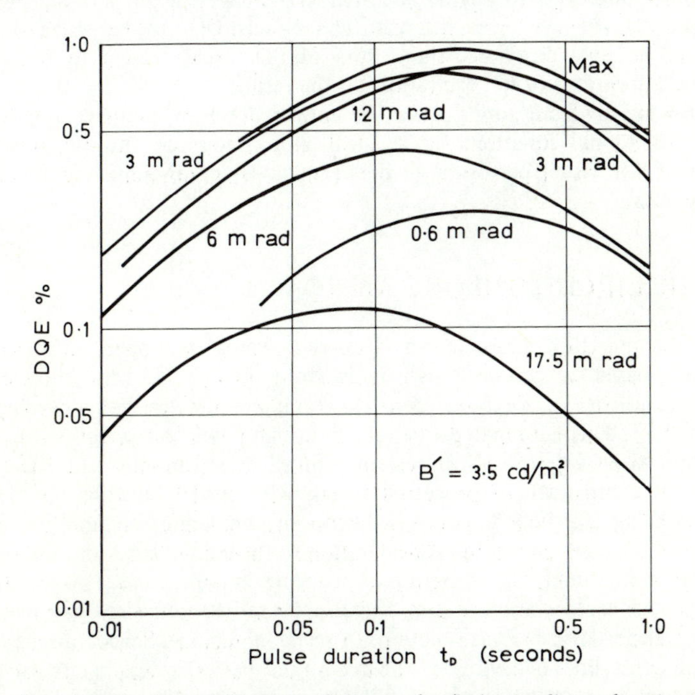

Fig. 6.5. Detective Quantum Efficiency of the eye for flash stimuli as a function of flash duration for various angular sizes. (Sizes given as diameters in mrads). (Reproduced from Jones[22] by courtesy of the Journal of the Optical Society of America).

uncertainty in the proper value of κ to be used in calculating quantum efficiency. He assumed that an ideal device has a threshold such that, if the amplitude of the power is above the threshold, it is concluded that a signal is present. Otherwise it is concluded that just noise is present. Thus, in order to be precise in computation of the quantum efficiency (from a rearrangement of Equation 6.10), the value of κ must correspond to the false alarm and detection probabilities in the determination of threshold (and must hence be related to d' in Section 3.4.2).

Jones calculated the DQE of the human visual system from Blackwell and McCready's experimental data for brief exposures of disc stimuli in one of four successive 2.5 s intervals[3]. The results are shown in Figs. 6.4 and 6.5, the DQE being plotted respectively against target diameter and pulse duration. A value of $\kappa = 1.22$ was found appropriate for these data, which were reported at the 50% detection probability. It can be seen that the DQE appears optimal at intermediate sizes (around 1.75 mrad) and intermediate pulse durations (from 0.08 s at 17.5 mrad diameter to 0.35 s. at 0.6 mrad diameter). Schnitzler[23], in a very useful review article, attributes the reductions of DQE for small sizes and short pulse durations to spatial and temporal image spread, although without indicating the reasons for such spread. The reduced DQE for large sizes and long durations he puts down to 'the failure of the visual system to function as a perfect interpretor of the excitations of the retina'.

Following on from Jones' work, Schnitzler develops complex and comprehensive functional equations for an ideal photon counter, but still shows poor agreement for visual performance between model predictions and Blackwell's Tiffany data.

6.3 THE CIRCUIT THEORY APPROACH

Whilst the quantum approach to vision was being developed, an alternative approach, based on certain known facts about the eye and wide knowledge of the characteristics of components of electronic circuits, was being developed by Schade[24-27]. This culminated in a very thorough possible description of the eye and brain as an electro-optical system coupled to a computer, the latter being capable of comparison of processed signals with stored information[28]. Whilst this modelling is primarily concerned with the luminance channel of vision it does not shirk from giving due consideration to the colour sense and the separate rod and cone functions. A series of working system constants are derived, defining such factors as stage gain, signal/noise ratios, equivalent pass bands and system transfer efficiency as functions of scene luminance. Schade discusses how to predict threshold behaviour of the eye for bar patterns, simple circular targets and random external noise, although in the case of the latter it is admitted that little data were available against which to check predictions.

Schade's model, then, assumes the eye to be comprised essentially of

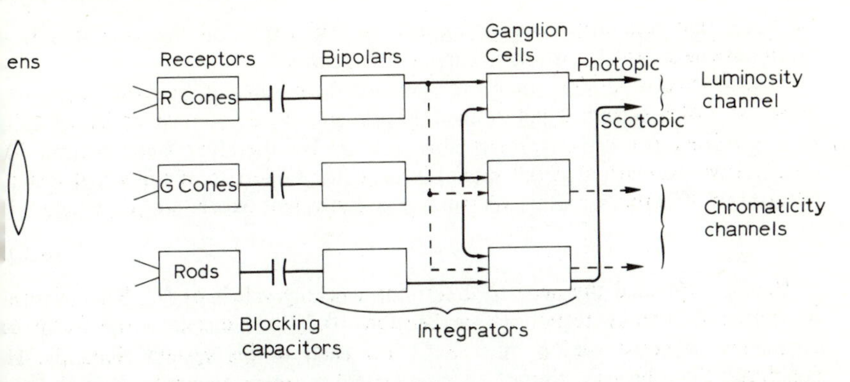

Fig. 6.6. Block diagram of the components of the visual system according to Schade. Noise is assumed to enter at the receptor level. (Reproduced from Shade[28] *by courtesy of the Journal of the Optical Society of America).*

elements as shown in Fig. 6.6, viz., a lens, a photosensitive surface, photoreceptors converting the light into electrical signals and providing a primary stage gain, the bipolar and ganglion cells (see Chapter 2) acting as integrators of variable gain (by change of integration area) dependent on illumination level (citing Polyak[29] and de Groot and Gebhard[30]), and the cortex, which is assumed to act as a comparator and to provide feedback signals. Noise is assumed to be present at the receptor level and is represented by a noise generator. Stage coupling is assumed to be by blocking capacitors, thus permitting only transmission of AC signals to the brain in accordance with the findings of Riggs *et al*[31]. It is assumed by Schade that the AC signals are generated by the ocular tremor of the eye (see Section 2.6). A difficulty can be seen in reconciling this concept with peripheral performance and grouped rods (see Chapter 7). It will be noted from Fig. 6.6 that Schade's model assumes only two kinds of cones – red and green sensitive – the blue function being provided by rod vision (c.f. Section 2.4).

Schade analyses the supposed visual system by stages, and to assist in this form of approach he introduces the concept of a *statistical unit* (su), which is defined as a quantity counted statistically as an independent event. The important property of this unit is that it is not restricted to any particular number of quanta but is defined, at *any point* in a system, as that number of quanta or electrons which are generated due to a minimal number of quanta being absorbed at the primary photo surface. Thus, at any point in the neural net, one su may represent from zero to many powers of ten electrons, dependent on the gain factor and feedback up to that point. Following from this concept of the su, Schade defines a *statistical transfer ratio* μ_s where

$$\mu_s = \frac{\text{output su's}}{\text{input su's}} \leqslant 1$$

He notes that this ratio does not express an efficiency, but that such a ratio often designated the quantum efficiency of a process.

Schade next considers that the standard deviation and electrical signal-tc noise ratio of a system is determined, in principle, by a count of statistical unit in a specified (or effective) sampling area $\bar{a}$. He therefore states a need fo quantitative evaluation of the spatial integrating properties of the visual systen in terms of effective sampling apertures $\bar{a}$ or equivalent pass bands N_e, where

$$\bar{a} = N_e^{-2}. \qquad (6.12$$

Having expressed the need to determine working values of N_e, Schade turn to spatial frequency response (see Section 10.2) and discusses the forms o frequency response which must exist for each of his system elements. H concludes that the lens, tremor and the retinal receptor elements must all have 'normal' spectra, that is responses which are unity at zero frequency and fal with increasing frequency. Conversely his blocking capacitors will have ar inverse form of response, being zero at zero frequency. Now each response function may be represented by an equivalent pass band N_e and, Schade claims all N_e's of 'normal' form, that is having positive amplitude coefficients, may be added with good accuracy by the formula

$$N_{e(total)}^{-2} = N_{e(1)}^{-2} + N_{e(2)}^{-2} + \ldots + N_{e(n)}^{-2} \qquad (6.13$$

Thus the complete system response, except for the effect of the blocking capacitors, may be simply represented. However, when it comes to combinatio of the N_e's with positive and negative amplitude coefficients 'they must be added as spectra'. This seems to the author to defeat the object of using the N_e concept for the eye, since one cannot recover a unique spectrum from a combined $N_{e(total)}$. However, possibly one may argue, as argued by Békésy (see Section 6.5.4), that the absolute form of such a combined frequency response is not important, provided that its area is correct (see also Section 10.3).

Having discussed his model at some length, Schade proceeds to evaluate and tabulate N_e's and transfer characteristics of components of the visual system as a function of scene luminance from Blackwell's Tiffany data[2] for circular disc stimuli. He then cross checks his findings against certain threshold data for point stimuli and visual acuity tests. In interpreting Blackwell's data he highlights the need for inclusion of negative feedback to account for threshold trends and illustrates how successful it is when applied.

A discussion of the compatibility of computed characteristics and neuro-physiological observations follows in which Schade finds good agreement. Similarly he claims that his computed gamma functions are compatible with known characteristics of photoconductors. Finally the problem of perception of spatial noise in a scene is given thorough consideration, and formulae are developed to handle noise situations in the absence of signal. It is pointed out that the problem of noise perception in the presence of fine signal structure is highly complex and dependent on signal/noise spatial interactions. A highly significant point brought out is that, for noise perception, the 'contrast' of the noise is attenuated by system components in proportion to the *diameter* of the effective sampling area ($1/N_e$), as opposed to signal contrast reduction being in proportion to the *area* of effective sampling ($1/N_e^2$).

Whilst this treatment of Schade's is very thorough and attractive, it is not clear how we are to guarantee its applicability when extrapolated to the much wider range of complex stimulus situations met with in everyday life. Also, because of the very considerable variations in component N_e's and transfer characteristics with conditions in an ill-defined manner, use of this model would appear to be complicated and restricted in practice.

6.4 THE ELEMENT CONTRIBUTION THEORY

A further approach to the problems of threshold vision in plain fields has been that due to Graham and co-workers[32-34] elaborated by Blackwell and co-workers[35]. Their approach is aimed primarily at overcoming the limitations of the quantum theory at extremes of size (see Fig. 6.3) and for long exposures, whilst being compatible with known neurophysiological behaviour and the roughly constant σ/C_M of frequency of seeing curves discussed in Chapter 3. Blackwell[35] claims that the constancy of σ/C_M found in extensive studies by himself and his co-workers allows only a very few possible relationships between background luminance B' and sensory effect Q_s. The simplest of all – a linear relationship between B' and Q_s – is stated to be at complete variance with known neurophysiological behaviour, as also is the positively accelerated function implied by direct application of the decision theory of Tanner and Swets (see Chapter 3) to threshold trends. On the other hand a relationship which is variable with conditions is not considered attractive or readily verifiable. On the strength of these statements Blackwell suggests that the only likely predictable function is of the form

$$Q_s(p) = K - \frac{\gamma_p}{(B' + c)} \tag{6.14}$$

where $Q_s(p)$ is the sensory effect at a probability of detection p, γ_p is a constant associated with $Q_s(p)$, c and K are arbitrary constants.

K is then the maximum possible value of $Q_s(p)$ and c is a measure of the neural noise in the absence of stimulation. It is further claimed by Blackwell that the relationship between $\Delta\bar{B}/B'$ and B', where $\Delta\bar{B}$ is the mean threshold luminance increment, may be obtained by differentiating Equation 6.14 to yield

$$\frac{\Delta\bar{B}}{B'} = K(1 + c/B') \tag{6.15}$$

The validity of Equation 6.15 as a predictor of threshold increment with

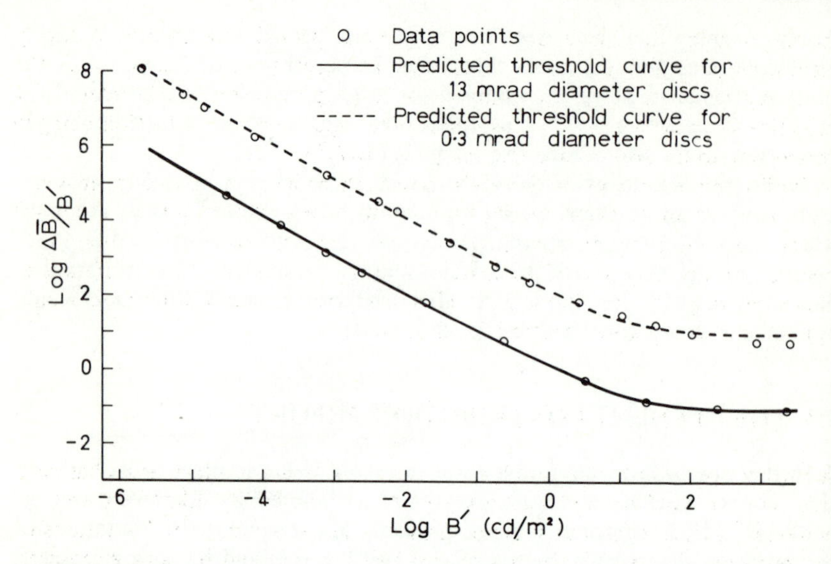

Fig. 6.7. Predictions of threshold data for 0.3 mrad and 13 mrad diameter circular disc stimuli of Blackwell and Law[36] *using the spatial element contribution theory. (Reproduced from Blackwell*[35] *by courtesy of the Journal of the Optical Society of America).*

variations of background luminance is confirmed for circular stimuli by predicting the experimental results of Blackwell and Law[36] for 0.3 mrad and 13 mrad circular stimuli presented for 0.01 s exposure. The fits, shown in Fig. 6.7, are seen to be good on the whole, but with a tendency for errors at high background luminances. Data of Blackwell and Law, presented in terms of $\Delta\bar{B}$ against B', show a constancy of $\Delta\bar{B}$ at background levels below about 0.035 cd/m² for these presentation conditions (c.f. Section 4.2).

Of course, in the luminance domain the quantum theories provide almost as adequate a fit to simple threshold data as does Blackwell's theory. It is in the size domain, and for long exposure times, where severe inadequacies in the quantum theories appear to exist.

In order to attempt to overcome the problem of size effects, Graham and co-workers[32] assumed an empirical relationship between the magnitude of threshold and the size of circular stimuli, where the neural effects q_s of various elements of area contained within a stimulus of radius r_s were assumed to sum in accordance with a power law of distance r from the element to the centre of the stimulus. Thus

$$Q_s = k_1 q_s \int_0^{2\pi} \int_0^{r_s} r\,\mathrm{d}r \frac{\mathrm{d}\theta}{r^z} \qquad (6.16)$$

where Q_s is the total sensory effect, r is the separation of the element from the centre of the stimulus, k_1 is an arbitrary constant. The threshold is then considered to be inversely proportional to Q_s.

Graham *et al*[32] found $z = 1.49$ for foveal cones, and in Graham and Bartlett[33] and Brown[34] the theory was extended to include rectangles. They recognised that the theory must break down as $r \rightarrow 0$ and suggested that, for small values of r, the function must approach a finite limit.

The true spatial element contribution theory is an outgrowth of Graham *et al*'s work and was first stated by Blackwell and Austin in 1952[37]. It was further developed by Kincaid, Blackwell and Kristofferson[38] and finally expounded in full by Blackwell in 1963[35]. It is suggested by Blackwell that it may be thought of in terms of topographical mappings of receptor cells on cortical cells. The concept then is that any one receptor excites many cortical cells and similarly each cortical cell is excited by many receptors, the two mapping functions being symmetrical and similar. Then the assumption is made that, at a particular neural site, a criterion level of Q_s is required in order for a detection to take place. If then the resultant sensory effect at such a neural site is the additive effects of various receptor cells weighted in accordance with the distance separating each from that contributing the maximum sensory effect, the excitation potential V_e for a given shape and size will be given as

$$V_e = \int_0^{2\pi} \int_0^{r_s(\theta)} r\psi(r)\,\mathrm{d}r\,\mathrm{d}\theta \qquad (6.17)$$

where $\psi(r)$ is called the spatial element contribution coefficient.

Derivation of $\psi(r)$ is most conveniently obtained from threshold data for circular stimuli where, since $V_e \propto 1/\Delta\bar{B}$

$$\frac{1}{\Delta\bar{B}} = \frac{2\pi}{k} \int_0^{r_s} r\psi(r)\,\mathrm{d}r, \qquad (6.18)$$

the value of $\psi(r)$ being arbitrarily set to unity when $r = 0$. Blackwell proceeds to present derived functions $\psi(r)$ for various exposure times, various background luminances and also specific monochromatic illumination conditions. A typical set of functions for various background luminances are shown in Fig. 6.8 where it will be seen that the 'summation area' increases in width as the luminance reduces. It will be seen from Fig. 6.9 that 'summation area' is also strongly a function of colour of light. (Data for both figures are from Smith *et al*[39]). All these 'summation' trends are seen by the present author to bear a remarkable resemblance to the spread functions of the refraction optics of the eye as discussed in Chapter 2 (a broadening spread function due to opening pupil at low light levels and a blur spread function due to residual defocus for blue light). However, Blackwell does not refer to the quality of the refraction optics and it is not obvious how such a similarity can be related to the area integration theories.

It is admitted by Blackwell that little attempt had been made to predict thresholds for non-circular stimuli except for limited rectangular stimuli. In this

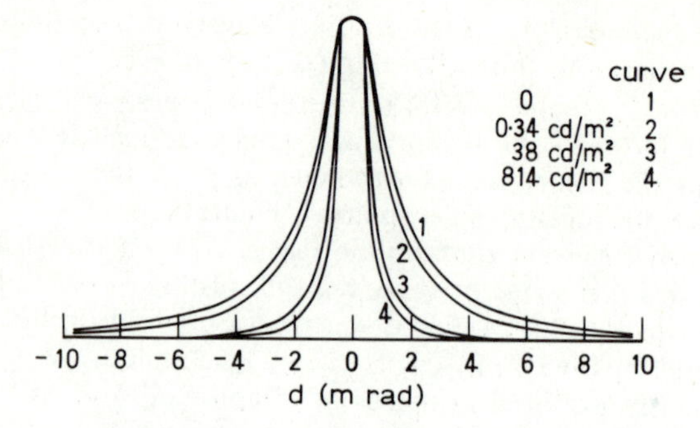

Fig. 6.8. A typical set of spatial element contribution functions for an 0.1 s exposure and various adaptation levels. (Reproduced from Blackwell[35] by courtesy of the Journal of the Optical Society of America).

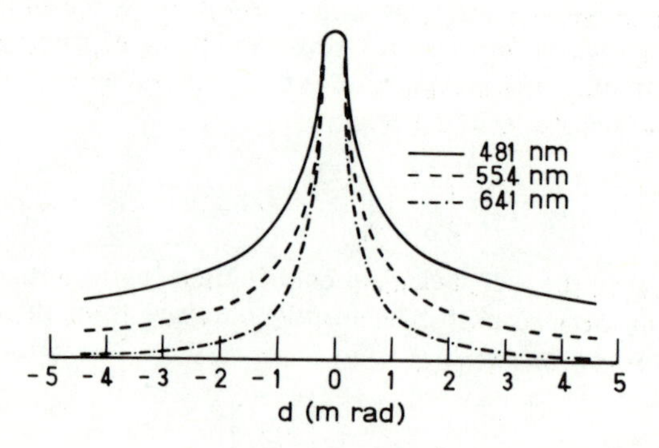

Fig. 6.9. Spatial element contribution functions for various monochromatic illumination situations. (Reproduced from Blackwell[35] by courtesy of the Journal of the Optical Society of America).

latter case the predictions for long thin rectangles were regularly too high. Certain attempts have, however, been made to predict thresholds for pairs of point stimuli as a function of separation. It has been found that in such a case the thresholds are predicted well when the two element contribution functions overlap. For larger separations good predictions are obtained by assuming that in this case there are two chances of seeing one stimulus, thus increasing the probability according to multiple chance statistics.

As far as temporal behaviour is concerned Graham and colleagues[40,41] suggested, as a result of Hartline's studies on the *Limulus* eye[42] (see Section 6.5.5), that the effects of stimulation accumulate for a specific storage interval and then suddenly discharge. This results in predicted thresholds obeying Bloch's (or Bunsen and Roscoe's) law

$$\Delta\bar{B} \times t_D = K$$

for short exposure times with a sudden transition to $\Delta\bar{B} = K$ at the end of a storage interval. Such a concept is not in very good agreement with experimental results for exposure times of more than about 0.05 s (see Fig. 4.7) and presents difficulties in practical application. As an alternative, Blackwell and colleagues proposed a temporal element contribution function similar in concept to the spatial one previously described. It assumes that each temporal event (or incremental portion of a temporal event) produces a temporal after-effect due to the various lags which must exist in the neural networks. Then, as with the spatial function, the overall effect is the integral of incremental components weighted relative to their time of occurence.

The first major studies were carried out by Clark and Blackwell[7] who experimented with the presentation of single pulses of variable pulse length and double pulses of pulse length 0.0025 s but with variable separation time. They showed that the quantum theory and Graham's theory only provided a good fit at short exposure times for single pulses and for small separations of double pulses. The Blackwell theory, although providing very good fits at short exposure times including prediction of absolute levels for single pulses using a temporal element contribution function derived from double pulse data, equally does not provide a good prediction for exposure times in excess of about 0.2 s and pulse separations of more than 0.05 s. A major improvement in fits for both single and double pulses is achieved by invoking the multiple chance concept again as for the widely spatially separated points. With such allowances for accumulation of probabilities with multiple chances and an assumption of a CNS (central nervous system) scanning mechanism which cycles six times a second, each scan erasing data from each previous scan, Blackwell shows good predictive capability with his model. It is not clear, however, how the six cycle scanning is reconciled with the mean glimpse rate of around 3 per second or indeed how

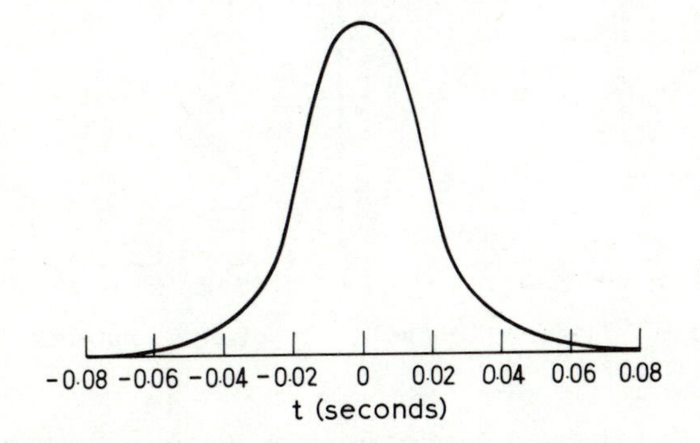

Fig. 6.10. A typical temporal element contribution function. (Reproduced from Blackwell[35] by courtesy of the Journal of the Optical Society of America).

such a total signal erasure is supposed to take place. A typical temporal element contribution function as used for predictive modelling by Blackwell is shown in Fig. 6.10.

Overall the Blackwell theories certainly provide a considerably better fit to a range of experimental threshold data for circular targets in plain fields and for certain pulse presentations than do quantum theories. However, there is little evidence that they are equally successful in handling more complex stimulus presentations and they, together with the quantum theory approaches, seem incompatible with border effects such as the Mach effect (see Section 2.9 and Section 6.5 below).

6.5 THE INHIBITION THEORIES

Whilst several schools of researchers had been attempting to fit various sets of threshold data in relatively simple fields by developing models based on direct utilisation of stimulus energy, a completely different approach was being followed by physiologists as a result of the startling findings of Ernst Mach in the late 1800's (see Ratliff[43] for a complete translation of six of Mach's main papers). In general Mach discovered that, if he viewed any pattern which had

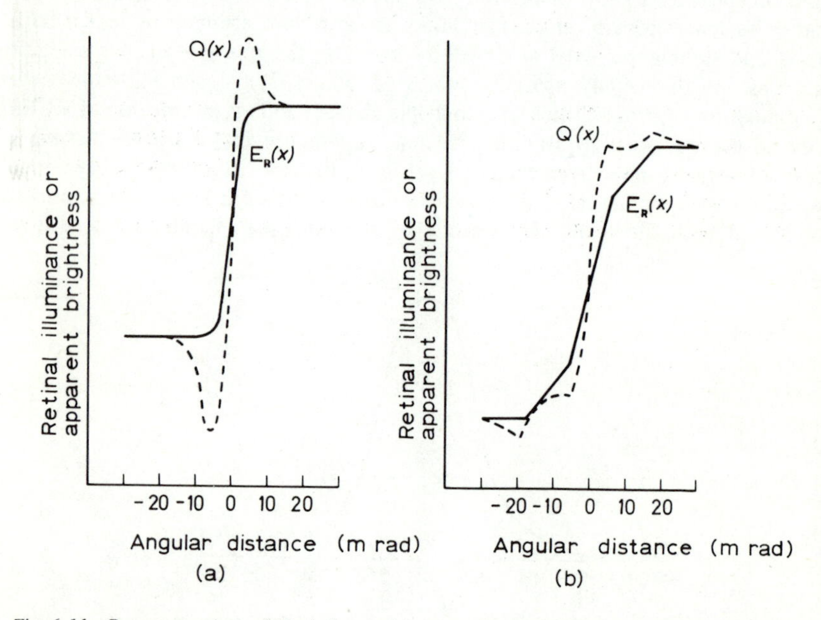

Fig. 6.11. Demonstration of Mach Bands (a) at an edge. (b) at discontinuities in illuminance gradient. $E_R(x)$ *≡ illumination,* $Q(x)$ *≡ response. (Reproduced from Fry[44] by courtesy of the American Academy of Optometry).*

sudden changes in luminance profile, he observed strange bright or dark bands in the vicinity of the discontinuity. The most familiar of these – the usually quoted Mach bands – appeared at or near a luminance step, and took the form of an accentuation of the light and dark portions of the scene adjacent to the luminance step (see Fig. 6.11(a)). Such a phenomenon can be relatively simply modelled by considering the important attribute of the scene to be the 1st differential of luminance. However, Mach also found that bright and dark bands, together with other distortions, occurred when there was a sudden change of luminance *gradient* in the viewed scene (Fig. 6.11(b)). Such a finding was also confirmed by Fry[44]. The various findings of Mach and others have led a series of workers to attempt to model the neural processes of vision in order to predict these edge and illuminance gradient effects, such modelling being strictly suprathreshold as opposed to the threshold models in earlier sections. A very thorough summary of six of the more important of these models is to be found in Ratliff[43]. A short summary is given below, including certain added observations by the present author.

6.5.1 Mach (1865 onwards) – Reciprocal Inhibition

Mach himself, having noted the strange distortions mentioned above, questioned whether the relationship between local stimulus intensity I and neural response Q_s might be expressed in terms of the second derivative of the two dimensional luminance distribution in the original stimulus. He thus proposed the possible relationship

$$Q_s = I - m\left(\frac{\mathrm{d}^2 I}{\mathrm{d}x^2} + \frac{\mathrm{d}^2 I}{\mathrm{d}y^2}\right) \tag{6.19}$$

where m is a constant.

Ratliff points out that his equation cannot, of course, be taken at face value since infinite values of $\mathrm{d}^2I/\mathrm{d}x^2$ or $\mathrm{d}^2I/\mathrm{d}y^2$ do not result in infinitely bright or infinitessimally narrow bands. However, what Ratliff does *not* comment on is that, at the retina, infinite values of $\mathrm{d}^2I/\mathrm{d}x^2$ and $\mathrm{d}^2I/\mathrm{d}y^2$ never occur due to the degrading effects of the refraction optics of the eye and retinal diffusion. Hence Mach's original equation is rather more possible than Ratliff implies.

In order to refine his modelling, and at the same time take account of retinal effects rather than stimulus characteristics, Mach next postulated that the observed brightness distortions could only be explained on the basis of reciprocal action of neighbouring areas of the retina. He expressed this reciprocal action in a number of ways, of which Ratliff describes one which he considers most in keeping with modern electrophysiological evidence. This is such that there is an assumed reciprocal action between any two retinal points g and j separated by distance x which is determined by a function $\phi(\mathrm{r})$.

Then one may say that

$$Q_g = I_g \frac{I_g \Sigma_j \phi(x_{jg}) \cdot \Delta a_j}{\Sigma_j I_j \phi(x_{jg}) \cdot \Delta a_j} \tag{6.20}$$

where Δa_j is the area of a receptor at j, I_g and I_j are the intensities of the stimulus at g and j respectively and $\phi(x_{jg})$ is the value of $\phi(r)$ for separation gj.

Mach assumed that the function $\phi(r)$ would probably be of exponential form as shown in Fig. 6.12.

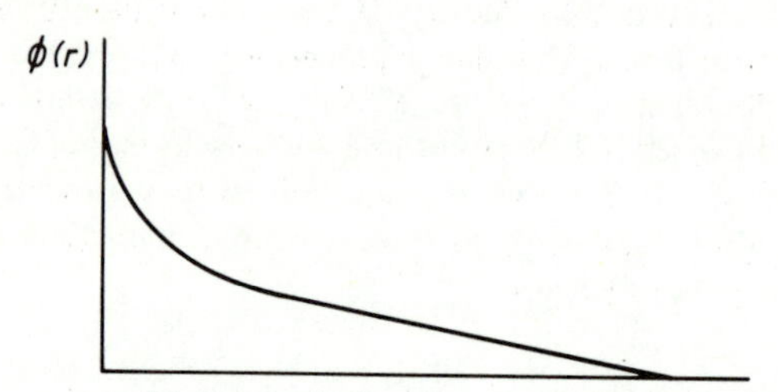

Fig. 6.12. The form of Mach's interaction function $\phi(r)$. (Reproduced from Ratliff[43] by courtesy of Holden–Day Inc., San Francisco).

Such a function will yield a value of Q_g equal to I_g for uniform areas of illumination, since then the numerator

$$I_g \Sigma_j \phi(x_{jg}) \cdot \Delta a_j$$

and the denominator

$$\Sigma_j I_j \cdot \phi(x_{jg}) \cdot \Delta a_j$$

are equal. Conversely, for regions where I_j is less than I_g there will be a contribution to Q_g greater than unity and for regions where I_j is greater than I_g the contribution will be less than unity. It is very easy to see that, for a region of uniform slope as in Fig. 6.11(b), the contributions to the response at the central parts of the slopes from below and above will average out to unity, whilst at or near sudden changes of slope there will be a tendency to distortion where magnitude will depend on the rate of change of the slope. Hence such a function as Equation 6.20 broadly predicts the observed visual behaviour and is effectively a second differential model. Ratliff points out that the numerator is a constant if $\phi(x_{jg})$ is a constant function, whilst the denominator is the sum of a series of values I_j times a second constant. Thus one may rewrite Equation 6.20 as

$$Q_g = I_g \frac{I_g K_e}{\Sigma_j I_j k_{jg}} \tag{6.21}$$

where K_e may now be considered as an excitatory constant and k_{jg} as a separation-dependent inhibitory constant.

It will be seen in Chapter 13 how this concept of excitatory and inhibitory influences has been recently used to explain threshold trends for structured fields.

6.5.2 Fry (1948) – Electrical Inhibitory Fields

Fry's model[44] is based on assumed inhibitory electric field effects as a result of two electrophysiological studies by Granit[45,46]. Granit discovered an inhibitory component in the retinal action potential* and Fry assumed that the contribution to such a potential from one stimulated point would be graded with distance from the point. Then, since each point in an extended retinal stimulus will set up its own graded field, the total field distribution will be the sum of component influences and each local response may be expected to be a simple local component field sum. This is different to Mach's model in that there is now no assumed neural interaction, the inhibition potentials being directly controlled by the retinal illuminance distribution. Fry assumes that the form of component field distribution function is gaussian – viz.

$$\frac{\Delta V}{\Delta a} \propto E_R \exp\left(-\frac{d_s^2}{\sigma^2}\right) \tag{6.22}$$

where σ expresses the extent of the inhibitory field, Δa is the area of a stimulated element and ΔV is the PD at any point at a distance d_s from the point of stimulation. From experimental data he finds that $\sigma \approx 6$ mrad.

The total PD at any point on the retina is then $\int \Delta V/\Delta a$ or

$$V_g \propto \int_{-\infty}^{+\infty}\int_{-\infty}^{+\infty} E_R(x,y)\cdot \exp\left(-\frac{x^2}{\sigma^2}\right)\exp\left(-\frac{y^2}{\sigma^2}\right)\cdot dy\, dx \tag{6.23}$$

If the response (Q_g) of the underlying neuron at g is assumed to be proportional to log $E_R(g)$ and inversely related to V_g, then

$$Q_g \propto \frac{E_R(g)}{(1+V_g)} \tag{6.24}$$

Fry compared the appearance of various complex patterns with Q_g's predicted and found good agreement. In particular his model correctly predicts the bands at the discontinuities in Fig. 6.11(b).

*The retinal action potential is a potential difference existing between the front and back surfaces of the retina which is generated in response to illumination level.

6.5.3 Huggins and Licklider (1951) – Double Differentiation and Weighting Function Model

Huggins and Licklider[47] proposed a neural inhibitory model which assumed basically first and second differentiation of stimulus distribution by retinal neural units. The mechanisms assumed for the first and second differencing are as shown in Fig. 6.13. For 1st differencing, adjacent pairs of neural units are assumed to have gains of +1 and – 1 respectively. If, then, the outputs of pairs are combined, the result is a 1st difference signal which is sensitive to polarity of the illuminance gradient. Second differencing is then assumed to be achieved by triple grouping with units having gains of +1, – 2 and +1 respectively. Alternatively, if *every* 1st order unit output is combined with *both* adjacent outputs, and these adjacent outputs combined again, then the result is effectively the same, whereupon both 1st and 2nd differences can be directly obtained from each group of 3 retinal units. Having postulated retinal mechanisms for 1st and 2nd differencing, Huggins and Licklider go on to discuss weighting functions necessary to smooth outputs and recover a close

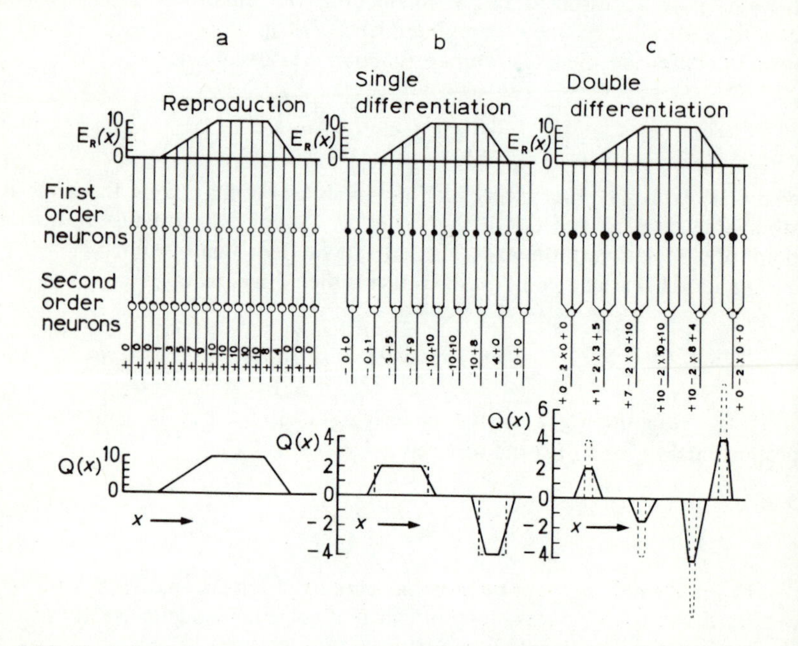

Fig. 6.13. Hypothetical neural mechanism for (a) reproduction, (b) differentiation, (c) double differentiation of stimulus distribution, as proposed by Huggins & Licklider. (Reproduced from Huggins and Licklider[47] by courtesy of the American Institute of Physics).

approximation to the input E(x). They define the neural response Q(x) as

$$Q(x) = K \int_{-\infty}^{+\infty} E(x - \xi) \cdot w(\xi) \cdot d\xi \tag{6.25}$$

where $w(\xi)$ is the weighting function for local interactions. They find, based on the luminance distribution of the *original viewed scene*, that the weighting function $w(\xi)$ requires to have a central positive lobe and surrounding negative lobes. After allowing for the spread due to the refraction optics of the eye (which Huggins and Licklider did not do), the residual weighting function is most likely to be mainly a negative function, thus resembling Mach's inhibition function $\phi(r)$ and Fry's graded field distribution function.

6.5.4 Békésy (1960) – 'Neural Unit' and Superposition Integral

Békésy proposed, as long ago as 1928, that contrast effects similar to Mach bands were probably not restricted to the visual system. It has since been demonstrated that lateral inhibition is present in hearing (e.g. Galambos and Davis[48] and Katsuki *et al*[49]) and also in skin sensation (e.g. Mountcastle and Powell[50]). As a result Békésy[51] proposed a simple summation/inhibition model of the form shown in Fig. 6.14. He claims this is equivalent to a continuous function, in that it predicts Mach bands of strength dependent on the magnitude of the illumination gradient and on the rate of change of gradient. However, it is much easier to define dimensions for a simplified rectangular form. In terms of measured data, Békésy has determined equivalent widths and strengths of the component parts of the function. He claims that the sizes of e and i are not critical, as long as the ratio e/i is held constant and the width of e is small compared with the width of i.

Ratliff points out that Békésy's and Huggins and Licklider's models are basically similar when $e/i = 1$, but that Békésy's model is more versatile since one achieves a better prediction of the Mach bands by employing a value $e/i > 1$.

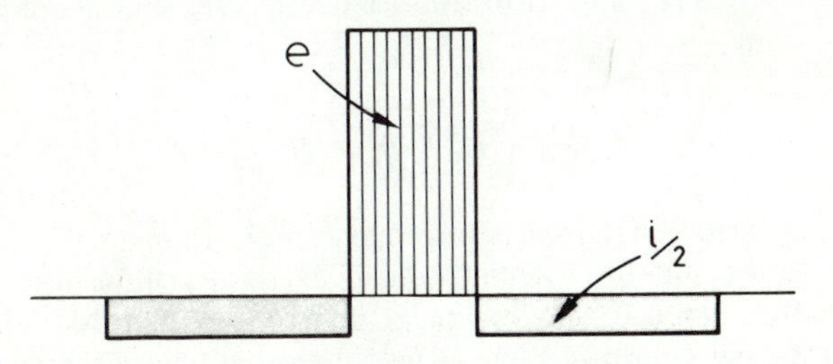

Fig. 6.14. Békésy's rectangular summation/inhibition function. (Reproduced from Ratliff[43] by courtesy of Holden–Day Inc., San Francisco).

6.5.5 Hartline and Ratliff (1954 and later) – Recurrent Inhibition

Hartline and Ratliff[52-54] mainly based their modelling on their studies of the compound eye of *Limulus* (the horseshoe crab). The *Limulus* eye contains about 1 000 ommatidia (little eyes) each of which appears to function as a receptor unit[55]. Hence it can be arranged that detailed illumination distribution of the 'retina' can be controlled exactly. Their findings are that each ommatidium mutually inhibits its neighbours. Hence, unlike other models of the present group, the magnitude of local inhibition depends on the *response* of neighbouring ommatidia rather than on their level of stimulation. Thus excitatory and inhibitory effects act simultaneously and must be described by simultaneous equations. The full treatment is too complex to summarise here and the reader who is interested is referred to Ratliff's book or the original papers. The inhibitory influence is believed by Ratliff, on indirect evidence of Kirschfeld and Reichardt[56], to be gaussian. Although a good case is made for this model for *Limulus*, there is an obvious danger in direct extrapolation to primate visual systems.

6.5.6 Taylor (1956) – Recurrent Inhibition

Taylor[57] formulated a model on similar lines to Hartline and Ratliff based on electrophysiological experiments on vertebrates. His model is based specifically on one of several special types of ganglion cell in the cat retina – one which has an 'on' centre, and an 'off' surround receptive field (as found by Kuffler[58]). He assumed that interaction can be specified in terms of two factors k and s which denote the magnitude and spread of interaction respectively (assumed a rectangular function). He also assumed that the frequency of response pulses was roughly proportional to log I. Then the interactions can be expressed as a series of simultaneous equations, one for each unit of the system. For a single row of units infinitely long (in order to ignore end effects) the response is then given as

$$Q_g = e_g + k \sum_{g-d_s}^{g+d_s} Q_j \tag{6.26}$$

where e_g is the excitation, normalised so that $e_g = Q_g$ when $k = 0$.

In order to account for a summation of excitatory influences observed at low illumination levels[59], Taylor suggests that k could be dependent on illumination, taking a positive value at low intensities (giving summation), being zero at some intermediate intensity and becoming negative at high intensities (giving inhibition). Ratliff points out that this ability to vary inhibition and supress entirely as a function of illumination level is unique to Taylor's model.

6.5.7 Discussion

Whilst all the inhibition models have unique features they are all very similar in concept and all imply that the most important areas of a viewed scene are those *adjacent to changes in luminance.* In the detection domain Lamar, Hecht, Schlaer and Hendley[60,61] found that the best way to explain their threshold variations for rectangles as a function of aspect ratio (see Section 4.5) was to assume that the main contribution to the stimulus was from a ribbon one minute of arc wide just inside the perimeter. In the field of image evaluation subjective quality is found to be closely correlated with acutance and MTFA which are both effectively measures of edge sharpness (see Section 10.3). In real life many scenes contain structured target objects and interactions of background structure with the target. Yet again studies in recent years on stabilised retinal images (e.g. Riggs *et al*[31] and Ditchburn and Ginsborg[62]) have consistently shown that when discontinuities are prevented from moving on the retina the image of the scene fades out leaving a sensation of a uniform field brightness. All the foregoing has led the author to the belief that edge gradients *in the retinal image* are the all important features of the visual scene, and that any comprehensive modelling of the visual process for detection must be based on these retinal edge gradients.

6.6 VISTARAQ

Before leaving this historical survey of modelling of thresholds of detection, it is considered important to draw the reader's attention to one other model – that recently proposed by Owen[63].

Owen starts his formulation from the Weber-Fechner law that physical sensation is proportional to log stimulus. The retinal responses are assumed to obey this law, thus implying that the probability of detection in a single glimpse must be a function of log B', log C and log n where B', C and n are the background luminance, the contrast and the number of retinal receptors excited respectively. It is next deduced from Blackwell's data that the probability of detection depends on the product

$$(\log B'/B'_0 \cdot \log C/C_0 \cdot \log n/n_0)$$

where B'_0, C_0 and n_0 represent minimum values of their respective variables which can result in a detection decision. Finally it is assumed that the probability of detection of a given stimulus is normally distributed for variations of this product of logarithms about a mean, which Owen terms the *threshold product.* Now it has been shown by Owen that between the 10% and 90% values log normal and normal distribution curves cannot be discriminated between within experimental accuracies. This model is thus not at variance with Blackwell's normal distribution (Section 3.1) and has the added virtue that it

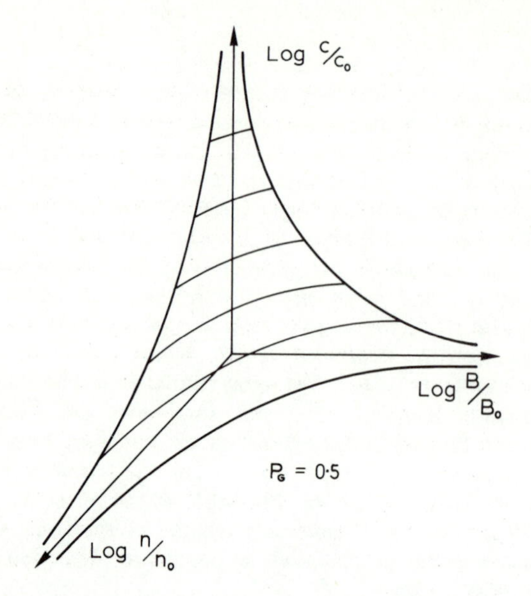

Fig. 6.15. The cubic surface for 50% probability of detection. (Reproduced from Owen[63] *by permission. Copyright Controller H.M.S.O., London, 1975).*

gives zero probability with zero stimulus – not the case with an assumed normal distribution as mentioned in Section 3.1.

For a given probability of detection, then,

$$\log \frac{B'}{B'_0} \cdot \log \frac{C}{C_0} \cdot \log \frac{n}{n_0} = \text{constant} \tag{6.27}$$

Thus a constant probability of detection is represented by cubic surfaces in the three dimensional space enclosed by orthogonal axes of $\log B'/B'_0$, $\log C/C_0$ and $\log n/n_0$ as shown in Fig. 6.15. In particular, where contrast is the threshold variable, as is usual, we may state the threshold product for 50% probability as

$$\log \frac{B'}{B'_0} \cdot \log \frac{\epsilon}{\epsilon_0} \cdot \log \frac{n}{n_0} = b \tag{6.28}$$

where ϵ_0 is now the minimum possible 50% threshold contrast.

In practice, calculation of the value of n depends on the type of target presented, it being necessary to allow for retinal image blur when targets are not completely resolved. Also it has been found necessary in practice to vary n_0 to some extent, dependent on the type of detection task. For most simple targets and foveal cone vision it is claimed that a value of $n_0 = 1$ is required. For peripherally viewed targets n_0 increases progressively to 7, whilst for rod vision a value of $n_0 = 125$ has been found appropriate. Log B'_0 has been deduced from Blackwell's data as −3.0 for photopic vision and −12.5 for scotopic vision, where B'_0 is in cd/m^2.

Minor variations of b are found necessary as a function of B' for background levels less than about 3 cd/m^2, whilst log ϵ_0 is to a small extent dependent on background level, particularly for single glimpse viewing at high background levels. For $B' > 3$ cd/m^2 Owen suggests a general formula

$$\log \epsilon = \log \epsilon_0 + \frac{12}{\log n\,(\log B' + 3.0)} \tag{6.29}$$

for single glimpse viewing. A multiple glimpse formula is then derived from Equation 6.29 of the form

$$\log \epsilon = \log \epsilon_0 + \frac{7}{\log n\,(\log B' + 3.0)} \tag{6.30}$$

for extended viewing. Trend curves of log ϵ_0 as functions of log B' for both formulae are provided by Owen[63].

This model, whilst not pretending to *explain* the visual process other than by allowing for retinal image quality (Section 2.3) and the variation of retinal receptor spacing with retinal position (Section 2.4), does provide a remarkably good fit to threshold trends for simple objects over the entire range of photopic vision (e.g. Fig. 6.16). Also, being essentially simple in form, it is admirably

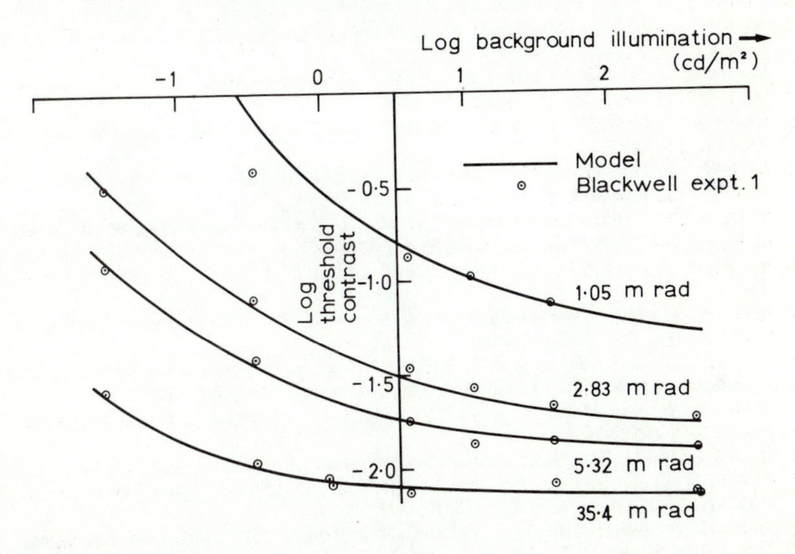

Fig. 6.16. Model predictions of threshold contrast as a function of background luminance, as measured for limited time viewing of disc stimuli by Blackwell. (Reproduced from Owen[63] by permission. Copyright Controller H.M.S.O., London, 1975).

suited to computerised predictions, including allowance for external effects such as atmospherics. Owen himself has developed a versatile computer program based on this threshold model which has wide capabilities. (Equations 6.27 to 6.30 are reproduced from Owen[63] by permission. Copyright Controller, H.M.S.O. 1975).

REFERENCES

1. Ronchi, L. (1971). 'Ricco's Law, *'Atti della Fondazione Giorgio Ronchi,* **26**, 751
2. Blackwell, H. R. (1946). 'Contrast Thresholds of the Human Eye', *J. Opt. Soc. Am.,* **36**, 624
3. Blackwell, H. R. and McCready, D. W. (1958). 'Foveal Contrast Thresholds for Various Durations of Single Pulses', University of Michigan Rep. No. 2455–13–F
4. Taylor, J. H. (1969). 'Factors underlying Visual Search Performance', Scripps Institution of Oceanography Rep. No. 69–22
5. Piper, H. (1903). 'About the Dependence of the Light Stimulus Value of an Object on its Area for Large Angular Subtense', (in German). *Z. Psychol.* **32**, 98
6. Velden, H. A. van der (1944). 'About the Minimum Light Quanta necessary for sensing by the Human Eye', *Physica,* **11**, 179
7. Clark, W. C. and Blackwell, H. R. (1960). 'Relations Between Visibility Thresholds for Single and Double Pulses', University of Michigan Rep. No. 2144–343–T.
8. Blondel, A. and Rey, J. (1911). 'On the Perception of Short Light Pulses at their Limit of Visibility', (in French). *J. de Physique,* **1**, 5th Ser. 530, 643
9. Blondel, A. and Rey, J. (1912). 'The Perception of Lights at their Range Limits', *Trans. I.E.S.,* London, 7, 625
10. Taylor, J. H. (1972). 'Air-to-Ground Visibility of Lights at Low Background Levels', AGARD Conference Proceedings No. 100 (Ed. H. F. Huddleston), London, November, p. B8–1
11. Barnes, R. B. Von and Czerny, M. (1933). 'Can a photon Shot Effect be observed with the Eye?', (in German). *Z. Physik,* **79**, 436
12. Hecht, S., Schlaer, S. and Pirenne, M. H. (1942). 'Energy, Quanta and Vision', *J. Gen. Physiol.,* **25**, 819
13. Rose, A. (1942). 'The Relative Sensitivities of Television Pick-up Tubes, Photographic Film and the Human Eye', *Proc. I.R.E.,* June, 293
14. Vries, H. de (1943). 'The Quantum Character of Light and its Bearing upon Threshold of Vision, the Differential Sensitivity and Visual Acuity of the Eye', *Physica,* **10**, 553
15. Bruscaglioni, R. (1942). 'The Fundamental Equation of Vision', (in Italian), *Ottica,* **6**, 326
16. Rose, A. (1948). 'The Sensitivity Performance of the Human Eye on an Absolute Scale', *J. Opt. Soc. Am.,* **38**, 196
17. Connor, J. P. and Ganoung, R. E. (1935). 'An Experimental Determination of Visual Thresholds at Low Values of Illumination', *J. Opt. Soc. Am.,* **25**, 287
18. Cobb, P. W. and Moss, F. K. (1928). 'The Four Variables of Visual Threshold', *J. Frank. Inst.* **205**, 831
19. Hecht, S. (1942). 'The Quantum Relation of Vision', *J. Opt. Soc. Am.,* **32**, 42
20. Brumberg, E. M., Vavilov, S. I. and Sverdlov, Z. M. (1943), 'Visual Measurements of Quantum Fluctuations', *J. Phys. USSR.,* **7**, 1
21. Jones, L. A. and Higgins, G. C. (1946). 'Photographic Granularity and Graininess', *J. Opt. Soc. Am.,* **36**, 203
22. Jones, R. C. (1959). 'Quantum Efficiency of Human Vision', *J. Opt. Soc. Am.,* **49**, 645
23. Schnitzler, A. D. (1973). 'Analysis of Noise-required Contrast and Modulation in

Image-detecting and Display Systems', in *Perception of Displayed Information*, L. M. Biberman, (Ed.) Plenum

24. Schade, O. H. (1951). Image Gradation Graininess and Sharpness in Television and Motion Picture Systems. Pt. I. Image Structure and Transfer Characteristics', *J. Soc. Mot. Pic. Tele. Eng.*, **56**, 137
25. Schade, O. H. (1952). 'Image Gradation Graininess and Sharpness in Television and Motion Picture Systems. Pt. II. The Grain Structure of Motion Picture Images – an Analysis of Deviations and Fluctuation of the Sample Number', *J. Soc. Mot. Pic. Tele. Eng.*, **61**, 181
26. Schade, O. H. (1953). 'Image Gradation Graininess and Sharpness in Television and Motion Picture Systems Pt. III. The Grain Structure of Television Images', *J. Soc. Mot. Pic. Tele. Eng.* **61**, 97
27. Schade O. H. (1955). 'Image Gradation Graininess and Sharpness in Television and Motion Picture Systems Pt. IV. A + B Image Analysis in Photographic & Television Systems (Definition and Sharpness)'. *J. Soc. Mot. Pic. Tele. Eng.*, **64**, 593
28. Schade, O. H. (1956). 'Optical and Photoelectric Analog of the Eye', *J. Opt. Soc. Am.*, **46**, 721
29. Polyak, S. (1941). *The Retina*, University of Chicago Press
30. Groot, S. G. de and Gebhard, J. W. (1952). 'Pupil Size as determined by Adapting Luminance', *J. Opt. Soc. Am.* **42**, 492
31. Riggs, L. A., Ratliff, F., Cornsweet, J. C. and Cornsweet T. N. (1953). 'The Disappearance of Steadily Fixated Visual Test Objects', *J. Opt. Soc. Am.*, **43**, 495
32. Graham, C. H., Brown, R. H. and Mote, F. A. (Jr.) (1939). 'The Relation of Size of Stimulus and Intensity in the Human Eye: I. Intensity Thresholds for White Light', *J. Exptl. Psychol.*, **24**, 555
33. Graham, C. H. and Bartlett, N. R. (1939). 'The Relation of Size of Stimulus and Intensity in the Human Eye: II. Intensity Thresholds for Red and Violet Light', *J. Exptl. Psychol.*, **24**, 574
34. Brown, R. H. (1947). 'Minor Studies from the Psychological Laboratory of Clark University. XXXIV' Complete Spatial Summation in the Peripheral Retina of the Human Eye'. *Am. J. Psychol.*, **60**, 254
35. Blackwell, H. R. (1963). 'Neural Theories of Simple Visual Discriminations', *J. Opt. Soc. Am.*, **53**, 129
36. Blackwell, H. R. and Law, O. T. (Jr.) (1958). University of Michigan Eng. Research Inst. Rep. 2455–9–F
37. Blackwell, H. R. and Austin, G. A. (1952). 'Spatial Summation in the Central Fovea', *J. Opt. Soc. Am.*, **42**, 873
38. Kincaid, W. M., Blackwell, H. R. and Kristofferson, A. B. (1960), 'Neural Formulation of the Effects of Target Size and Shape upon Visual Detection', *J. Opt. Soc. Am.*, **50**, 143
39. Smith, S. W., Blackwell, H. R. and Cutchshaw, C. M. (1958), University of Michigan Eng. Research Inst. Rep. 2144–346–T
40. Graham, C. H. and Margaria, R. (1935). 'Area and Intensity-Time Relation in the Peripheral Retina', *Am. J. Physiol.*, **113**, 299
41. Graham, C. H. and Kemp, E. H. (1938). 'Brightness Discrimination as a Function of the Duration of the Increment in Intensity', *J. Gen. Physiol.*, **21**, 635
42. Hartline, H. K. (1934). 'Intensity and Duration in the Excitation of Single Photoreceptor Units', *J. Cell and Comp. Physiol.*, **5**, 229
43. Ratliff, F. (1965). *Mach Bands*, Holden-Day Inc., San Francisco
44. Fry, G. A. (1948). 'Mechanisms subserving Simultaneous Brightness Contrast', *Am. J. Optom. & Arch. Am. Acad. Optom.*, **25**, 162
45. Granit, R. (1933). 'The Components of the Retinal Action Potential and their Relation to the Discharge in the Optic Nerve', *J. Physiol.*, **77**, 207

46. Granit, R. (1946). 'The Distribution of Exitation and Inhibition in Single-fiber Responses from a Polarised Retina', *J. Physiol.*, **105**, 45
47. Huggins, W. H. and Licklider J. C. R. (1951). 'Place Mechanisms of Auditory Frequency Analysis', *J. Acoust. Soc. Am.*, **23**, 290
48. Galambos, R. and Davis, H. (1944). 'Inhibition of Activity in Single Auditory Nerve Fibres by Acoustic Stimulation', *J. Neurophysiol.*, **7**, 287
49. Katsuki, Y., Watanabe, T. and Suga, N. (1959), 'Interaction of Auditory Neurons in Response to Two Sound Stimuli in Cat', *J. Neurophysiol.*, **22**, 603
50. Mountcastle, V. B. and Powell, P. S. (1959). 'Neural Mechanisms subserving Cutaneous Sensibility, with Special Reference to the Role of Afferent Inhibition in Sensory Perception and Discrimination', *Bull. Johns Hopkins Hosp.*, **105**, 201
51. Békésy, G. V. (1960). 'Neural Inhibitory Units of the Eye and Skin, Quantitative Description of Contrast Phenomena', *J. Opt. Soc. Am.*, **50**, 1060
52. Hartline, H. K. and Ratliff, F. (1954). 'Spatial Summation of Inhibitory Influences in the Eye of *Limulus*', *Science*, **120**, 781
53. Hartline, H. K. and Ratliff, F. (1957). 'Inhibitory Interaction of Receptor Units in the Eye of *Limulus*', *J. Gen. Physiol.*, **40**, 357
54. Hartline, H. K. and Ratliff, F. (1958). 'Spatial Summation of Inhibitory Influence in the Eye of *Limulus*, and the Mutual Interaction of Receptor Units', *J. Gen. Physiol.*, **41**, 1049
55. Ratliff, F. and Hartline, H. K. (1959). 'The Response of *Limulus* Optic Nerve Fibres to Patterns of Illumination on the Receptor Mosaic', *J. Gen. Physiol.*, **42**, 1241
56. Kirschfeld, K. and Reichardt, W. (1964). 'The Processing of Stationery Light Signals in the Complex Eye of *Limulus*', (in German). *Kybernetik*, **2**, 43
57. Taylor, W. K. (1956). 'Electrical Stimulation of some Nervous System Functional Activities', in *Information Theory*, Colin Cherry (Ed.) Academic Press
58. Kuffler S. W. (1953). 'Discharge Patterns and Functional Organisation of Mammalian Retina', *J. Neurophysiol.*, **16**, 37
59. Beitel, R. J. (1936). 'Inhibition of Threshold Excitation in the Human Eye', *J. Gen. Psychol.*, **14**, 31
60. Lamar, E. S., Hecht, S., Schlaer, S. and Hendley, C. D. (1947), 'Size, Shape and Contrast in the Detection of Targets by Daylight Vision. I. Data and Analytical Description', *J. Opt. Soc. Am.*, **37**, 531
61. Lamar, E. S. Hecht, S., Hendley, C. D. and Schlaer, S. (1948). Size, Shape and Contrast in Detection of Targets by Daylight Vision. II. Frequency of Seeing and the Quantum Theory of Cone Vision'. *J. Opt. Soc. Am.*, **38**, 741
62. Ditchburn, R. W. and Ginsborg, B. L. (1952). 'Vision with a Stabilised Retinal Image', *Nature*, **170**, 36
63. Owen, G. P. (1970). 'Visual Target Acquisition. Single Glimpse Probability of Detection', D.O.A.E. Memo. 7047

7 Modelling of Vision, 2: A Versatile Physically-based Model

It can be seen from the survey of the more comprehensive models of vision covered in Chapter 6 that there is a wide divergence of opinion as to the basic functions, even for the simplest of detection tasks. The effect of stimulus differential energy is variously considered to be predominently associated with the centre of the stimulus, as in the element contribution theory, or the edge, as in the inhibition theories. Transmission of data is implied to be by D.C. coupling in some cases, by A.C. coupling in others, particularly by Schade. The temporal storage of data is also variously described. Whilst some models explicitly imply a signal/noise situation, others do not. The inhibition theories do not appear to be related to threshold performance, nor the threshold models to known edge phenomena. Some models in their most refined forms do not appear to be able even to predict the full range of thresholds for simple circular stimuli, whilst others require complex empirical variations of various constants in order to provide a good prediction of these same simple data. Only the element contribution theory is positively shown to have the ability to predict multiple chance situations and few of the models seem to include an intentional allowance for the imperfections of the retinal image.

If all that one required to know about visual performance was concerned with detection of simple, isolated objects in plain fields, use of one or more of these models would suffice for all situations. However, vision in real life usually is not as simple as that. Many scenes are highly structured, many objects have textured surfaces, many objects have multiple contrasts against their immediate background (indeed what is the meaning of contrast when an object of interest has areas of different luminance or is seen against a non-uniform immediate background?[1]) In addition natural objects often have 'frilled' edges (e.g. trees and bushes) which tend to blur the outline (see for instance Chapter 11). Yet again some objects are characterised by a line profile with no solid centre (as in any outline drawing) or are merely a discontinuity in the field of view, unbounded on one or more sides and hence incapable of being represented as an area (e.g. a horizon, particularly between sea and sky).

In order to supplement existing models, where there is a difficulty in treating such conditions, an attempt was made by the author at modelling visual performance based on known general physical properties of the eye and an assumed logical form of simple, first stage processing by the central nervous system (i.e. processing aimed at detecting existence of local stimuli rather than recognition of any detail). The outcome was the progressive development of a model of the visual process which is presently able to predict, using a simple set of constants, a very wide range of simple detection thresholds, to relate many sets of isolated data one to another, and by its very nature to allow extrapolation into the complex regimes of practical vision[2-7]. As will be seen later, many facets of the model are similar to facets of the models discussed in Chapter 6, but facets of several different models are brought together.

This chapter is devoted to a summarised development of the model and its application to various simple visual threshold situations discussed in the preceding chapters. Its applicability to more complex viewing situations will be discussed in Chapters 8, 9, 12 and 13. It should be noted that, although the reader is referred to previous publications for detail information on model development, certain data contained in these previous publications are now believed to be in error. In such cases the information believed to be erroneous is superceded by data contained herein. It should also be stressed at this point that, whilst this model is believed to be compatible with the general form of neural network in the retina, it *does not* purport to explain all the detailed workings of the neural networks either at the retina or in the cortex, but only to provide what might be termed an overall information transfer function between retinal images and the decision making part of the brain. Should the reader be interested in pursuing the detailed workings of the neural networks he is referred to Pirenne[8], Polyak[9], Alpern[10], Brindley[11], Cornsweet[12] and Walter[13] as a start to an appreciation of the probable detailed workings of the eye and brain. Such studies are not, however, considered by the author to help very much with general appreciation of threshold behaviour. Rather, from the author's own experience, they tend to cloud the issue with excessive detail.

7.1 BASIC CONCEPTS

Before proceeding to the progressive development of the model, it is necessary to define the various concepts on which it is based. These are, briefly, as follows:—

(1) *Retinal image quality*

That the retinal image is degraded due to the spread function of the refraction optics of the eye, as measured by such workers as Westheimer and Campbell[14], Campbell and Gubisch[15] and Flamant[16], and also due to retinal diffusion[17,18]. The retinal receptors thus receive a degraded impression of the scene viewed, the degradation being somewhat dependent on the working pupil diameter (see Section 2.3). In addition the retinal image for normal viewing is presumed to be degraded further by effects of involuntary tremor and drift during a given glimpse (see Section 2.6). Fig. 7.1 illustrates a typical illuminance cross section through the foveal retinal image of an extended object – showing the softened profile due to the spread function of the refraction optics of the eye and retinal diffusion, together with tremor and drift. Fig. 6.1 shows the composite spread function which produces such a softening of the profile.

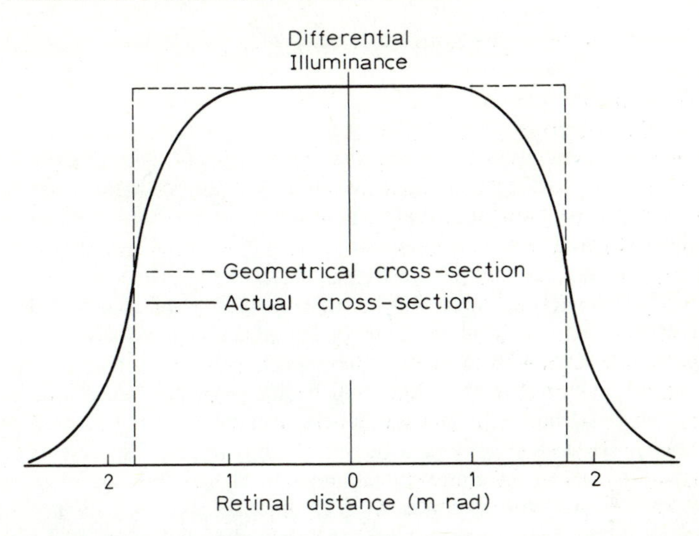

Fig. 7.1. Typical retinal illuminance cross-section for a 3.5 mrad diameter disc stimulus (showing degradation of borders).

(2) *Form of data transmission*

That the primary method of data transmission from the retinal photoreceptors to the brain is effectively by difference signals[2,6,7]. In this sense the model is basically similar to Schade's model (Section 6.3). However, rather than assuming tremor to be the source of the difference signals, these are assumed to be generated in a manner similar to that proposed in certain inhibition models (Section 6.5), particularly that of Huggins and Licklider[6]. It will be shown later (Section 7.5) that an assumption that tremor is the source of differencing appears to be incompatible with peripheral threshold trends. On the other hand a neural differencing concept is directly relatable to the difference processing in the neural networks proposed by some workers in order to explain colour vision behaviour (e.g. Hurvich and Jameson[19] and Guth and Lodge[20] – also Section 2.4). It can also readily be related to the inhibitory surround responses measured by Werblin in the bipolar cells of *Necturus*[21-23]. The fact that Mach, Huggins and Licklider, and Fry all found a need for second differentiation to explain all the supra-threshold Mach phenomena (Section 6.5) is not considered to be at variance with the present threshold model since, if second differencing is produced by two differencing processes in series, at first difference thresholds the second differences will generally be sub-threshold.

Apart from being an economical way of transmitting data from a very complex sense organ, it will be realised that the above assumptions potentially allow for prediction of thresholds of many forms of complex stimuli (e.g. blurred outlines, the horizon, line drawings, multiple contrast stimuli, etc.).

(3) *Noise environment*

That the detection process takes place in the presence of random noise both within and external to the visual system. It is postulated that one must expect there to exist different sensitivities in different elements of the neural network, with the result that there will exist, at any one time, what might be described as spatial noise between the various receptor channels. Such noise must equally be present in difference signals. Also, due to the matrix structure of the retina, there will be additional first difference 'noise' as a result of the 'lay' of local parts of an image contour with respect to the matrix structure. In addition it is to be expected that each receptor channel will exhibit temporal noise, which will be increased by 'noise' due to the interaction of tremor, drift and the matrix structure with the contour image. The noise environment is 3 dimensional (2 dimensions of space and one of time), as opposed to the conventional one dimensional noise environment of, for instance, electronic systems. At high luminances and with textureless scenes the internal noise may be assumed to be the total noise. However, for low luminances the incoming image will be affected by quantum noise as assumed by Rose and Jones (Section 6.2), whilst, for textured surfaces, additional spatial noise must be allowed for (see Section 12.7).

(4) *Decision Process*

That the detection decision is made on the strength of a processed 'intelligence' signal at the brain, this in itself inevitably having temporal noise associated with it (e.g. Ronchi *et al*[24]). Such a 'decision' signal may well be associated with the 'decision' or 'expectancy' wave described by Walter[25,26]. It would also be at this level that the decision theory of Swets *et al*[27] (see also Chapter 3) may be expected to apply.

(5) *Strength of 'intelligence'*

That the 'intelligence' available to the brain for simple detection is the number of associatable component signals transmitted in the presence of noise, its strength being measured in terms of the number of statistical comparisons which can be made between the associatable signal components. If the signal strengths at the receptor difference level and decision level are S_D and S_B respectively, then we may thus say that

$$S_B \propto \frac{S_D \cdot n(n-1)}{2!} \tag{7.1}$$

where n is the number of associatable signal components.

But, for constant temporal and input noise, S_B may be assumed to be a constant. Therefore

$$S_D \propto \frac{1}{n(n-1)} \tag{7.2}$$

(6) *Confidence level*

That the detection decision is made at a confidence level related to the strength of the 'intelligence' reaching the brain. This confidence level will itself be dependent on such factors as state of briefing, state of learning and pay-off (punishment and/or reward against results). This factor is thus largely psychological by nature and must be in turn related to quantum absorption (Section 6.2).

(7) *Saccadic effects*

That the retinal image is normally displaced to a new position on the retina at each saccade, thus providing a largely uncontaminated new set of difference signal components in the presence of a *different* set of noise components for simple stimuli. Such saccadic motion, considered in conjunction with the first differencing, also automatically accounts for the fact that the visual processes largely 'wipe the image clean' at each glimpse (c.f. Blackwell's 6c/s scanning – Section 6.4).

This overlaying of signal on different spatial noise structure must lead to a gradual suppression of spatial noise, eventually leading to an essentially spatial noise free situation with consequent improvement of threshold.

The gradual suppression of inter-receptor noise is illustrated in Fig. 7.2, where typical (statistical) sets of neural signals for 5 threshold glimpses at a square object of 10 min of arc side length are shown, together with the progressive superposition of the 'significant' signal data. It can be seen that, whilst only 20% of the contour is defined on any one glimpse in this situation, and it is hard to relate the rather random signal spikes to one another, after 5 glimpses the profile association is becoming strong. This

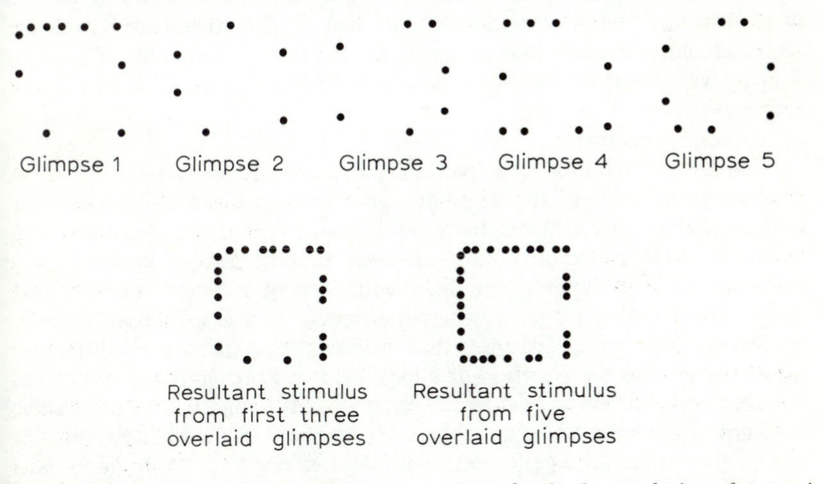

Fig. 7.2. Illustration of the progressive suppression of noise by overlaying of successive glimpses.

statistical multiple chance improvement is essentially as implied by Blackwell (Section 6.4).

(8) *Neural response thresholds*

That all noise functions, both internal and external to the eye, are Poissonian, and that all neural processes require a finite detectable difference in the absence of noise in order to operate (c.f. Cornsweet[12], Chapter 6). This leads to a series of basic functions of the form

$$F(N) = k_1 N^{-0.5} + k_2 \tag{7.3}$$

where k_1 is a measure of the uncertainty associated with the particular noise function and k_2 is the finite difference necessary when N tends to infinity in order that a response shall be achieved.

(9) *Local adaptation state*

That the local adaptation levels of the retina are controlled by a simple exponential lateral coupling as proposed by Mach, Fry and others (see Section 6.5) as part of the necessary 'inhibition' to account for Mach bands. Such lateral cross coupling would not be expected to be significant when considering detection of simple stimuli in plain fields due to the low differentials of retinal illuminance. It *would* become significant, however, where strong scene contrasts occur (see Chapter 13). This lateral cross-coupling, together with transmission of difference signals, is in keeping with the Retinex theory of Land and McCann[28], which subjectively explains the way in which we interpret uniformity of reflectivity of surfaces in the presence of shading, etc.

(10) *Receptor independence*

That the individual types of retinal receptors operate independently as far as luminance thresholds are concerned – that is, that difference signals are obtained only between like receptors for luminance thresholds. This is in keeping with Schade's modelling (Section 6.3) and also the opponent colour vision models (Section 2.4).

(11) *Receptor grouping*

That thresholds at a local part of the retina are defined by effective receptor group spacings at that point, cones being assumed to be effectively individually coupled to the optic nerve, at least in the several degrees around the fovea. Rods, on the other hand, are assumed to be grouped in clusters as a main source of incremental low light sensitivity. It is further assumed that retinal image quality (as measured at the receptors) falls off proportionately to the receptor group spacings, thus maintaining constancy of difference signal strengths as a proportion of a local luminance difference at all parts of the near foveal retina at least (see Section 7.5). Although there is no positive evidence of this known to the author, such a trend in image quality is possible due to the marked changes in retinal diffusion effects away from the fovea as anticipated by Ohzu and Enoch[17].

(12) *Spatial noise level*

That the spatial noise in the retina is such that, at threshold, a 30% change in 'strength' of stimulus difference components is sufficient to change probability of a component signal response at the bipolar cells from virtually zero to virtually 100%. This is in keeping with the findings of Silverthorn[29], who has found that a reprocessing of standard deviations for individual sessions of Taylor's single glimpse threshold data[30] yields a strikingly stable lower limit of 0.1 C_M, this being independent of stimulus size and observer where C_M is the mean contrast threshold for the session. This reprocessing of Taylor's frequency of seeing data has already been discussed in Chapter 3, and is shown in Fig. 3.2. It must be concluded that the spatial noise for a single glimpse is unlikely to be larger than that associated with an RMS of 0.1 C_M, the rest of the conventional frequency of seeing spread being associated with temporal variations in threshold over longer periods than a single glimpse, possibly as a result of variability in ganglion integration[7].

With the foregoing in mind, an inspection of the line spread function of Fig. 6.1 will show that, if the peak region is at or just above component threshold, the regions displaced by more than about one fifth of a millirad from the peak will effectively be lost in noise for sharp discontinuities. Thus, at least for foveal detection of simple, symmetrical stimuli, one might expect the *only* region of importance to be the region of *maximum* illuminance gradient, covering no more than 2 receptor pair widths across the illuminance contour. It was on the basis of this assumption that the initial model was constructed[2,3]. For more complex stimuli such an assumption is now realised to be oversimplified and the statistical probability of component response must be elicited[6,7]. This will be discussed further in Section 7.7 and further confirmatory evidence of the effective suppression of softer parts of profiles will be presented in Chapter 12.

(13) *Temporal response*

That the total visual system behaves like a capacitive electrical network with a charge time constant (τ) of the order of 0.05 s and a discharge time constant of the same order or rather longer. This is in keeping with much experimental data showing evidence of a summation or integration time (e.g. Ronchi and Longobardi[31], Schweitzer and Troelstra[32]), as well as the empirical findings of Blackwell[33] (see also Section 6.4) and Schnitzler[34]. It should perhaps be stated at this point that such an assumption about the overall visual system does not in any way conflict with the known physiological data about detail transmission of information in the neural networks by pulse trains. That pulses in such trains are associated with much shorter time constants than 0.05 s is a necessity for useful performance of a pulse coded transmission system (see Section 2.11 and Pirenne[35] for instance) and has indirect implication on the behaviour of the visual system to repetitive flash stimuli. However, for single light pulse and continuous forms of presentation it is considered entirely adequate to deal only with the total system response.

The foregoing led the author to propose a basic equation of vision for constant luminance detection thresholds of simple stimuli of the form

$$F(\epsilon) = K_1 F(n) + \delta \qquad (7.4)$$

where $F(\epsilon)$ is a function of the profile shape and 'strength' in the retinal image (ϵ being the psychometric contrast $\Delta B/B'$), $F(n)$ is a function of a number of differential signals available to the brain as discussed in (5), K_1 is a constant for a given background luminance, experimental set-up and probability of detection, being associated with retinal and neural noise, δ is a constant for a given background luminance and probability of detection, being determined by the level of intelligence at the brain on which the observer is prepared to make a decision. By inference δ is also dependent on the number of glimpses in a multiple glimpse situation and on the task (see (6) and (7)).

It has been found possible to define the various functions and constants in this equation in terms of physically known properties of the eye, the quantum theory and decision theory concepts, such as to provide a very widely applicable predictive model of visual detection in plain fields, where the only major indeterminancy is in the value of δ, depending as it does on psychological factors as discussed in (6). There is, in addition, a smaller uncertainty in the value of K_1, dependent on the particular experimental set up. For a given form of experimentation, fair approximations to δ and K_1 can usually be made from the variety of classical experimentation available. It will be tentatively shown in subsequent chapters that the foregoing model also appears to have much wider ranging predictive capabilities for structured scenes and simple recognition tasks.

7.2 THE BASIC FOVEAL MODEL

Since the majority of critical visual tasks are carried out by foveal inspection for extended periods, the visual model was first propounded for foveal detection[2]. We shall follow the same procedure here and subsequently extend it to other viewing domains. It should be noted that such a restriction immediately limits us to photopic luminance levels, since the fovea contains no rod receptors[36] (see also Section 2.4). Many of the succeeding paragraphs will also be restricted to photopic luminance levels, but Section 7.11 is devoted to a short discussion of a potential parallel modelling approach to scotopic vision.

Now for foveal detection of simple circular stimuli of a size significantly greater than that of the spread function it can be shown[3] that $F(\epsilon)$ in Equation 7.4 may be given as

$$F(\epsilon) = \log_e \left[\frac{(K_2 + K_3)\epsilon + 1}{K_3 \epsilon + 1} \right] \qquad (7.5)$$

where K_2 is a slope constant defined as the ratio of the mean differential illuminance between two adjacent receptors at the maximum slope region of the illuminance profile to the differential illuminance defining ϵ, and K_3 is a constant defining the position on the profile at which the maximum slope occurs, it being defined as the ratio of the difference between illuminance at the maximum slope and that of the background to the differential illuminance defining ϵ. Equation 7.5 may be approximated by

$$F(\epsilon) \approx K_2 \epsilon \tag{7.6}$$

for threshold situations where $(K_2 + K_3)\epsilon$ is very much less than unity (since $\log_e(1 + x) \approx x$ for $x \ll 1$). This approximation is found to be valid over the majority of situations met in photopic viewing, and considerably eases calculations and appreciation when some of the extensions to the model are considered (but see Section 7.6).

For large circular stimuli and constant background luminance one may thus write

$$\log_e \left[\frac{(K_2 + K_3)\epsilon + 1}{K_3 \epsilon + 1} \right] \approx K_2 \epsilon = \frac{K_1}{n(n-1)} + \delta \tag{7.7}$$

δ being chosen to suit single glimpse or long viewing times as appropriate[7].

In the above equation K_2 may be taken as 0.163 approximately for naked eye viewing. Equation 7.7 is also approximately correct for application to other than circular stimuli, providing that all dimensions are large compared to the spread function and that the stimuli have simple profiles. For convenience, n may be taken as the number of receptor pairs around the perimeter of the stimulus image (see Lavin and Overington[3]).

For small stimuli it is necessary to consider what happens when one convolutes a spread function with an object profile of similar size to itself or smaller. In the limit (the point source situation) the resultant image profile will be of the same dimensions as the spread function and of a maximum slope proportional to the *area* of the subtended object. Thus the LHS of Equation 7.7 must be modified to a form

$$F(\epsilon) = \log_e \left[\frac{(K_2 + K_3) c_v{}^2 \alpha^2 \epsilon + 1}{K_3 c_v{}^2 \alpha^2 \epsilon + 1} \right] \tag{7.8}$$

where c_v is a constant relating the maximum slope of the spread function to that of the edge of the image of an extended object and has a value of 2.05 approximately for naked eye viewing and α is the diameter of the unresolved 'point' object (in mrad).

At the same time the number of retinal receptor pairs involved in the detection becomes constant at a value n_0 defined by the spread function shape (approximately nine). Thus the RHS of Equation 7.7 becomes constant and the

LHS approximates to

$$F(\epsilon) = K_2 c_v{}^2 \alpha^2 \epsilon \quad (\propto \alpha^2 \epsilon)$$

a statement of Ricco's Law that area x contrast is a constant (see Section 6.1). This transition from Equation 7.7 for small objects also provides an explanation for the apparent fall off of quantum efficiency for small objects as discussed in Section 6.2.

For intermediate sizes between the large stimulus and the point source it is necessary to provide transition formulae which define the progressive collapse of the stimulus profile and the progressive approach to a fixed contour length defined by n_0. It has been found that both these trends may be adequately modelled, at least for circular targets, by rewriting Equation 7.7 as [3]

$$K_2 Z = \frac{K_1}{n^2} + \delta \tag{7.9}$$

where $Z = c_v{}^2 \alpha^2 \epsilon / (c_v{}^4 \alpha^4 + 1)^{1/2}$ and n is given as $15.4\ (\alpha^2 + 0.34)^{1/2}$ (ref. 3).

The ability of Equation 7.9 to model infinite viewing time threshold curves is illustrated by Fig. 7.3, where two of Blackwell's high luminance, infinite viewing time curves[37] (see also Section 4.1) are predicted simply by choosing values of K_1 to fit the Ricco's law region and a common value of δ.

For non-circular targets of low aspect ratio it is normally adequate to model as an equivalent circular target if the profile is simple and sharp. Equally for targets of large aspect ratio it is usually adequate to model in terms of an

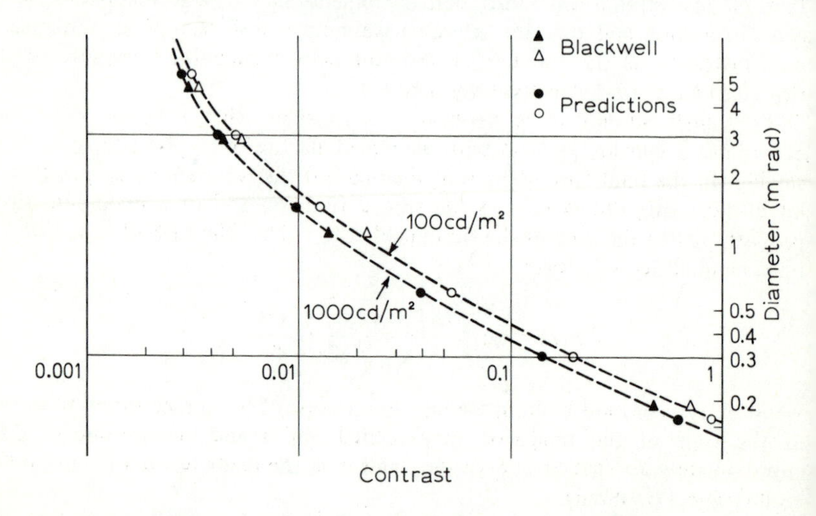

Fig. 7.3. Illustrating the predictive capability of the maximum gradient model for disc stimulus thresholds at high luminance. (Reproduced from Overington[4] by courtesy of the Advisory Group for Aerospace Research and Development of NATO).

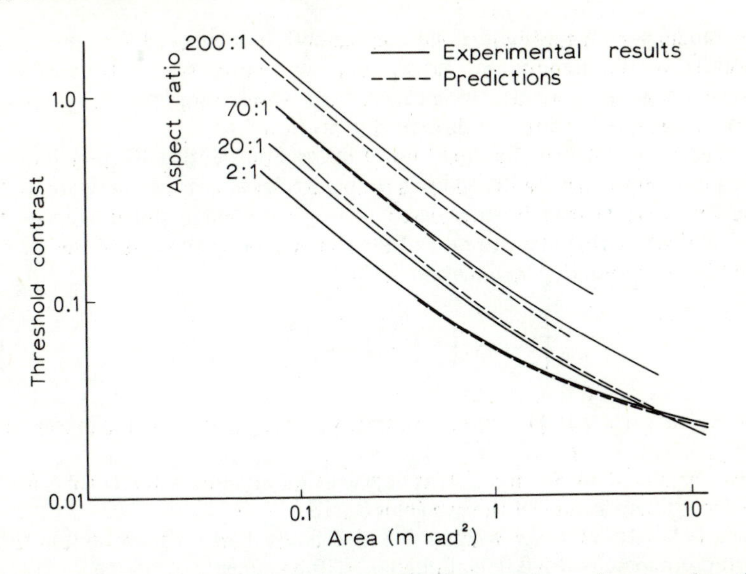

Fig. 7.4. Prediction of Lamar et al *thresholds for a variety of rectangular stimuli. (Reproduced from Overington and Lavin*[2] *by courtesy of Optica Acta).*

equivalent rectangle of dimensions x and y, whence

$$Z = \frac{(c_v')^2 \cdot x \cdot y \cdot \epsilon}{[(c_v')^2 \cdot x^2 + 1]^{1/2} [(c_v')^2 \cdot y^2 + 1]^{1/2}}$$

and

$$n = 9.9[(x + y)^2 + 0.83]^{1/2}$$

In the above $c_v' = 2c_v/\sqrt{\pi}$.

The predictive capability of this version of the model is shown in Fig. 7.4, where the entire range of rectangular thresholds due to Lamar, Hecht, Hendley and Schlaer[38] (see also Section 4.5) are predicted using one pair of values for K_1 and δ.

7.3 THE EFFECTS OF PRESENTATION TIME

It would appear from Fig. 4.1 that there is a fairly marked difference in visual performance depending on whether one has one glimpse or a long time for studying an object. Referring back to Section 4.7 it will also be recalled that experimental results appear to indicate that threshold is inversely proportional to presentation time for very short exposures, becoming rather uncertain at longer presentation times, with a discontinuity around 0.25 – 0.5s. How then do we model such a threshold trend completely? It is believed that the answer lies jointly in the assumption that the complete visual system behaves as an electrical charge network and that noise is progressively suppressed by successive overlay of signal information as described in Section 7.1.

Looked at in this light the build up of information during the period of one fixation or glimpse may be likened to a simple electrical charging network – that is, the total charge may be considered to be dependent exponentially on the length of flow of charging current and the charging time constant of the system. This leads to a reordering of Equation 7.9 to

$$\left[1-\exp\left(-\frac{t_D}{\tau}\right)\right]K_2'\epsilon=\frac{K_1}{n^2}+\delta \tag{7.10}$$

where K_2' is such that $[1-\exp(-0.3/\tau)]K_2'\equiv K_2$ and t_D is the presentation time ($t_D \ngtr$ a glimpse).

Also, as stated in Section 7.1, it appears that an approximate value for the integration time constant of the eye (τ) is 0.05 s.

Since $(1-\exp x)\to -x$ when $x \ll 1$ it follows that, for presentation times very short compared with 0.05 s, Equation 7.10 becomes

$$t_D K_2'\frac{\epsilon}{\tau}=\frac{K_1}{n^2}+\delta \tag{7.11}$$

which, for a given size, gives $\epsilon \propto 1/t_D$ (Bloch's Law) (Section 6.1). For long presentation times

$$\left[1-\exp\left(-\frac{t_D}{\tau}\right)\right]\longrightarrow 1$$

and, at $t_D = 1/3$ s (the end of an average glimpse time),

$$\left[1-\exp\left(-\frac{t_D}{\tau}\right)\right]=[1-\exp(-6)]\approx 1$$

In other words, at the end of an average glimpse time this model predicts that almost all of the possible accumulation of information from a fixation has been achieved. There would thus be little point in natural fixations being much longer than 1/3 s on average. The above is also a possible explanation of the apparent fall off of quantum efficiency for long pulses (c.f. Section 6.2).

Let us now turn our attention to the multiple glimpse situation. Here we have suggested the concept of gradual suppression of noise by an overlaying of signal information on a set of random noise patterns. Then, if we have m glimpses, it is suggested that the suppression of noise should be represented as a cumulative probability function of the form

$$\Phi_m = 1 - [1 - x^{-0.5}]^m \tag{7.12}$$

where Φ_m is the residual noise after m glimpses.

It appears from study of experimental data that this suppression of noise may be largely accounted for by considering δ to be composed of two terms, one related to the absolute, noiseless threshold and the other being synonymous with x in Equation 7.12. For high photopic luminances this 'noise' component of δ appears to be approximately double the absolute, noiseless component, yielding infinite viewing time values of δ approximately one third of those for single glimpse viewing. It is not at this time clear whether there should be two similar components of K_1, owing to a very considerable variability of K_1 from one set of experimental data to another (e.g. Taylor[30] and Blackwell and McCready[39]).

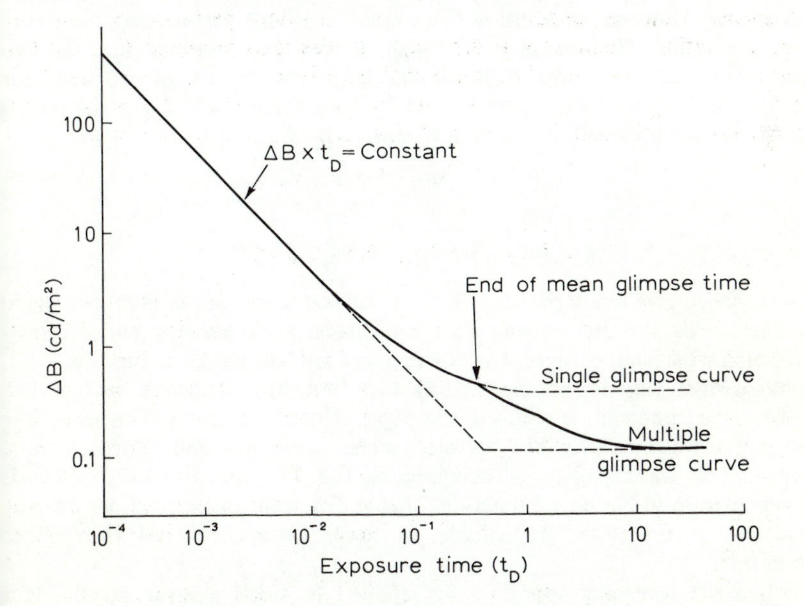

Fig. 7.5. Prediction of an entire temporal threshold curve, including both short presentation time and multiple glimpse viewing.

Figure 7.5 shows the composite threshold prediction as a function of presentation time for single and multiple glimpses for one target size. The discontinuity at 1/3 s will be blurred (as Clark and Blackwell[40] found) in a practical situation, owing to the fact that the glimpse durations exhibit a considerable spread. It is interesting to note that the improvement for multiple glimpses is virtually complete after 10–15 s, thus predicting Blackwell's findings[37] that viewing times of 15 s and greater could be considered to be effectively infinite.

7.4 THE EFFECTS OF SCENE LUMINANCE

Thus far the modelling in this chapter has been restricted to a particular adaptation level, constants being chosen to suit. Obviously if a model is to be in any sense general in its application there should be a simple law relating it to physical factors which change with scene luminance. Let us, then, look at such factors.

Now firstly it might be assumed that, as adaptation level changes, the effect is a change of gain in the neural networks (c.f. Schade, Section 6.3). Then, considering an analogy with electronic amplifiers, the higher the gain the more noisy the output. We might thus expect the performance to be controlled partially by a component inversely proportional to the square root of the retinal illuminance. However, in addition there must be a finite performance limit even when the retinal illuminance is very high. It was thus proposed that the two constants in the basic foveal model should each be composed of two terms (of the form $k_1 E_R{}^{-0.5} + k_2$, where k_1 and k_2 are constants and E_R is the retinal illuminance in trollands). It was found[5] that such a form of K_1 and δ, where

$$K_1 \approx 15.4 E_R{}^{-0.5} + 0.48$$

and

$$\delta \approx 1.25 \times 10^{-3} E_R{}^{-0.5} + 4 \times 10^{-4}$$

would indeed give a very presentable fit to the major part of the photopic region of Blackwell's infinite viewing time thresholds, even ignoring minor effects associated with pupil changes. The fits achieved are reproduced as Fig. 7.6.

Minor discrepancies (not significant) were thought to be due to the fact that as the scene luminance is reduced, the pupil diameter increases. This leads to a non-proportional relationship between scene luminance and retinal illuminance – a fact taken account of in computing Fig. 7.6 – but also inevitably leads to degradation in the spread function. This in turn leads to expectations of small losses in performance, particularly at scene luminances less than about 3.5 cd/m^2.

When the foregoing approach was applied to single glimpse thresholds it became apparent that the luminance functions were not all embracing. In an attempt to model the data of Blackwell and McCready[39] it appeared that a reciprocal law of the form $(k_1 E_R{}^{-1} + k_2)$ was more applicable for both δ and K_1[41]. This of course relates directly to Blackwell's finding of Equation 6.15 for short time viewing. The discrepancy between the functions for short and long time viewing may, it is suggested, also account in part for the discrepancy between Rose's and Bruscaglioni's quantum vision models (Section 6.2). No positive explanation of the difference is offered at this time, but it is tentatively suggested that the multiple glimpse situation may possibly progressively suppress some of the effects of quantum noise in a similar manner to the suppression of spatial noise due to inter-receptor differences discussed previously, thus effec-

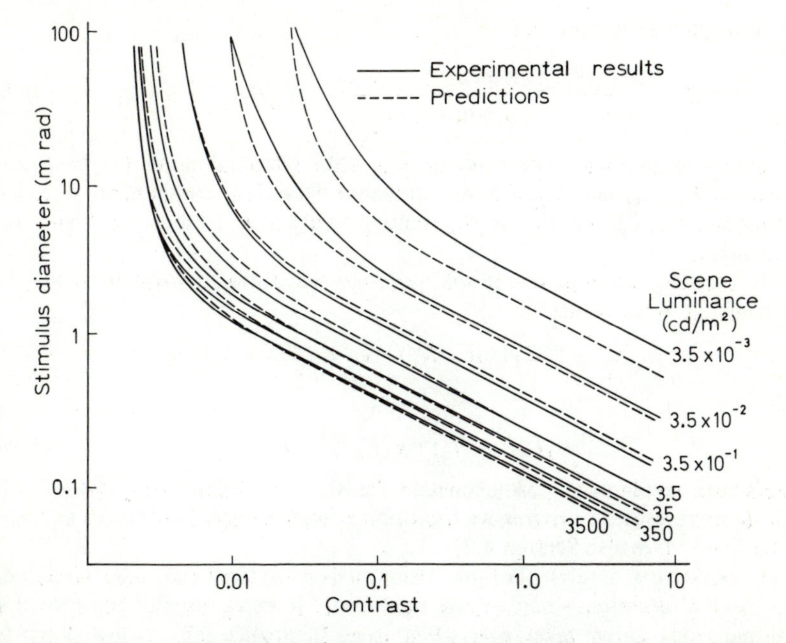

ig. 7.6. Prediction of the entire infinite viewing time thresholds for photopic luminances or disc stimuli as found by Blackwell. (Reproduced from Overington[5] by courtesy of the ournal of the Optical Society of America).

vely introducing a $\sqrt{E_R}$ factor for infinite viewing times. However, regardless f the reason for the differences, the two forms of K_1 and δ for single glimpses nd infinite viewing time respectively do provide useful and accurate threshold end models. In addition to the difference of δ for single glimpse and infinite ewing discussed in Section 7.3 it is necessary to increase δ substantially for free hoice viewing (as might be expected from confidence considerations). The etailed form of this increase has not been established at this time, but looks to e of the order of 5 for conversion from forced to completely free choice. 'ariations of K_1 of the order of 2 or 3 appear necessary to account for different xperimental conditions.

Strictly speaking it is also necessary to consider an effective luminance due to e temporal internal system noise at very low luminance levels, since internal oise gives rise to a dark current[42] even in the absence of external stimulus. uch a system noise should be introduced into the left hand side of Equation 7. To account fully for scene luminance (with due regard for variation of oread function with pupil diameter) Equation 7.7 should then read

$$K_2 \frac{(\Delta B - N_v)}{B'} = \frac{(K_B E_R{}^{-1} + K_0)}{n(n-1)} + (\delta_B E_R{}^{-1} + \delta_0) \qquad (7.13)$$

for single glimpse viewing and

$$K_2 \frac{(\Delta B - N_v)}{B'} = \frac{(K_B E_R{}^{-0.5} + K_0)}{n(n-1)} + (\delta'_B E_R{}^{-0.5} + \delta'_0) \qquad (7.14)$$

for extended viewing, where N_v is the RMS temporal noise in the neural networks; K_B, δ_B and δ'_B are the luminance dependent components of K_1, δ and δ'; and K_0, δ_0 and δ'_0 are the limiting values of K_1, δ and δ' at high scene luminances.

For constant stimulus size it will be found that these two equations may be rearranged into the forms

$$K_2(\Delta B - N_v) = k_1 + k_2 E_R \qquad (7.15)$$

and

$$K_2(\Delta B - N_v) = k_1 E_R{}^{0.5} + k_2 E_R \qquad (7.16)$$

which both condense to ΔB = constant for very low luminances, $\Delta B \propto E_R$ at high luminances and a progressive transition at intermediate luminances as found by Barlow[42] (see also Section 4.2).

In considering natural photopic vision, it is considered that only rarely will one meet a situation where N_v is significant, it being usually the case that scotopic (rod) vision takes over when scene luminance falls so low. Such an allowance for N_v in scotopic vision is, however, very necessary.

7.5 PERIPHERAL THRESHOLDS

Although most critical visual tasks are executed using foveal vision, one must not ignore the importance of peripheral vision. It is, after all, peripheral vision which provides our basic orientation and also cues us for objects of interest on which we should concentrate our foveal vision. Furthermore, in a search situation – as will be dealt with in Chapter 8 – peripheral vision becomes all important.

Considered in terms of the visual model being discussed in this chapter, three factors might be expected to contribute to peripheral threshold variations. Firstly, if the optical spread function changes markedly with retinal image position, one would expect a change in K_2. Secondly, the local retinal receptor spacing would be expected to have two effects at the same time – a lowering of the number of receptor pairs around a given contour, and an increase in the level of the available stimulus due to the larger separation in retinal receptor pairs providing an increase in K_2.

Now experimental work on local threshold performance by a number of workers has shown threshold trends which appear to be closely related to the distribution of retinal receptors as measured by Østerberg[43] and Polyak[44]. For instance Kishto[45] has shown considerable variation in photopic threshold round certain annular regions of the retina centred on the fovea, which he has elated to non-uniform distributions of retinal cones (as for instance in the icinity of the blind spot). Equally Yves le Grand[46] draws attention to a near dentity between the rate of decrease of visual acuity and the rate of increase of etinal cone receptor spacing as the angle from the fovea is increased, at least out o an eccentricity of several tens of m radians. Taylor[47] has shown peripheral hreshold performances for disc stimuli with decreasing luminance levels to change gradually from a strong foveal optimum at high photopic levels to a pronounced optimum at some 0.25 rad from the fovea at scotopic levels, where od vision predominates. In the light of these findings it was assumed that, possibly, the fall-off of optical quality and the increase in separation of receptor pairs might balance out. Performance would then be simply related to the fall-off of linear receptor density at angles away from the fovea as measured by Østerberg.

It was found that a convenient and adequate description of the average linear etinal cone density, at least in the region out to some 0.25 rad from the fovea, was the function

$$n_\theta = n_F \left(\frac{180\theta}{\pi} + 1 \right)^{-1/2} \qquad (7.17)$$

where n_F and n_θ are the number of cones in a given angular distance in the centre of the fovea and at an angle θ radians from the fovea respectively. Using his cone density function, the basic foveal threshold equation was extended to

$$K_2 \epsilon = \frac{K_1(180\theta/\pi + 1)}{n^2} + \delta \qquad (7.18)$$

where δ is appropriate to the viewing time (Section 7.2).

Using Equation 7.18 the comprehensive peripheral threshold data of Taylor[30] for one high luminance level (260 cd/m^2) were predicted with the esults shown in Fig. 7.7. This was considered a good validation of the adequacy of this form of model out to at least 0.2 rads from the fovea (with the obvious exception of the region around the blind spot).

The suggestion that optical quality might fall off strongly with increasing eccentricity is at variance with claims by Le Grand[46] and Van Meeteren[48], whilst being at least tentatively possible due to retinal diffusion according to Ohzu and Enoch[17]. No really positive data appear to exist on this matter, but an assumption that image quality falls off roughly in proportion to local retinal matrix coarseness would appear to be a desirable property of the visual system in order to retain any sort of balance between resolution and image noisiness. If the local optical quality is excessively good relative to the local receptor unit spacing, it can be shown that the resultant is either very noisy or very inefficient imagery. This optimal matching between matrix size and optical image spread is considered for electronic matrix detectors by Lavin and Quick[49]. It may

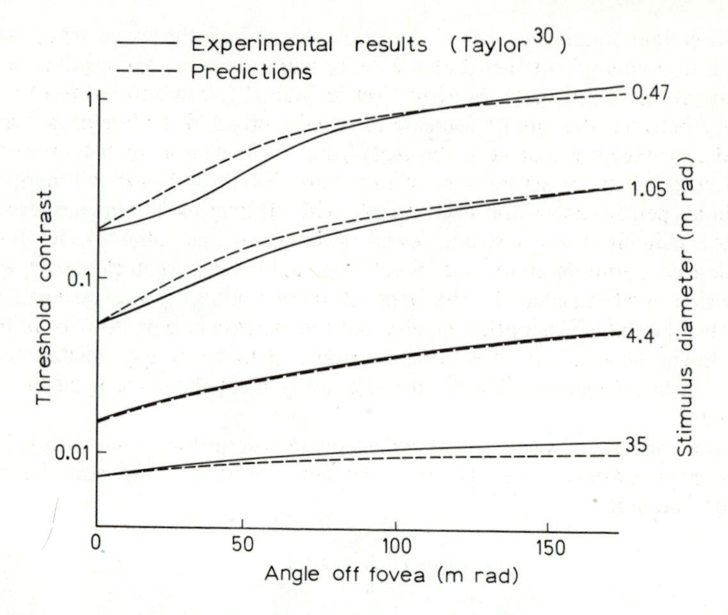

Fig. 7.7. Prediction of peripheral thresholds for disc stimuli at moderate luminance.

similarly be reasoned that, with the increased receptor grouping in the far periphery, the idea proposed by Schade (Section 6.3) that ocular tremor is the source of AC neural signals is hardly compatible with efficient signal processing.

Support for the concept of optimal balance between optical quality and matrix coarseness may be found when one realises that, even in the most primitive eyes such as that of *Limulus*, the effective spread function (in this case the angular coverage of each ommatidium – see Section 6.5.5) is matched to the angular spacing between ommatidia to much the same extent as the foveal spread function is matched to the foveal receptor spacing in the human eye[50].

7.6 SIGN OF CONTRAST

Most experimental work on detection thresholds is carried out on positive contrast stimuli for convenience, it being very difficult to provide large, uniform areas of high luminance with small negative contrast stimuli superimposed. However, much recognition work is carried out using essentially negative contrast stimuli, and some real world detection situations also involve negative contrasts (for instance the detection of aircraft against the sky is usually a negative contrast situation – see Chapter 17). As discussed in Section 4.13 there has been conflicting evidence suggesting on the one hand that positive and

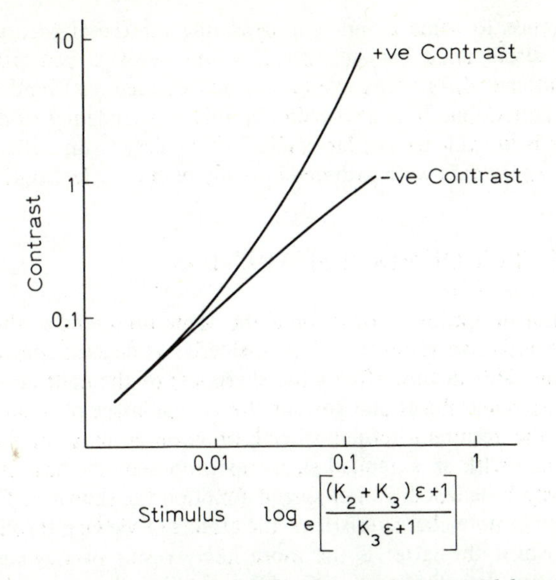

Fig. 7.8. Illustration of the theoretical differences to be expected between positive and negative contrast stimuli. (Reproduced from Overington[2] by courtesy of Optica Acta).

negative contrast stimuli of equal contrast are equally detectable (e.g. Blackwell[37]) and on the other hand that high negative contrast stimuli are more easily detectable than equal positive contrast stimuli (e.g. Achurch[51], Clark *et al*[52] and Hawkins and Church[53]). It is instructive to refer back to the full form of the foveal vision model (Equation 7.7) at this point. Whereas for low contrasts we found that the approximation $K_2\epsilon$ was adequate for the left hand side of the equation, if the contrast is not low it is found not only that the simple approximation does not hold but also that the predicted thresholds for negative and positive contrast stimuli of equivalent size are markedly different. In Fig. 7.8 the values of

$$\log_e\left[\frac{(K_2 + K_3)\epsilon + 1}{K_3\epsilon + 1}\right]$$

for positive and negative contrasts are plotted against contrast for constant K_2 and K_3. It will be seen that, for equal sensation to that for a high negative contrast of (say) -0.8, a positive contrast of $+2.9$ is required. In practice such a condition of high ϵ and constant K_2 and K_3 will only apply at very low luminance levels (cf. the low luminance differences between increment and decrement thresholds found by a number of workers – e.g. Patel and Jones[54], Short[55] and Cohn[56]). At high luminance levels the effective value of K_2 will be modified by small size corrections (see Equation 7.9) whilst K_3 will tend to zero, these trends to some extent compensating for the effects of large ϵ an reducing the discrepancy between positive and negative contrast thresholds However, significant differences would still be expected, as found by Patel an Jones, Short and Cohn. It is also believed that the tendency to differences a high contrasts is in evidence in Blackwell's Tiffany data[37] on critical inspection but that insufficient data were gathered at really high negative contrasts.

7.7 THE EFFECT OF EDGE SHARPNESS

If the eye is not in optimum focus, or if the scene presented to the eye is itsel imperfect, the effective result may be considered as degradation of the retina spread function. This in turn affects the sharpness of the contour of the retina image of an extended object and softens the retinal image of small objects. It i possible that the resultant retinal spread function is of a number of forms ranging from one with an extended skirt superimposed on a basically naked ey core to a grossly broadened overall spread function (as shown in Fig. 7.9). Th former of these is more characteristic of the effects of viewing through aberrate optical aids, whilst the latter is the more likely result of viewing a degrade image on a screen. In either case, since the area under the spread function mus

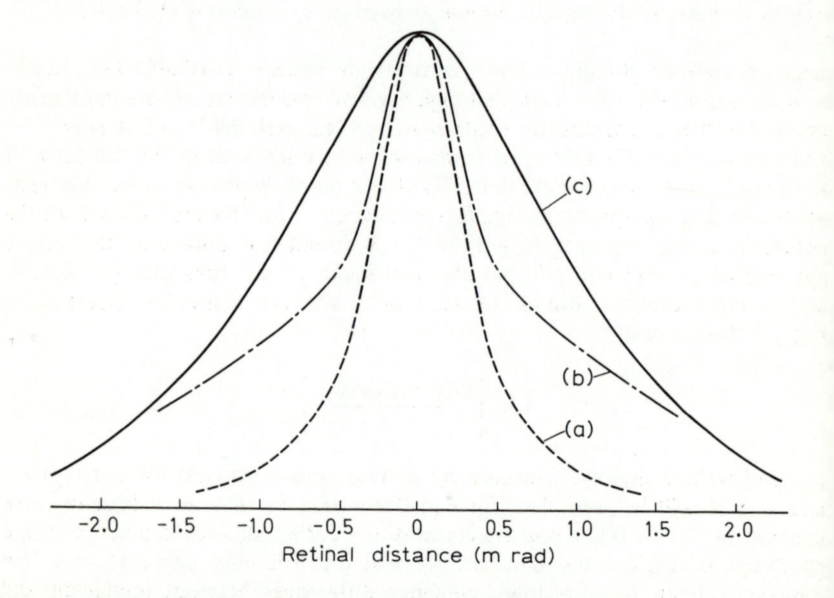

Fig. 7.9. Illustrating extreme forms of additional blurring of the retinal image due t influences external to the eye. Curve (a): typical naked eye spread function, Curve (b) predominantly broadened skirt whilst retaining a sharp central core, Curve (c): overal broadening of the entire spread function.

e unity, the central peak height will be reduced from that of the naked eye. This will result in a reduction of maximum edge gradient for extended objects, which is effectively a reduction in K_2 in Equation 7.7. In addition, in the case of the generally broadened spread function, instead of the peak region of the spread function effectively straddling only two or three receptors, it may increase to straddle several receptors.

7.7.1 Effect of the broadened spread function

Let us consider what the above means in terms of the basic vision model. Firstly, in all cases it is necessary to consider an operator on K_2 which, for extended objects, may be thought of as the ratio of the effective peaks of the (absolute) line spread functions for the total system and the naked eye.

This leads to the modified equation

$$\log_e \left[\frac{\left\{ \dfrac{K_2 \cdot A_T(x)_M}{A_e(x)_M} + K_3 \right\} \epsilon + 1}{K_3 \epsilon + 1} \right] \approx K_2 \cdot \frac{A_T(x)_M}{A_e(x)_M} \cdot \epsilon = K_1 F(n, \theta) + \delta \tag{7.19}$$

where $A_T(x)_M$ and $A_e(x)_M$ are the maximum values of the line spread functions of the total system and the eye respectively, both averaged over one receptor spacing. For small objects it is also necessary to replace c_v in Equation 7.9 by c_s, which looks after the transition from edge gradient to spread function for the complete system. In severe cases it will be necessary, in addition, to adjust the limit value n_0 for small objects. For any system of known performance it is possible to determine $A_T(x)_M/A_e(x)_M$, c_s and the modified n_0 with little difficulty.

Considering only extended objects it would thus appear, from Equation 7.19, that threshold contrast should be inversely proportional to $A_T(x)_M/A_e(x)_M$. To check this, an attempt was made to predict the blurred data of Fry[57]. These blurred borders were of considerable length and of a gaussian blur form. The predictions (shown in Fig. 7.10) were found to be very good. It was, however, of some concern that, with very degraded borders, the effective maximum slope region straddled several receptors and yet it appeared that one difference signal only was sufficient. A critical study of the formula showed that, for such long borders and infinite viewing times, the term $K_1 F(n)$ in Equation 7.19 was small compared with δ. Thus, even if other parts of the contour did contribute to the total stimulus in the case of blurred edges, they would have little effect on the value of $(K_1 F(n) + \delta)$.

In order to study the detail behaviour of the receptors across degraded edges it was considered necessary to study the effect of quality for various sizes of

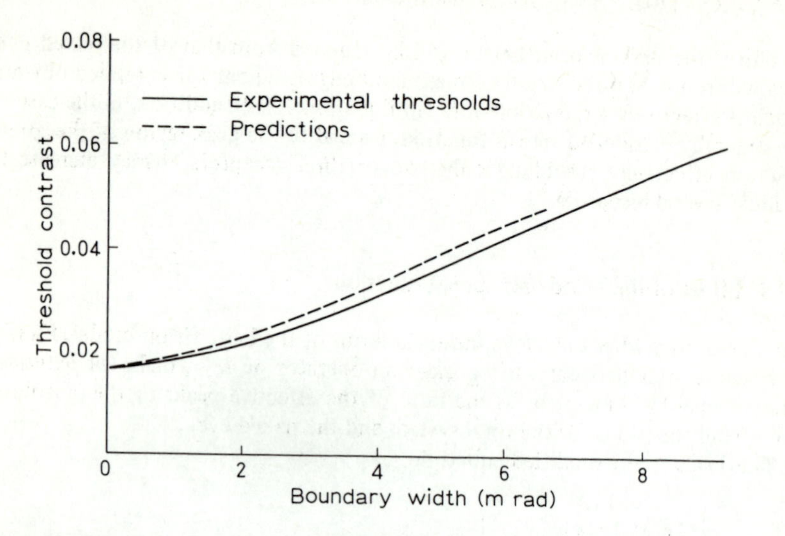

Fig. 7.10. Prediction of the blurred border thresholds as reported by Fry. (Reproduced from Overington[5] by courtesy of the Journal of the Optical Society of America).

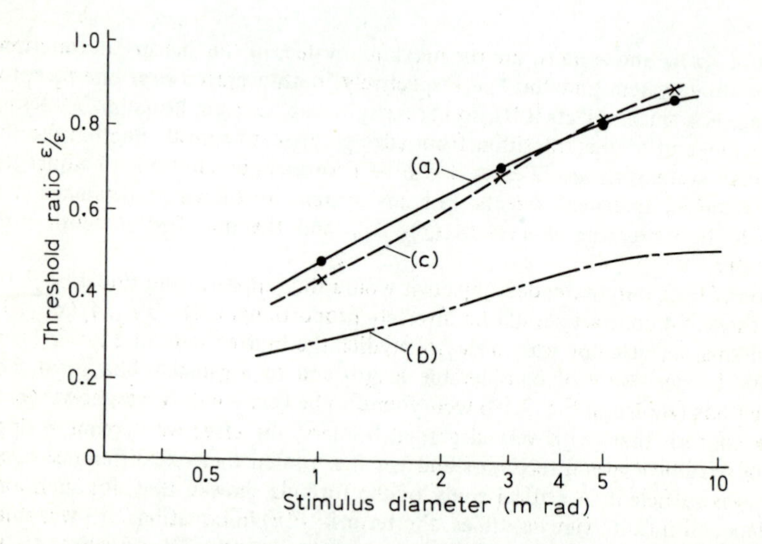

Fig. 7.11. Illustrating the difference between the thresholds for sharply image disc stimuli (ε) and for controlled blurred disc stimuli (ε') as a function of stimulus size. Curve (a): observed trend. Curve (b): predictions using a simple maximum gradient concept. Curve (c): predictions making probabilistic allowance for contributing components away from the geometrical edge. (Data from Crowther and Overington[7]).

stimuli at a short presentation time. To this end a controlled experiment was carried out at BAC(GW), where circular stimuli of diameters between 1 mrad and 44 mrad were presented for 1/3 s (single glimpse) in optimal focus (effectively visually perfect) and at a set of controlled defocus conditions[7,58]. It was found that, for all stimuli tested, the degradation in threshold as blur increased was much less than predicted by the assumption that only one ring of receptors was contributing to the neural response (Fig. 7.11). In fact, for large stimuli, which should have approximated to the long blurred border situation, there was very little degradation of threshold with increasing blur. Detailed inspection of Ogle's data for blurred stimuli[59,60] also revealed far less effect of blur for large stimuli than evident in the case of blurred borders. There thus appeared to be effects of both size and method of presentation which must be taken into account.

7.7.2 Effect of neural interaction

For a possible explanation of the foregoing findings it is necessary to look critically at the scale of probable inhibitory functions in the neural units of the retina (Chapter 2). Now it has already been assumed, as a basic component of the present vision model, that the bipolar cell response is essentially restricted to small groupings of adjacent receptors. However, what of the ganglion cells which are fed from the bipolars? Accepting, as discussed in Chapter 2, that it is the X ganglions which are likely to be associated with cone vision, one may infer from the literature that the central excitatory field of typical ganglions associated with photopic vision behaves essentially as an integrator of bipolar cell responses[7]. The size of this central excitatory field for human beings may be estimated by comparing the visual systems of the cat and man. It has been found that the spatial frequency for peak contrast sensitivity of cats[61] is about one sixth of that of man (Section 4.14). Also it has been found that the X ganglion cells of the cat vary in size of central excitatory field from about 9 mrad to several tens of mradians, the size of field being roughly related to distance from the fovea[61]. One may then speculate that the central excitatory field of foveal ganglion cells in man will be in approximately the same ratio to the smallest excitatory field found in the cat as the ratio of peak frequencies of contrast sensitivities for man and the cat. Thus the diameter of the excitatory integration may be expected to be about 1.5 mrad. Now, for simple, sharply imaged stimuli, this integration means that the outputs of some 6 or 7 bipolar cells along the contour are summed by each ganglion cell along the contour, reduced summations occuring for ganglion cells displaced from the contour. Such a summation will retain the basic sensitivity of the neural networks only to the peak of the contour at threshold. However, for blurred stimuli, owing to the additional probabalistic contributions of bipolar cell responses away from the peak gradient of the contour, the effective strength of ganglion cell signal will be increased compared to that for the simple row of stimulated bipolar cells. This has the effect of compensating to some extent for the reduction of K_2 due to the softening of the spread function and means that an additional factor η_G must be introduced into the vision equation. Equation 7.19, in its approximate form, then becomes

$$K_2 \cdot \eta_G \cdot \frac{A_T(x)_M}{A_e(x)_M} \cdot \epsilon = K_1 F(n, \theta) + \delta \tag{7.20}$$

where η_G takes the value of unity for sharp stimuli.

Attempts at predicting the results of the BAC(GW) blurred disc experiment using the foregoing concept yielded a set of thresholds in very close agreement with the experimental results. It was also found possible, using the same approach, to model the thresholds for sharp and blurred diamond stimuli closely. The predictions for the blurred disc stimuli are shown in Fig. 7.11 whilst the complete results and a detailed description of the theory are to be found in Overington[6] and Crowther and Overington[7].

7.7.3 Effects of presentation time

It remains to attempt an explanation of the marked differences between the long time viewing effects of blur, which appear to be large, and the short time viewing effects, which are small. It is suggested that a possible explanation for the difference is once again to be found in the effects of noise. An essential assumption of the predictions of the BAC(GW) blur experiment is the generation of bipolar cell responses probabalistically in the presence of noise. If it is assumed that long time viewing effectively suppresses this noise, then one would expect that the effect of blur for long time viewing would be due to the peaks of the respective spread functions, as modelled in Equation 7.19 and found to fit experimental data obtained with long viewing times.

7.8 DYNAMIC THRESHOLDS

Thus far all modelling has been concerned with a static viewing situation (that is, the only motion has been that due to involuntary eye movements). In practice many objects of interest are moving. How, then, can the effect of motion on thresholds be modelled?

Let us study the retinal imagery for a moving object of simple shape. For such studies it is convenient to think of simple differences in illuminance of two adjacent receptors as producing the neural signals, to a first approximation. Two distinct situations may exist as shown in Fig. 7.12.

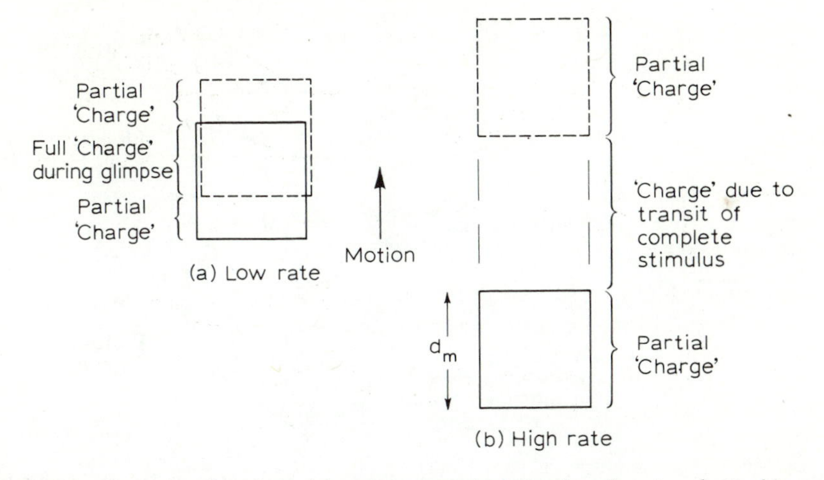

Fig. 7.12. Illustrating the relative 'charge accumulation' situations for rates of retinal image motion less than and greater than an image width in one glimpse.

7.8.1 Low rates

In Fig. 7.12(a) is shown the situation where there is only a little motion relative to the dimension of the stimulus which is swept. Here the main effect is to blur out the edges of the object being viewed into an illuminance ramp which may cover several rows of receptors. Such a situation will frequently exist in detection of slow moving objects. It will also be found in a pursuit tracking situation, where the object is moving fast and the task is one of detail inspection whilst trying to follow its motion with the eyeball. In such situations there will be two forms of accumulation of neural data. Firstly there will be differential signals generated on the end ramps due to the differential time of passage of the edge of the stimulus across the adjacent receptors along the line of image motion. Such differential signals will, to a first approximation, be limited to the duration of transit of the edge from one receptor to the next. At the same time differential signals will be generated from receptors *across* the line of motion and at the edges of the stimulus, the duration of such signals being the entire glimpse. The two situations are illustrated in Fig. 7.13. Which of the two sets of signals is significant in a given situation would seem to depend on the rate of motion, the swept dimension of the target and the sharpness of the image, since these factors will control both the number and strength of the two forms of signal. Whilst some attempts to model this regime rigorously have been made, there are to date little experimental data against which to check predictions. Hence, since the situation is one of great complexity, it will not be pursued further in this book.

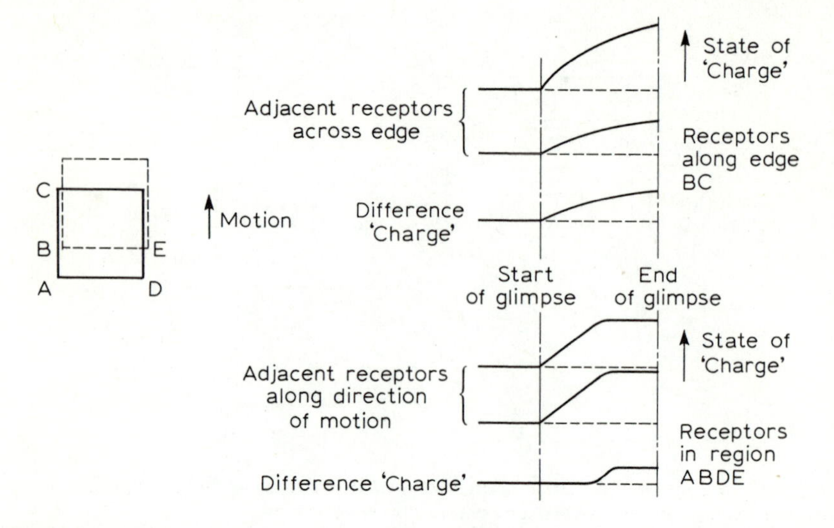

Fig. 7.13. Illustrating the two forms of possible signal generation associated with very low rates of retinal image motion.

7.8.2 High rates

In Fig. 7.12(b) is shown the second well defined situation – where the motion during a glimpse or exposure is considerably greater than the swept dimension d_m(mrad). In this case the situation approximates to the short presentation time conditions of Equation 7.10, but with the difference that it is only the receptors along the two swept edges of the stimulus which are to be considered in general (except when d_m is less than 0.5 to 0.8 mrad when, dependent on quality, it *may* be necessary to consider *all* receptors swept). Now each receptor which is traversed by the edge of the stimulus during the glimpse or exposure will be presented with a pulse of energy of duration d_m/v, where v is the velocity of the retinal image across the retina (mrad/s). Thus in Equation 7.10 we may put $t_D = d_m/v$. Also we may define n as

$$\frac{2(v \cdot t_D - d_m)}{s_R},$$

where s_R is the appropriate retinal receptor spacing, if we assume only maximum signals are utilised.

Then we may write the threshold equation for high rate motion as

$$\left[1 - \exp\left(-\frac{d_m}{v\tau}\right)\right] K_2'\epsilon = \frac{K_1 s_R{}^2}{4(v \cdot t_D - d_m)^2} + \delta$$

or

$$\left[1 - \exp\left(-\frac{d_m}{v\tau}\right)\right] K_2'\epsilon = \frac{K_1\left(\frac{180\theta}{\pi} + 1\right)}{95.5(vt_D - d_m)^2} + \delta \qquad (7.21)$$

For very high rates, where $d_m \ll v$ and

$$\frac{K_1\left(\frac{180\theta}{\pi} + 1\right)}{95.5(vt_D - d_m)^2} \ll \delta$$

Equation 7.21 simplifies markedly to

$$\frac{d_m}{v\tau} K_2'\epsilon \approx \delta$$

or

$$\epsilon \approx \frac{v\tau\delta}{K_2' d_m} \qquad (7.22)$$

Limited experimental data which satisfy the requirements for Equation 7.22 to hold have been studied at B.A.C.(GW) and the predictions of the equation have been found to hold well.

It should be noted that the stimulus shown in Fig. 7.12 is of simple rectangular form. For more complex figures, including in this case circles, it must be expected that the variation of swept dimension across the stimulus will result in dynamic effects analagous to edge softening. Thus due account must always be taken of object shape *and* direction of motion relative to the shape when attempting to model motion thresholds.

7.9 THE SCOTOPIC REGIME

To this point all the modelling in this chapter has been concerned with photopic or cone vision. What about scotopic or rod vision? At this point in time little work has been carried out at BAC(GW) to extend the concepts of the modelling described to rod vision. However, striking similarities exist between anomalous performance in photopic and scotopic vision – particularly in receptive field studies (Chapter 13) – except for the fact that the angular scale of effects is much increased. It is believed by the author that a possible, but as yet unproven, explanation might lie in an assumption that rod vision behaves in a similar way to cone vision, except for two important differences. Firstly, whereas cones are assumed to be individually coupled to the neural networks (at least near the fovea), rods are known to be grouped together in considerable numbers (e.g. Enroth-Cugell and Robson[61]). Hence each group of rods must be treated essentially as a single receptor, and processes involving differencing and adaptation might be expected to be on a much expanded scale. Secondly, whilst it seems likely from physiology, and suits our photopic model, that many of the ganglion cells coupled to cones are of the X type – which are sensitive to continuous flow of energy[61] – it seems equally likely that rod vision operates, at least partially, through Y ganglions, which respond only to transients of energy flow (Section 2.4). These two differences must be expected to result in scotopic vision having significant differences from photopic vision in its spectrum of behaviour, in addition to the scale effects due to rod grouping.

7.10 COLOUR THRESHOLDS

By implication all the modelling discussed in the last two chapters has assumed that target and background were neutral in colour. Is there an adequate way to model colour thresholds? Again, as with scotopic thresholds, little study of the application of the presently described modelling to the colour regime has been carried out. However, again it is believed by the author that it is likely that colour thresholds are determined by very similar processes to the luminance thresholds – but in this case by analysis of maximum difference signals between adjacent *unlike* receptors. Such processing would retain the basic simplicity of the luminance threshold modelling whilst, at the same time, fitting well with the observations of Land and McCann[28] on the possible scene reconstruction by the brain which allows us to perceive colours correctly even when parts of an area are considerably distorted due to coloured reflection effects. It is also immediately and fully compatible with the opponent colour theories of Hurvich and Jameson[19] and Guth and Lodge[20]. Furthermore it permits at least some measure of explanation of the astonishing false colour sensations demonstrated by Land[63–65], who found that almost full colour sensations could be produced by viewing, in register, two specially prepared black and white transparencies, each illuminated by a monochromatic light source *provided that the registration was good* (see Section 2.4). A small amount of relative displacement destroyed the illusion of colour, thus implying strongly that the sensations were produced only by edge differences.

7.11 DISCUSSION

In this chapter a widely versatile, yet simple, model of the visual processes for threshold detection of simple objects in plain fields has been developed. It will be shown in Chapters 12 and 13 that such a simple model is also able to predict detection thresholds in a variety of complex field situations and certain simple recognition thresholds. The model will also be further developed in Chapter 8 to permit threshold modelling in an empty field search situation and in Chapter 10 as an aid to the interpretation of instrumentally measured image quality of optical aids.

REFERENCES

1. Overington, I. (1972). 'Some Aspects of the Variation and Measurement of Contrast of Targets in Field Trials' in *Proceedings of the NATO/APOR Conference on Field Trials and Acquisition of Tactical Operational Data*, Vol. 1, p. 137
2. Overington, I. and Lavin, E. P. (1971). 'A Model of Threshold Detection Performance for the Central Fovea', *Optica Acta*, **18**, 341
3. Lavin, E. P. and Overington, I. (1972). 'Visual Modelling', Annex E of *Final Report on the Third Visual Studies Contract*, BAC (GW) Ref. L50/196/1535
4. Overington, I. (1972). 'Modelling of Random Human Visual Search Performance based on the Physical Properties of the Eye', in *AGARD Conference Proceedings No. 100*, (Ed. H. F. Huddleston), p. B2–1, London
5. Overington, I. (1973). 'Interaction of Vision with Optical Aids', *J. Opt. Soc. Am.*, **63**, 1043
6. Overington, I. (1974). 'An Exploratory Study into the Various Observed Complex Functional Characteristics of Vision and their Compatability with a Unified Simple Modelling', BAC (GW) Ref. ST12386
7. Crowther, A. G. and Overington, I. (1975). 'Experiments on the Detection of Blurred Targets', BAC (GW) Rep. No. ST10840
8. Pirenne, M. H. (1967). *Vision and the Eye*, Chapman and Hall
9. Polyak, S. ((1941). *The Retina*, University of Chicago Press
10. Alpern, M. (1969). Chapter 1–8 in *The Eye*, (Ed. H. Davson), Vol. 3, Academic Press
11. Brindley, G. S. (1970). *Physiology of the Retina and Visual Pathway*, 2nd edn., Edward Arnold
12. Cornsweet, T. N. (1970). *Visual Perception*, Academic Press
13. Walter, W. G. (1953). *The Living Brain*, Duckworth & Co., London
14. Westheimer, G. and Campbell, F. W. (1962). 'Light Distribution in the Image Formed by the Living Human Eye', *J. Opt. Soc. Am.*, **52**, 1040
15. Campbell, F. W. and Gubisch, R. W. (1966). 'Optical Quality of the Human Eye', *J. Physiol.*, **186**, 558
16. Flamant, F. (1955). 'Study of the Distribution of Light in the Retinal Image of a Slit', (in French), *Revue d'Optique*, **34**, 433
17. Ohzu, H. and Enoch, J. M. (1972). 'Optical Modulation by the Isolated Human Fovea', *Vision Research*, **12**, 245
18. Ohzu, H., Enoch, J. M. and O'Hair, J. C. (1972). 'Optical Modulation by the Isolated Retina and Retinal Receptors', *Vision Research*, **12**, 231
19. Hurvich, L. M. and Jameson, D. (1957). 'An Opponent-process Theory of Colour Vision', *Psych. Review*, **64**, 384
20. Cuth, S. L. and Lodge, H. R. (1973). 'Heterochromatic Additivity, Foveal Spectral Sensitivity, and a New Colour Model', *J. Opt. Soc. Am.*, **63**, 450
21. Werblin, F. S. and Dowling, J. E. (1969). 'Organisation of the Retina of the Mud Puppy, *Necturus Maculosus:* II. Intracellular Recording', *J. Neurophysiol.*, **32**, 339
22. Werblin, F. S. (1971). 'Adaptation in a Vertebrate Retina: Intracellular Recording in *Necturus*', *J. Neurophysiol.*, **34**, 228
23. Werblin, F. S. (1973). 'The Control of Sensitivity in the Retina', *Scientific American*, 71

24. Ronchi, L., Fontani, M. and Pedata, F., (1972). 'Spontaneous Brain Activity as Background Noise', *Atti della Fondazione Giorgio Ronchi,* **27**, 363
25. Walter, W. G. (1967) 'Slow Potential Changes in the Human Brain Associated with Expectancy, Decision and Intention', in *The Evoked Potentials, Electroenceph. Clin. Neurophysiol. Suppl. 26.*, (W. Cobb and C. Morocutti. Eds.) Elsevier, 123.
26. Walter, W. G. (1967). 'Electric Signs of Expectancy and Decision in the Human Brain', in *Cybernetic Problems in Bionics,* (Ed. Oestericher), Gordon and Breach, 361
27. Swets, J. A., Tanner, W. P. (Jr) and Birdsall, T. G. (1961). 'Decision Processes in Perception', *Psych. Revue,* **68**, 301
28. Land, E. H. and McCann, J. J. (1971). 'Lightness and Retinex Theory', *J. Opt. Soc. Am.,* **61**, 1
29. Silverthorn, D. G. and Garland, N. (1972). 'A Further Account of Progress in Modelling the Visual Acquisition of Ground Targets from the Air', BAC (GW) Human Factors Study Note Series 7, No. 17
30. Taylor, J. H. (1962). 'Contrast Thresholds as a Function of Retinal Position and Target Size for the Light-adapted Eye', *Proceedings of the NAS-NRC Vision Committee*
31. Ronchi, L. and Longobardi, G. (1971). 'Luminance-Time Relationship in Binocular Vision', *Atti della Fondazione Giorgio Ronchi,* **26**, 239
32. Schweitzer, N. M. J. and Troelstra, A. (1963). 'An Analysis of the b-wave in the Human ERG', *Vision Research,* **3**, 213
33. Blackwell, H. R. (1963). 'Neural Theories of Simple Visual Discriminations', *J. Opt. Soc. Am.,* **53**, 129
34. Schnitzler, A. D. (1973). 'Image-detector Model and Parameters of the Human Visual System', *J. Opt. Soc. Am.,* **63**, 1357
35. Pirenne, M. H. (1967). *Vision and the Eye*, Chap. 5, Chapman and Hall
36. Pirenne, M. H. (1967). *Vision and the Eye*, Chap. 2, Chapman and Hall
37. Blackwell, H. R. (1946). 'Contrast Thresholds of the Human Eye', *J. Opt. Soc. Am.,* **36**, 624
38. Lamar, E. S., Hecht, S., Hendley, C. D. and Schlaer, S. (1948). 'Size, Shape and Contrast in Detection of Targets by Daylight Vision: II Frequency of Seeing and the Quantum Theory of Cone Vision', *J. Opt. Soc. Am.,* **38**, 741
39. Blackwell, H. R. and McCready, D. W. (1952). 'Foveal Detection Thresholds for Various Durations of Target Presentations', *Minutes and Proceedings of NAS-NRC Vision Committee*, AGSIL/53/4405, 249
40. Clark, W. C. and Blackwell, H. R. (1960). 'Relation Between Visibility Thresholds for Single and Double Pulses', University of Michigan Rep. No. 2144–343–T
41. Overington, I. (1974). 'An Investigation into the Reasons for the Degraded Thresholds obtained in the Size Probability Experiment', BAC (GW) Rep. No. ST10961
42. Barlow, H. B. (1957). 'Increment Thresholds at Low Intensities considered as Signal-noise Discriminations', *J. Physiol.,* **136**, 469
43. Østerberg, G. (1935). 'Topography of the Layer of Rods and Cones', *Acta Ophthal.*, **13**, Suppl. 6
44. Polyak, S. (1936). 'Minute Structure of the Retina in Monkeys and in Apes', *Arch. Ophthalmol,* (Chicago), **15**, 477
45. Kishto, B. N. (1970). 'Variation of the Visual Threshold with Retinal Location', *Vision Research,* **10**, 745
46. Le Grand, Y. (1967). *Form and Space Vision,* (Translation by Millodot, M. and Heath, G. G.), Indiana University Press
47. Taylor, J. H. (1969). 'Factors underlying Visual Search Performance', Scripps Institution of Oceanography Rep. No. 69–22
48. Meeteren, A. Van (1973). 'Calculations on the Optical Modulation Transfer of the Human Eye for White Light', Institute for Perception TNO, Rep. IZF1973–2
49. Lavin, H. and Quick, M. (1974). 'The OTF in Electro-optical Imaging Systems', in *Proceedings of the SPIE, Vol. 46, Image Assessment and Specification*, 279

50. Ratliff, F. and Hartline, H. K. (1959). 'The Response of *Limulus* Optic Nerve Fibres to Patterns of Illumination on the Receptor Mosaic', *J. Gen. Physiol.*, **42**, 1241
51. Achurch, I. C. (1963). 'Initial Analysis of Data from the Contrast Recognition Experiments', Assessment Group Memo. No. 154, BAC (GW) Bristol
52. Clark, M. I., Herdan, B. L., Lloyd-Bostock, P. M. R. and Tyrwhitt-Drake, B. G., (1966). 'Visual Studies – Studentship Experiment', Optics Group, BAC (GW) Bristol
53. Hawkins, K. and Church, N. T. (1969). 'Contrast Sign Dependence', Study Note No. 2 of *Research into Factors affecting the Detection of Aircraft through Optical Sights*, BAC (GW) Ref. L50/20/PHY/186/1059
54. Patel, A. S. and Jones, R. W. (1968). 'Increment and Decrement Visual Thresholds', *J. Opt. Soc. Am.*, **58**, 696
55. Short, A. D. (1966). 'Decremental and Incremental Visual Thresholds', *J. Physiol.*, **185**, 646
56. Cohn, T. E. (1974). 'A New Hypothesis to explain why the Increment Threshold exceeds the Decrement Threshold', *Vision Research*, **14**, 1277
57. Fry, G. A. (1965). 'The Eye and Vision', in *Applied Optics and Optical Engineering*, Vol. 2, (Ed. R. Kingslake), Academic Press
58. Hedges, S. L. (1974). 'An Experiment to Investigate the Effects of Optical Quality on the Detection of Circular Targets', BAC (GW) Rep. No. ST11165
59. Ogle, K. N. (1961). 'Foveal Contrast Thresholds with Blurring of the Retinal Image and Increasing Size of Test Stimulus', *J. Opt. Soc. Am.*, **51**, 862
60. Ogle, K. N. (1961). 'Peripheral Contrast Thresholds and Blurring of the Retinal Image for a Point Light Source', *J. Opt. Soc. Am.*, **51**, 1265
61. Enroth-Cugell, C. and Robson, J. G. (1966). 'The Contrast Sensitivity of Retinal Ganglion Cells of the Cat', *J. Physiol.*, **187**, 517
62. Rushton, W. A. H. (1965). 'Visual Adaptation', *The Ferrier Lecture, Proc. R. Soc. B.*, **162**, 20
63. Land, E. H. (1959). 'Experiments in Colour Vision', *Scientific American*, **May**
64. Land, E. H. (1959). 'Colour Vision and the Natural Image, Pt. I', *Proc. Nat. Acad. Sci.*, **45**, 115
65. Land, E. H. (1959). 'Colour Vision and the Natural Image, Pt. II', *Proc. Nat. Acad. Sci.*, **45**, 636

8 Rudimentary Search Modelling

In the preceding chapters the processes of acquisition have been limited to the detection of objects in a known position in the visual field. Whilst such knowledge is a necessary foundation from which to develop an appreciation of the practical functions of vision, the visual task is frequently more complicated. If detection of the presence of an object somewhere in the visual field is the task one is concerned with – as, for instance, in the task of detecting the presence of an aircraft in the sky or of a vehicle in open country – then the important difference between this task and basic detection threshold studies discussed so far is the need to *search* for the object over an extended field. The search process is discussed at length by Koopman[1], Krendel and Wodinsky[2] and Bloomfield[3], amongst others.

During a search process the eye scans the scene in a series of jumps, dwelling for a fraction of a second on each area. Such a series of jumps and dwells are the search equivalent of the involuntary saccades and inter-saccadic intervals described for fixated vision in earlier chapters. In an empty field situation – that is, where the background scene contains little structure – the jumps, or voluntary saccades, are usually considered to be random (e.g. Koopman[1] and Krendel and Wodinsky[2]). Conversely, for structured backgrounds, they are often found to be concentrated in areas of high interest, as might be expected (e.g. Enoch[4] and Williams[5]). Each dwell period is conventionally referred to in search modelling as a *glimpse.* The glimpse frequency is usually of the same order as that for involuntary saccades – that is, about three per second on average – although there is some evidence that, for tasks which are *felt* to be difficult, this frequency drops to around one per second (e.g. Enoch[4]). At each and every glimpse there exists a probability either that the gaze will be centred on the object – yielding a foveal single glimpse viewing situation – or that the object may be seen anywhere within a larger or smaller region of peripheral vision as determined by the viewing conditions. As shown in Section 4.8, the retinal position on which the object falls on a particular glimpse will modify greatly the chances of seeing the object on that glimpse. Furthermore, it seems in practice that, if there is a reasonably high confidence of having seen an object in the peripheral field at a given glimpse, the next glimpse will be drawn towards that object, thus strengthening the effect of the stimulus and allowing confirmation or rejection as being the object of interest.

In this chapter we shall be concerned with the rudimentary search situation – the detection of an object in an unknown position in a plain field. Other forms of search – and in particular that of searching a structured field for a particular object amongst similar (or confusible) objects – will be dealt with in Section 13.3.

8.1 VISUAL LOBE CONCEPTS

In order to attempt to study empty field search it is first necessary to find some way of defining the relative probability of perceiving an object at various points in the visual field. This in turn requires that we combine, in some way, the trends of 50% thresholds as discussed in Section 4.8 with the frequency of seeing concepts discussed in Chapter 3. Such a combination process leads to a set of probability curves which are in some ways analagous to the illuminance cross-sections of the retinal images of a group of progressively smaller disc stimuli due to interactions of the spread function with the geometric shape[6,7]. For instance, for small targets and suprathreshold foveal single glimpse situations the probability will remain at essentially 100% out to some retinal eccentricity, after which the probability will fall off progressively, much as the illuminance of a degraded image falls off. However, the shape of the function must be expected to be determined both by the size and shape of the object and by the retinal receptor density function derived from Østerberg (Section 2.4).

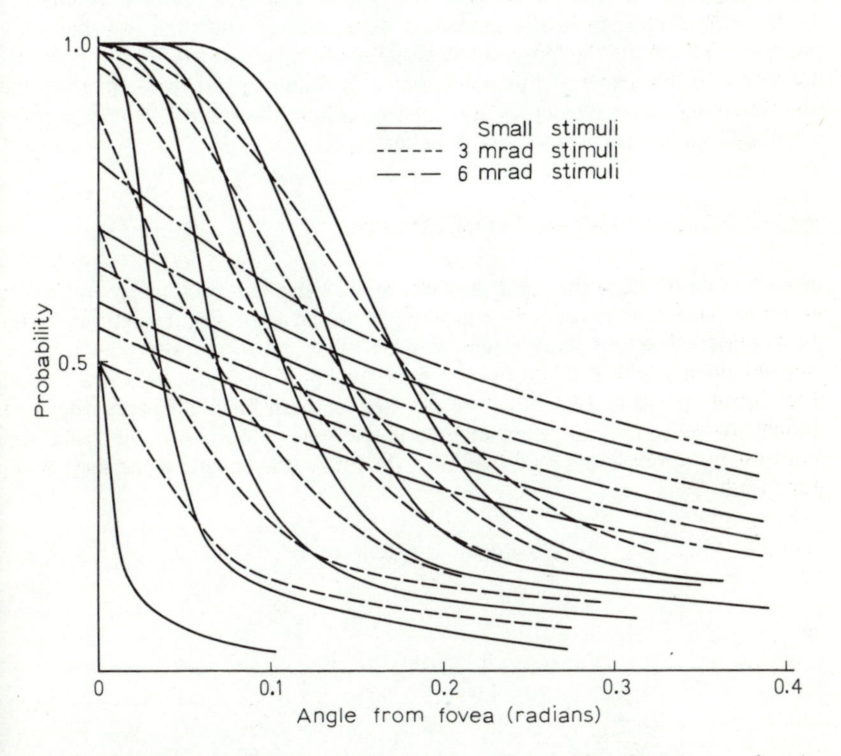

Fig. 8.1. Variation of single-glimpse probability p_G with retina l eccentricity as a function of stimulus contrast (groups of curves are shown for various visual lobe radii at p_G = 0.5).

The result is that, for large objects, the change of probability with retinal position is likely to be very small. This is in keeping with the peripheral threshold trends (see Section 4.8). As examples of the form of such probability maps, the peripheral probabilities computed from experimental threshold data and assumed frequency of seeing spreads for a series of contrasts are shown for large and small circular stimuli in Fig. 8.1. Such probability cross-sections are known as visual lobes.

The computation of visual lobes is relatively complicated. In an attempt to simplify practical usage of visual lobe concepts it has been proposed by a number of people (e.g. Owen[8] and Davies[9,10]) that it is adequate, for most purposes, to define each particular visual lobe as a cylindrical function with an angular diameter equal to the angle at which, for a given situation, there exists a 50% probability of single glimpse acquisition. Although it is difficult to provide positive proof of the universality of such an approximation, the idea enjoys wide acceptance. It is believed by the author, from certain sensitivity modelling, that such an approximation is frequently entirely adequate in practice, despite the soft visual lobe profiles illustrated in Fig. 8.1. For this reason, in order to retain some simplicity in the modelling developed in this book, we shall only consider this simplified, approximate visual lobe concept at any length. It should be noted by the reader at this point that it is common practice to refer to the rigorous visual lobe concept as 'soft-shelled', whilst the approximate form to be developed further is known as 'hard-shelled'.

8.2 SINGLE GLIMPSE PROBABILITY

In order to determine the total probability of seeing an object in a given time in terms of hard-shell visual lobe concepts, it is first necessary to determine the probability of seeing in a single glimpse. This in turn must obviously be dependent on the size of the field to be searched. It must also be dependent on the size of the visual lobe. Now the size of the visual lobe for images which are effectively static on the retina may be determined, for a given object size and contrast, by a reordering of Equation 7.18. From this equation one may write, for threshold,

$$\left(\frac{180\theta_v}{\pi}+1\right)=\frac{(K_2\epsilon-\delta)n^2}{K_1}$$

or

$$\theta_v=\left\{\frac{(K_2\epsilon-\delta)n^2}{K_1}-1\right\}\frac{\pi}{180} \tag{8.1}$$

where θ_v is the radius of the hard shell visual lobe.

For images which are moving on the retina, although in principle a similar approach may be made, reference back to Equations 7.20 and 7.21 will show that the situation becomes very complex. Only when motion is very fast compared to stimulus size is the dynamic situation easy, and in such a situation Equation 7.21 implies a trend towards a condition where threshold is constant over the majority of the retina (i.e. an equal probability of seeing wherever the stimulus is presented). This latter statement should not, however, be taken as an absolute fact, but rather as an indication of the order of things.

Returning to the essentially static retinal image situation, it is necessary to determine the single glimpse probability for a given lobe size within a given field. Now the simple expression for the probability of seeing a target anywhere in a search area A_F, when the visual lobe area is a_v, is given as

$$p_G = \frac{a_v}{A_F} \tag{8.2}$$

However, the inadequacy of this expression is apparent from the fact that if $a_v > A_F$ then $p_G > 1$, which is not possible. The reason for this is, of course, that if $a_v > A_F$, a portion of the visual lobe of size $(a_v - A_F)$ is wasted outside the search area. A frequently used correction for this situation is to employ the expression

$$p_G = \frac{a_v}{(A_F + a_v)} \tag{8.3}$$

which has the correct limiting behaviour. However, it is shown by Davies[9] that, if one develops an exact solution, this should have the form

$$p_G = \left[\sqrt{\left(\frac{a_v}{A_F}\right)} - \frac{a_v}{4A_F}\right]^2 \tag{8.4}$$

which reaches $p_G = 1$ at $a_v = 4A_F$. It should be noted that this function only applies for

$$0 < a_v \leqslant 4A_F.$$

For $a_v > 4A_F$, $p_G = 1$.

An alternative way of expressing Equation 8.4 is in terms of rectangular or elliptical functions, when probability is given as

$$p_G = \left(\frac{x_v}{X_F} - \frac{x_v^2}{4X_F^2}\right)\left(\frac{y_v}{Y_F} - \frac{y_v^2}{4Y_F^2}\right) \tag{8.5}$$

where x_v and y_v are the two dimensions of the visual lobe and X_F and Y_F are the two dimensions of the search field. In the particular case of assumed circular visual lobes and search fields, by symmetry and inspection, one may conclude that

$$p_G = \left(\frac{\theta_v}{\theta_F} - \frac{\theta_v^2}{4\theta_F^2}\right)^2 \quad \text{for } \theta_v \leqslant 2\theta_F \tag{8.6}$$

and

$$p_G = 1 \quad \text{for } \theta_v > 2\theta_F$$

where θ_F is the radius of the search field.

The reader wishing to study the topic of single glimpse probability in depth is referred to the extended treatise by Davies[9].

8.3 CUMULATIVE SEARCH PROBABILITY

Having established the single glimpse probability it is necessary to consider how the cumulative probability of seeing builds up with increasing search time. Here we find ourselves with a situation similiar to that already invoked for build up of intelligence about the existence of a stimulus in the fixated viewing situation (Section 7.3). If the glimpses are truly independent (as they should be in searching an empty field), if the stimulus remains of constant strength, and if the foveal probability is unity, then the cumulative probability of detection in m glimpses may be defined (e.g. Koopman[1], Krendel and Wodinsky[2] and Weatherburn[11]) as

$$\Phi_m = 1 - \Pi(1 - p_G) = 1 - (1 - p_G)^m \tag{8.7}$$

Assuming an average glimpse rate of 3 per second this yields a family of curves as shown in Fig. 8.2. In practice the accumulations with time, when extrapolated back to zero probability, are found to follow curves closely resembling these theoretical curves, but with small time delays before any significant probability builds up (e.g. Krendel and Wodinsky[2] and Bloomfield[3]). This is, of course, due to the fact that response, in a multiple glimpse situation, tends to occur *at the end* of a particular glimpse, thus adding a delay time of the order of 1/3 s to the theoretical curves.

The above is entirely adequate for a situation where the single glimpse probabilities are constant with time (i.e. the stimulus does not change with time) and the foveal probability is unity. However, what of the many situations where one or both these conditions do not hold?

Let us first consider the situation where p_G is constant, but the foveal single glimpse probability is less than unity. In such a situation it seems wrong to consider the single glimpse value of q^{II} to be equatable to $(1 - p_G)$, since this then takes no account of the foveal probability p_f. At the same time, although Φ_m for multiple glimpses may be larger than p_f, it does not necessarily reach unity if p_f is small.

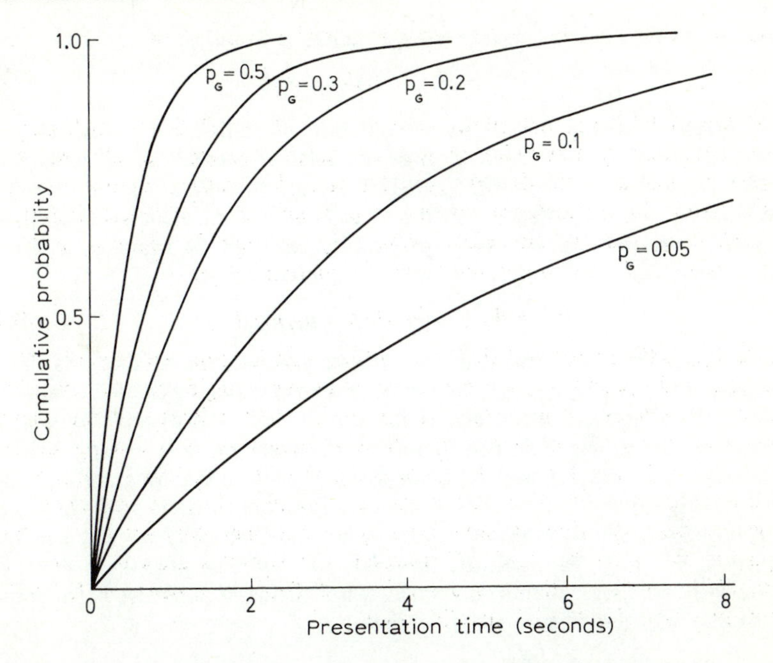

Fig. 8.2. Cumulative probability as a function of time for constant stimuli and various single glimpse probabilities (mean glimpse time 1/3 s).

In an attempt to allow for these two factors, some workers have suggested that Equation 8.7 should be written

$$\Phi_m = 1 - (1 - p_f p_G)^m, \quad m < M$$
$$\Phi_m = \Phi_f, \quad m \geqslant M \tag{8.8}$$

where M is the number of glimpses for the computed cumulative probability to reach the foveal probability for infinite viewing time Φ_f. Whilst such a function does provide for a reasonable value of

$$q = (1 - p_f p_G)$$

this being now related to a visual lobe size defined where probability is $p_f/2$, the discontinuity when Φ_m reaches Φ_f does not seem correct. Rather the present author believes that one should find a function which approaches the limit value of Φ_f asymptotically. Such a function would be arrived at by assuming a need to multiply all cumulative probabilities in an equation such as Equation 8.7 by the factor Φ_f, thus scaling the usual cumulative probability curve given by Equation 8.7 proportionately over its entire length. It is thus proposed that a suitable function for Φ_m for any constant strength stimulus would be

$$\Phi_m = \Phi_f - \Phi_f(1 - p_f p_G)^m \tag{8.9}$$

Although Equation 8.9 allows computation of cumulative probabilities for many situations as a function of time (in units of number of glimpses m), it cannot be used as it stands if the stimulus strength is varying. Such is frequently the situation in, for instance, viewing of or from moving vehicles, aircraft, etc. In such situations the cumulative probability can only be found by a step by step computation using a rigorous form of Equation 8.9, viz.

$$\Phi_m = \Phi_{fm} - \Phi_{fm}\Pi(1 - p_{fr}p_{Gr}) \tag{8.10}$$

where Φ_{fm} is the foveal probability for infinite viewing time which exists after m glimpses, and p_{fr} and p_{Gr} are the foveal and general single glimpse probabilities on the rth glimpse. Fortunately, if the search field is large and the stimulus only small, the values of Φ_f and p_f will usually reach unity before p_G becomes significant (c.f. Fig. 8.1 and Equation 8.6). If these conditions are combined with a fairly rapid stimulus growth through threshold then the errors in Φ_m due to ignoring contributions where Φ_f and p_f are less than unity are very small and Equation 8.9 may be used. If, however, the stimulus growth is slow, and particularly for large stimuli and small search fields, it is believed that serious errors may arise due to such simplification.

8.4 PRACTICAL APPLICATIONS

In order to illustrate both the application and the type of results coming out of the rudimentary search modelling discussed in this chapter, let us consider some practical examples.

8.4.1 Search for a static stimulus

Such a situation is applicable to search for particular stars in astronomy, search for signal lights in a poor visibility situation and search for isolated threshold objects in such spheres as microscopy. Since the stimulus is essentially static, the value of p_G remains constant. Φ_m is thus a simple function of time. However, the accumulation is very dependent on the values of p_G and the foveal single glimpse probability p_f. These in turn are dependent on stimulus size, shape and contrast, whilst, in addition, p_G is dependent on search field radius θ_F.

Practical measurements of the search time as a function of object size, object contrast and search field diameter for circular stimuli were carried out by Krendel and Wodinsky[12]. As an attempt to check the adequacy of the modelling of empty field search described in the previous section, a number of selected conditions from Krendel and Wodinsky's work were modelled by the present author[13]. It was found that good predictions were obtained for a variety of target sizes and contrasts with large search fields, but that there was considerable disagreement with small search fields (of the order of 0.12 rad diameter). However, this disagreement at small search fields is in general agreement with the findings of Enoch[14] that an increasing number of wasted glimpses occur during empty field search of fields less than 0.2 rad in diameter. It would thus appear that hard shell visual lobe search modelling is tentatively adequate in a practical static stimulus situation, provided that due account is taken of wasted glimpses for small search fields.

8.4.2 Search for a growing stimulus

An alternative and common search situation is that where the stimulus is growing in either size or contrast whilst search is progressing. In such a situation it is most likely that the cumulative probability is required not as a function of time but as either a direct or indirect function of the growth of the stimulus.

One particularly important practical case is in the acquisition of objects of interest from aircraft or of aircraft from the ground. In either of these situations the stimulus is growing rapidly in both size and contrast due to the rapid closure of viewing range. It is then most useful to compute Φ_m as a function of range. Such a computation may be accomplished by combining Equation 7.18 with Equation 8.10. Firstly, by defining ϵ and n in Equation 7.18 as functions of viewing range, a viewing range R is determined at which the foveal probability p_f is 50%. Then the viewing range is permitted to reduce by finite and convenient increments ΔR. For a median value of the first increment of reduction the hard shell visual lobe radius θ_v is computed from Equation 7.18. From this visual lobe and the effective search field radius θ_F, the value of p_G is computed (from Equation 8.6). A value of p_f is next computed by putting $\theta = 0$ in Equation 7.18. Knowing the rate of closure of range, the number of glimpses in this first range increment is determined. Then, from Equation 8.9, assuming constant probabilities in the small range increment, a value of Φ_m may be determined at a range $(R - \Delta R)$.

The above procedure may now be repeated for the next and subsequent increments of range reduction, except that at each subsequent stage the cumulative probability Φ_m includes not only the contribution $(1 - p_f p_G)_m$ for the increment but also the accumulation up to that point. In other words we may modify Equation 8.10 to read

$$\Phi_m = \Phi_f - \Phi_f(1 - p_f p_G)_m \times (\Phi'_f - \Phi'_m)/\Phi'_f \tag{8.11}$$

where $(\Phi'_f - \Phi'_m)/\Phi'_f$ refers to the accumulation up to the end of the previous range increment.

The above looks unpleasantly complicated with all the variable p_f's and Φ_f's to consider. Fortunately, as previously discussed, in many practical growth

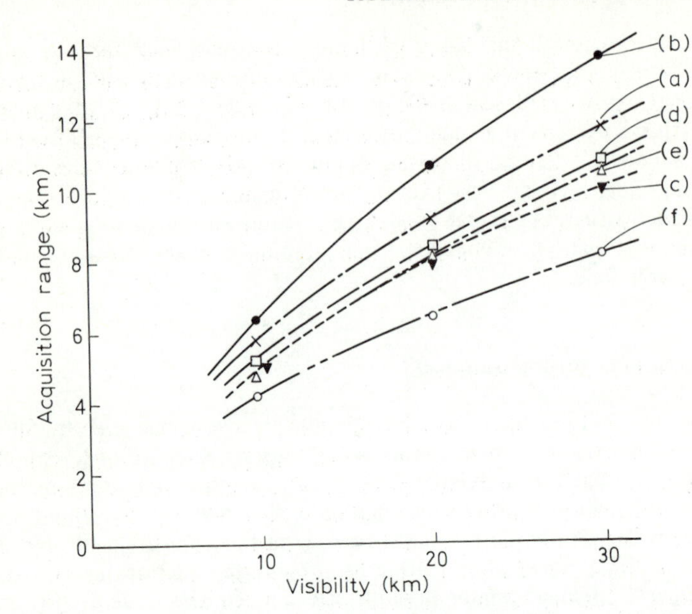

Fig. 8.3. Illustrating the prediction of variation of acquisition range with meteorological visibility for a target viewed through high power binoculars and approaching at 200 m/s. Curve (a): Reference conditions: $\theta_F = 0.435$ rad, Intrinsic contrast (C_o) = −0.9, Equivalent circle diameter in eye space (D_e) = 30 m, Field luminance B′ = 100 cd/m² (approx). Single parameter variations from reference conditions:– Curve (b): B′ = 1 000 cd/m² (approx). Curve (c): $C_o = -0.6$. Curve (d): $D_e = 23$ m. Curve (e): $\theta_F = 0.87$ rad. Twin parameter variation from reference condition (lower power binoculars):– Curve (f): $\theta_F = 0.11$ rad, $D_e = 7.2$ m. (Reproduced from Overington[13] by courtesy of the Advisory Group for Aerospace Research and Development of NATO).

situations involving moderate search, almost all the accumulation will occur after p_f and Φ_f have reached unity. Thus, for many practical purposes, it is adequate to consider Equation 8.11 to be replaced by

$$\Phi_m = 1 - (1 - p_G)_m(1 - \Phi'_m) \tag{8.12}$$

where Φ'_m is the accumulation to the end of the previous range reduction increment.

Some examples of theoretical and practical growth functions are to be found in Davies[9,10] and Overington[13], whilst a few computed examples illustrating effects of θ_F and stimulus parameters are to be found in Figs. 8.3 and 8.4. In order to compute such functions it was imperative that allowance be made for atmospheric effects on contrast. The form and modelling of these latter effects are discussed in Chapter 15.

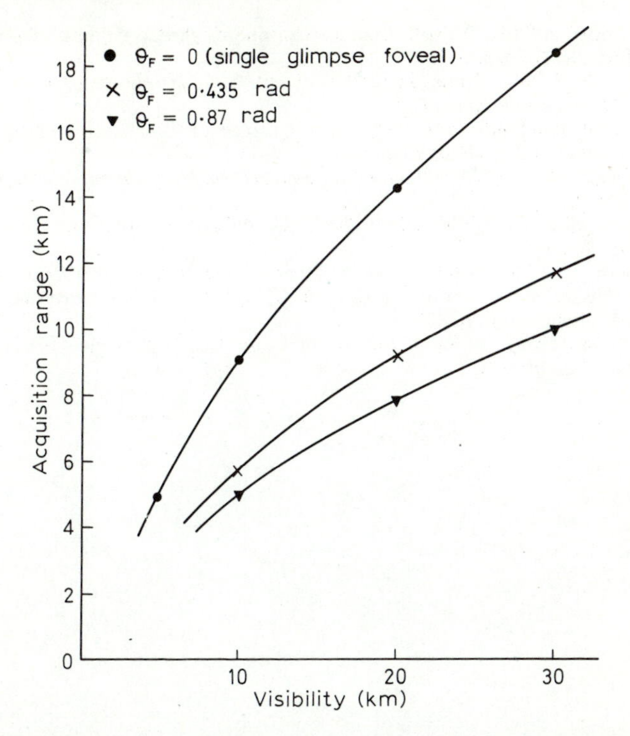

Fig. 8.4. Illustrating the sensitivity to search field size of the acquisition range/visibility function for a typical target approaching at 200 m/s and viewed through high power binoculars. (Reproduced from Overington[13] by courtesy of the Advisory Group for Aerospace Research and Development of NATO).

REFERENCES

1. Koopman, B. O. (1956). 'The Theory of Search, II. Target Detection', *Operations Research,* **4**, 503
2. Krendel, E. S. and Wodinsky, J. (1958). 'Visual Search in an Unstructured Visual Field', AFCRL–TR–59–51 Final Rep. *Franklin Institute Rep. No. F–A1851,* ASTIA No. AD211156
3. Bloomfield, J. (1970). *Visual Search*, PhD. Thesis, University of Nottingham
4. Enoch, J. M. (1959). 'Natural Tendencies in Visual Search of a Complex Display', in *Visual Search Techniques*, Publ. No. 172, National Academy of Sciences, National Research Council, pp. 92–93, 187–193, 251–252
5. Williams, L. G. (1967). 'A Study of Visual Search using Eye-Movement Recordings' *Honeywell Inc. Rep. No. 12009–IR2*
6. Overington, I. and Lavin, E. P. (1970). 'A Theory of Foveal Vision', App. 6 of *Final Report on Visual Studies II Contract*, BAC (GW) Ref. L50/20/PHY/196/1214
7. Overington, I. (1972). 'Some Aspects of the Variation and Measurement of Contrast of Targets in Field Trials', in *Proceedings of the NATO/APOR Conference on Field Trials and Acquisition of Tactical Operational Data*, Vol. 1, p. 137

8. Owen G. P. (1970). 'Visual Target Acquisition – Single Glimpse Probability of Detection', *DOAE Memo. 7047*
9. Davies, E. B. (1968). 'Visual Search Theory with Particular Reference to Air to Ground Vision', *RAE Tech. Rep. 68055*
10. Davies, E. B. and Smith, L. J. (1969). 'A Comparison of Visual Search Theory and RRE Experimental Data', *RAE Tech. Rep. 69057*
11. Weatherburn, C. E. (1961). *A First Course in Mathematical Statistics*, p. 22, Cambridge University Press
12. Krendel, E. S. and Wodinsky, J. (1960). 'Visual Search in Unstructured Fields', *J. Opt. Soc. Am.*, **50**, 562
13. Overington, I. (1972). 'Modelling of Random Human Visual Search Performance based on the Physical Properties of the Eye' in *AGARD Conference Proceedings No. 100*, (Ed. H. F. Huddleston), p. B2–1
14. Enoch, J. M. (1959). 'Effect of the Size of a Complex Display upon Visual Search', *J. Opt. Soc. Am.*, **49**, 280

9 General Considerations of Indirect Imagery

There is an increasing tendency in this age for us to need to use our eyes to view other than directly. By this is implied the very considerable employment nowadays of visual aids* of one form or another (binoculars, telescopes, microscopes etc.), and of such media as photography, television and image tubes. With all these devices it is rare that the image produced, and viewed as an object stimulus by the eyes, is perfect. In addition, particularly with certain forms of binocular optical aid, there can be problems at the man/aid interface. The purpose of this chapter is to draw attention to the differences between naked eye (or direct) vision and the various forms of indirect vision. The reader will be introduced to concepts and directed to other parts of this book and other literature for fuller study.

9.1 VISION THROUGH VISUAL AIDS

We shall start by considering the wide range of visual aids and some of the implications of viewing through them.

9.1.1 Forms of visual aids

The main forms of visual aid may be divided into three categories – telescopes (including binoculars), microscopes, magnifier systems. Each of these categories may be broken down further, introducing certain characteristic optical problems. Telescopic systems, for instance, may be monocular (i.e. viewed with one eye), binocular (i.e. having entirely separate optics for each eye) or bi-ocular (defined as having a common objective lens and twin eyepieces). These forms of visual aid are illustrated in Fig. 9.1. Microscopes may be either monocular or bi-ocular. Magnifier systems may again be monocular or bi-ocular. The primary effect of changing from a monocular to a bi-ocular or binocular system is one of increasing the signal/noise ratio and hence improving seeing capability (see Section 2.10). However, such an ideal may not always be achieved due to effects of binocular rivalry (Section 9.1.4). In addition for any visual aid, there must be potential problems of pupil eccentricity (Section 9.1.2) and mal-focus (Section 9.1.3), whilst for bi-ocular systems (other than magnifiers) there is a large loss of light due to the need for, at the very least, a beam splitter. This latter has the

*Visual aids are here defined as optical transmission systems which present an aerial image to the eye.

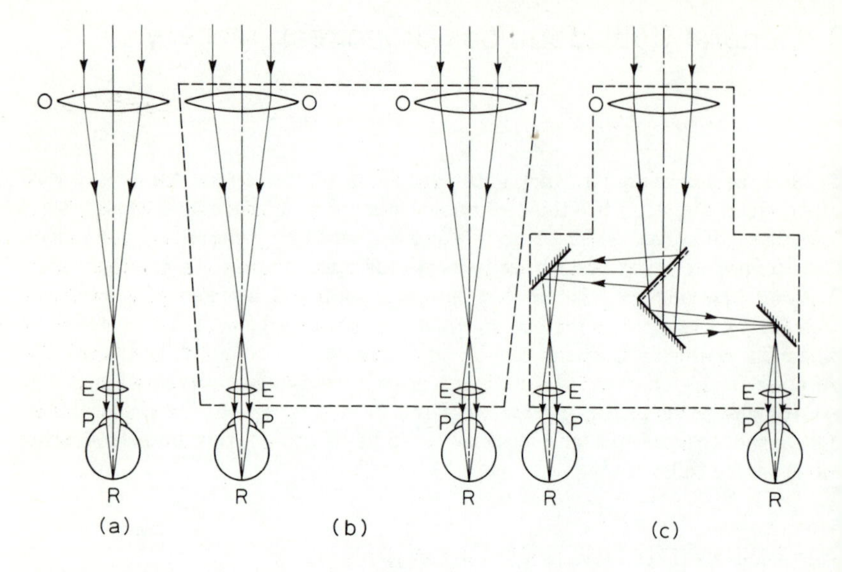

Fig. 9.1. Various types of visual aid. (a) Monocular, (b) Binocular, (c) Bi-ocular. O = objective, E = eyepiece, P = eye pupil, R = retina.

effect of reducing the effective scene luminance, which is not a serious problem with high luminance scenes but can be a problem as scene luminances are reduced (see Section 4.2). Finally, except in the very best of visual aids, exhibiting high quality blooming of lens surfaces against surface reflections, together with optical anti-reflection precautions on all mount surfaces, there will normally be significant veiling glare and a possibility of 'ghost' images (Section 9.1.6).

9.1.2 Eccentric pupils

Pupil eccentricity is defined as the extent to which the optical axis of the visual aid and the visual axis of the eye fail to coincide at the exit pupil of the equipment. It is illustrated in Fig. 9.2. Such pupil eccentricity is likely to exist in the use of monocular visual aids when the exit pupil is significantly larger or smaller than the eye pupil at the adaptation level of the scene viewed. It is also very likely to occur in the use of binocular and bi-ocular visual aids employing individual eyepieces due to errors in the setting of the eyepiece separation to match the interpupiliary distance (the distance between the visual axes of a person's two eyes – see Fig. 9.2). This interpupiliary distance can vary from person to person by as much as 10 mm and so it can be seen that, since typical

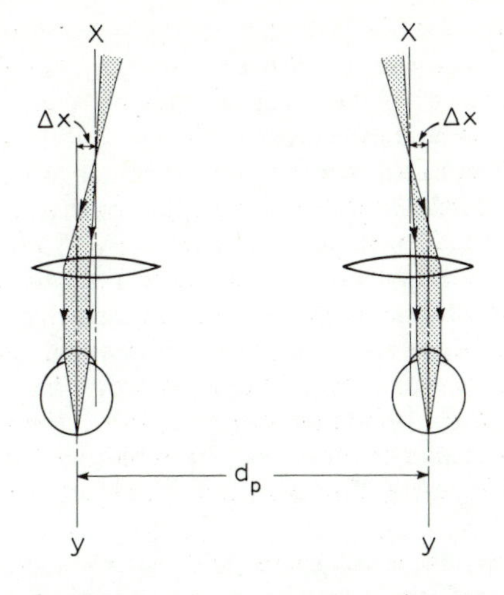

Fig. 9.2. Illustrating mismatch between the interpupiliary distance d_p *and the eyepiece separation for a binocular aid.* X–X *are the instrument optical axes,* y–y *are the eye's optical axes,* ΔX *is the pupil eccentricity.*

pupil diameters are 3–5 mm (see Section 2.2), considerable eccentricity errors can easily be introduced by other than the most careful setting of the separation between binocular eyepieces. An ideal eyepiece would have aberration balance such that the effective quality was unaffected by pupil eccentricity. However, such an eyepiece requires very special design (e.g. Kingslake[1] and Born and Wolf[2]) and is very rarely found on commercial visual aids (although one or two types have been designed for specialist use in the last few years, for instance by Pilkington Perkin Elmer of St. Asaph, Flintshire). The more conventional eyepieces have aberration balances such that, if pupil eccentricities of more than a few tenths of a millimetre are present, quite serious degradations in image quality can occur. Whilst such degradations can be detected readily when viewing a high contrast edge or high contrast fine detail, they are by no means so easily seen when viewing a more normal scene. Hence in normal field use of visual aids it is not particularly easy to find the optimum pupil position by observation but the effects of eccentricity on performance are, of course, still present.

9.1.3 The problem of focus

Mal-focus of visual aids is not necessarily a serious problem with monocular aids, since the eye's accommodative powers can compensate for quite wide variations of focus (see Section 2.8). However, just this ability to compensate for variations in focus settings makes it difficult for the unskilled observer to set the focus of a monocular system accurately. It has been found, by simple statistical tests at BAC (GW), that tolerances of the order of ± 1 dioptre on repeated settings of a monocular must be accepted as typical. With such variations, and the realisation that one of the more accepted means of setting binocular systems is to focus the oculars separately, it should be immediately apparent that it is possible to set the two oculars on binocular systems very differently. Such differences impose unnatural stresses on the two eyes, which are now required to focus at widely different ranges. The alternative method of focussing binocular aids – that of focussing the two oculars jointly – requires very considerable skill and an ability to focus both eyes to work equally. Since most people have a master eye, which dominates binocular visual behaviour very markedly[3], the most likely effect of mal-adjustment of binocular focus is that the subjective (or non-dominant) eye will cease to register an image, thus defeating the object of providing a binocular viewing system.

The focus problems discussed above, and the pupil eccentricities caused by errors in interpupiliary distance settings discussed in Section 9.1.2, may suggest to the reader that binocular instruments are hardly worth constructing. It is not the author's intention to give this impression – which is far from the truth – but merely to forewarn the reader of the pitfalls which must be guarded against and overcome if binocular equipments are to be used efficiently.

9.1.4 Binocular Rivalry

A further problem of indirect binocular or bi-ocular viewing is associated with convergence. In their natural state the muscles of the two eyes are so arranged that the bringing into play of the accommodation muscles automatically also brings into play the convergence muscles in order to fuse the images presented to the brain by the two eyes. Now if a binocular or bi-ocular visual aid is provided with focusing eyepieces then strictly there is only one final image distance setting which is fully compatible with a given alignment of the two oculars. If the two oculars are, in effect, parallel tubes then the only strictly correct setting is for final images at infinity, since it is only when viewing objects at infinity that the normal eyes are not in a state of convergence. In practice this matter is frequently overlooked, possibly because it is difficult in practice to become consciously aware of the potential effects of misalignment on performance. This is because the eyes have considerable tolerance on convergence[4] and the brain, being presented with two similar sets of data, will generally fuse the two at the expense of visual comfort. However, 'two wrongs never make a right', as an old saying goes, and some penalty must be paid for such compromises. What in fact is likely to happen is for performance to degrade – a fact which would only be realised by carrying out comparative tests under threshold conditions. To the author's knowledge little data are available on such threshold degradations.

9.1.5 Effects of direct coupling

A final, but not insignificant, problem associated with vision through visual aids is concerned with the nature of the interface. Most visual situations other than viewing through visual aids involve the naked eye looking at an object or image which may be considered to be incoherently illuminated – that is, the wavetrains of the light coming from any point of the object are of random phase and are not in any way related in phase to wavetrains of light from any other point on the object. However, in the case of viewing through visual aids, the coupling of the visual aids and the eye is by means of an aerial image. Now this aerial image is produced by complex interference of elementary parts of the wavefront emergent from the exit pupil of the visual aid and is dependent on the diffraction and aberration effects therein[5]. But, since the wavefront passing through the aerial image is not in any way destroyed before reaching the refraction optics of the eye, the effect of the eye's quality on this image must depend on the particular aberrations of the eye and the way in which they interact with those present in the wavefront due to the visual aid. Furthermore the diffraction effects will only be dependent on the visual aid pupil or the eye pupil, according to which is the smaller. The other will contribute nothing to the final quality. These problems are discussed in some detail by De Velis and Parrent[6] and the author[7].

9.1.6 Glare Effects (parasitic light)

All the quality effects discussed so far for visual aids have been essentially concerned with a softening of the boundaries between image areas of different illuminances. A further group of effects may be grouped under the general heading 'glare'. This group of effects are all due to light falling on the image plane in the wrong place, as opposed to local interference effects. Hence the use of the expression 'parasitic light' from the French term for glare effects, *lumiére parasitique.* This term describes the defect admirably.

The range of defects covered by the general term include the classical veiling glare (a uniform level of illuminance overlaying the image and reducing contrast)[8], flare light (being local glare close to the image of a bright object) and ghost images (being low intensity images of specific bright objects within or outside the field of view which are superimposed on the true image in the wrong place). For a detailed discussion of these and other minor glare factors the reader is referred to Martin[9].

9.1.7 Modelling of Vision through Visual Aids

Subject to being able to define the image quality at the retina, it becomes very easy, in principle, to extend the modelling described in Chapters 7 and 8 to the case of viewing through visual aids. All that is necessary, over and above the modelling of unsharp objects discussed in Section 7.7, and the search modelling of Chapter 8, is to allow for the effects of magnification, veiling glare and transmission. Now the only effects of magnification are to increase apparent object size and search field area, with attendant increase in apparent velocity in a dynamic situation. Equally the only basic effect of veiling glare is to reduce the apparent object contrast, whilst the effect of transmission losses is to reduce the apparent field luminance. These factors may be allowed for by writing

$$\log\left[\frac{\left\{K_2\cdot\eta_G\dfrac{A_T(x)_M}{A_e(x)_M}+K_3\right\}\eta_c\epsilon+1}{K_3\eta_c\epsilon+1}\right]\approx K_2\cdot\eta_G\frac{A_T(x)_M}{A_e(x)_M}\cdot\eta_c\epsilon=\frac{K_1{}'F(n)}{M^2}+\delta' \quad (9.1)$$

in place of Equations 7.19 and 7.20, and by writing

$$p_G=\left[\frac{\theta_v}{\theta_F{}'}-\frac{\theta_v{}^2}{4\theta_F{}'^2}\right]^2 \quad (9.2)$$

in place of Equation 8.6, where $(1-\eta_c)$ is the veiling glare for the viewing conditions, M is the magnification of the visual aid, $K_1{}'$ and δ' are the values of K_1 and δ for a retinal illuminance of TE_R where T is the transmission of the visual aid and E_R is the mean retinal illuminance for the scene luminance being viewed, $\theta_F{}'$ is the semi-field angle measured in image space, $\theta_F{}'$ being equal to $M\theta_F$, and θ_F being the semi-field angle in object space.

In Equation 9.1, as in Equation 7.19, $F(n)$ is, of course, the value associated with the angular subtense of the object at the viewing position, whilst

$$\frac{K_2\cdot\eta_G\cdot A_T(x)_M}{A_e(x)_M}$$

must take due account of multiple rows of receptor pairs contributing to the total ganglion signal strength for single glimpse and search situations as in Equation 7.20.

In practice one should actually allow for differential quality effects both for peripheral viewing and as a function of the angle of the sight line with respect to the optical axis. It is very rare that a visual aid is free from significant effects as a function of pupil eccentricity and viewing direction. However, to the author's knowledge very little is known about the relative effects of visual aid/eye relative performances as a function of viewing angle and pupil eccentricity. It is suggested, in the absence of such data, that Equations 9.1 and 9.2 could be used for predictions of thresholds and single glimpse probabilities, although one may need to allow for interactions of, for instance, relative effects of blur at various retinal eccentricities.

9.2 VIEWING OF DIFFUSE SECONDARY IMAGES

9.2.1 Forms of diffuse secondary images

Moving away from visual aids to other situations where secondary images are viewed, these may be of several forms. The simplest is where an image is formed on a fine structured diffusing surface, as when a photograph is projected on a finely structured screen, the image then being viewed from such a range that no grain structure is observed. Under such viewing conditions we have a secondary image most closely approximating a grainless image with imperfect definition due to the quality of the reproduction optics (camera and/or projection lens). It should be remembered that such a secondary image must *always* be degraded due to diffraction effects even if 'perfect' lenses are used!

More often photographic records will be viewed as projected on structured screens, (e.g. beaded screens) in which case a granular spatial noise structure is superimposed on the optical image. Equally, if photographs are studied under a microscope or in other direct viewing modes, even if the viewing optics are effectively perfect, the film grain structure will overlay the optical image data.

A very common form of image presentation nowadays is the television picture. Because of the nature of the electronics used to transmit and reconstitute this form of image it must, of necessity, have a 'line structure' associated with it. The picture will conventionally be reproduced in a series of horizontal lines[10]. Alternatively, in certain specialist situations, it may be reproduced as a series of vertical lines or even a helical line. All these presentations have in common the characteristic of high modulation one-dimensional noise of a specific spatial frequency, with the orthogonal dimension presenting data limited in quality by the TV camera lens and the band-width of the electro-optical system[10].

A final group of image presentation devices are the image convertors, image intensifiers and fibre optic displays. This group is all characterised by a fairly strong 2-dimensional noise overlaying the image presented. The form of this 2-dimensional noise is variable from device to device. In some image tubes the noise is in the form of random bright flashes[11]. In others (particularly the channel plate intensifiers[12]) the random flashes and the image are themselves superimposed on a matrix structure due to the channel plates. This latter matrix structure is also the predominent form of noise associated with fibre optic displays[13].

9.2.2 Photographic quality factors

The overall quality of a photographic image is complex, depending on the quality of the camera lens, the diffusion characteristics of the film emulsion in the camera[14], the chemical spread function of the film/developer combination (resulting in a form of degradation occurring in the development process)[15], the quality of any copying processing equipment and copy film, the projection or viewing optics and the quality of any viewing screen. In addition to the simple image degradation due to each of the above components there can be contributions of graininess from the original camera film, any copy film and any projection screen.

Since the images at each photographic emulsion stage are broadly diffuse they may be assumed to be incoherently coupled, hence avoiding the problems associated with direct coupling described in Section 9.1. In other words the quality factors at each stage and that of the eye may be considered to be additive. As such they may be dealt with directly, as discussed in Chapter 10. However, a word of warning is necessary at this point. The chemical spread function for certain film/developer combinations results in a so-called adjacency effect in the image which serves to enhance the local contrast across the image of a luminance discontinuity. This enhancement of contrast, which is dependent on the sharpness of the image on the film[15], constitutes a non-linearity in the optical system and adds difficulties to the process of image evaluation. It has the effect of subjectively sharpening up an image and as such is intentionally incorporated in many modern film emulsions. The importance, when considering visual acquisition from a photographic record, must depend very largely on how the eye operates. If, as the author believes, the all important image content is first differences of retinal image illuminance (Chapter 7), then it will be obvious that such edge sharpening can play a major part in acquisition.

The components of graininess may be defined in terms of one of a number of empirical functions which seek to relate the grain structure to a form of 'quality'. Perhaps one of the most notable of these is the Selwyn granularity coefficient[16] which is widely used as a measure of grain noise in the photographic industry. To date, to the author's knowledge, there has been nothing published relating such graininess to *objective* performance, although several authors have proposed relationships to *subjective* performance, e.g. Fry and Enoch[17] and Charman and Olin[18].

In photographic imagery it should be noted well by the reader that both quality and graininess are highly dependent on the density of the photographic image, both increasing with density up to a density of around 1.0, after which they tend to reduce again[19]. The relationship between quality and density for a typical film and development is shown in Figure 9.3. It is becoming a more common practice by film manufacturers to present curves of quality and graininess as functions of image density and development conditions in the film data sheets.

The role of glare in photographic systems can be an important one. It is not uncommon for significant simple veiling glare to be present in camera systems due to reflections within the camera body. In the author's experience this is particularly noticeable with miniature cine camera systems – possibly due to the close proximity of all potential sources of secondary reflection to each other and

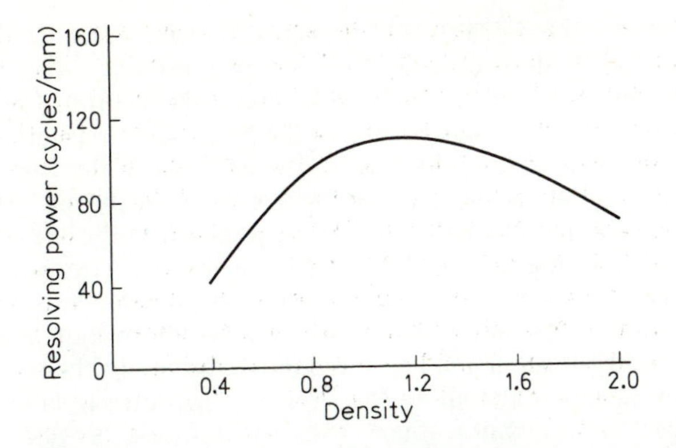

Fig. 9.3. The relationship between resolving power and density for a typical film and development (adapted from Mees and James (1966). The Theory of the Photographic Process, *Macmillan Co., New York).*

the film plane. Much more important is the reduction of contrast on projection due to all manner of sources of stray light. Such veiling glare on a projection screen can have very deleterious effects on material projected as illustrations, as many readers will have noticed. A slide prepared to suit a totally glare free projection facility may be totally useless in certain lecture theatres as a result of the severely reduced contrast whilst, conversely, a slide of high contrast, although admirable for lecture theatres having a high level of veiling glare, may be hurtful to the eyes in a projection facility which is totally glare free.

9.2.3 TV quality factors

When it comes to TV systems, as mentioned in Section 9.2.1 the predominant quality factors are the quality of the image formed on the photocathode of the TV camera tube by the camera lens and the characteristic one-dimensional line structure introduced by the line by line sampling of the photocathode image at the TV camera. However, in addition there must be, as for photographic systems, a form of image degradation by diffusion in the photocathode surface[20]. This diffusion is of an electrical origin (lateral leakage currents mainly) and is dependent on the 'drive' conditions for the TV system. This component of quality gets worse as the gain is increased to cope with lower signal levels and is also sensitive to the contrast settings[21]. In a practical system whether or not this quality factor is important depends largely on the quality of the imaging optics.

Additional quality factors introduced by the electronics – in addition to the line pattern – are the frequency transmission characteristics of the electronic circuits and their bandwidth (which control the quality of final image information *along* the TV lines[10]), the actual size and shape of the electron beam scanning spot which picks data off the photocathode[22] and the form of the scanning spot which reconstructs the image on the monitor. The interaction between the two scanning spot forms and the physical line separation can result in a variety of image distributions across the direction of the lines. Figure 9.4 endeavours to illustrate some of the distortions which can be introduced. At (a) we have a scanning spot of width considerably greater than the line separation. It can be seen that such a spot results in a grossly distorted reconstruction if the latter is produced from non-overlapping lines on the monitor, as is conventional. At (b) is shown an opposite situation, where some information is not scanned off the photocathode at all and hence cannot contribute to the reconstruction. At (c) we see an approximation to the ideal – a slight overlap of both electron beams, camera and monitor. Such an overlap would provide an optimal regeneration of information within the limits of the basic line frequency (e.g. Biberman[23]).

In an attempt to overcome the problems of viewing a picture obviously composed of a finite number of lines, two approaches have been tried. Firstly there is the use of random interlace*, where the two fields on the TV picture are allowed to drift slightly perpendicular to direction of scan, one with respect to the other[10]. It was hoped by this means largely to remove the effective line structure, since each complete picture would not overlay exactly in its entirity. Secondly there is the use of spot-wobble, where the monitor scanning spot is caused to oscillate perpendicular to the scan direction at high frequency and with an amplitude approximating to half the line spacing. This system is successful in removing the effective one-dimensional noise due to the scan lines, but can introduce its own brand of two-dimensional noise

Much discussion currently rages on whether one can measure the performance of a television system adequately in terms of conventional quality criteria. The author's feelings are that one can only apply conventional criteria with any guarantee of success to quality *along* scan lines. On the other hand, for the systems but also one important addition. Since the photocathode of a television both dimensions, provided that due account has first been taken of the distortions due to any excess scanning spot size on the photocathode.

Glare effects in television systems include those common to photographic systems but also one important addition. Since the photocathode of a television camera must be coated on the inside of a glass envelope, there exists the possibility of multiple internal reflections inside the glass of this envelope. If an intense bright spot appears in the image falling on the photocathode it can result in concentric rings of halation as shown in Fig. 9.5. Such halation rings quite commonly occur on commercial television when studio lights appear in the field of view. In severe cases such halation can result in complete obliteration of the

*Interlace in a TV system is achieved by scanning each alternate line and then filling in the missing lines on the next field (or presentation)[10].

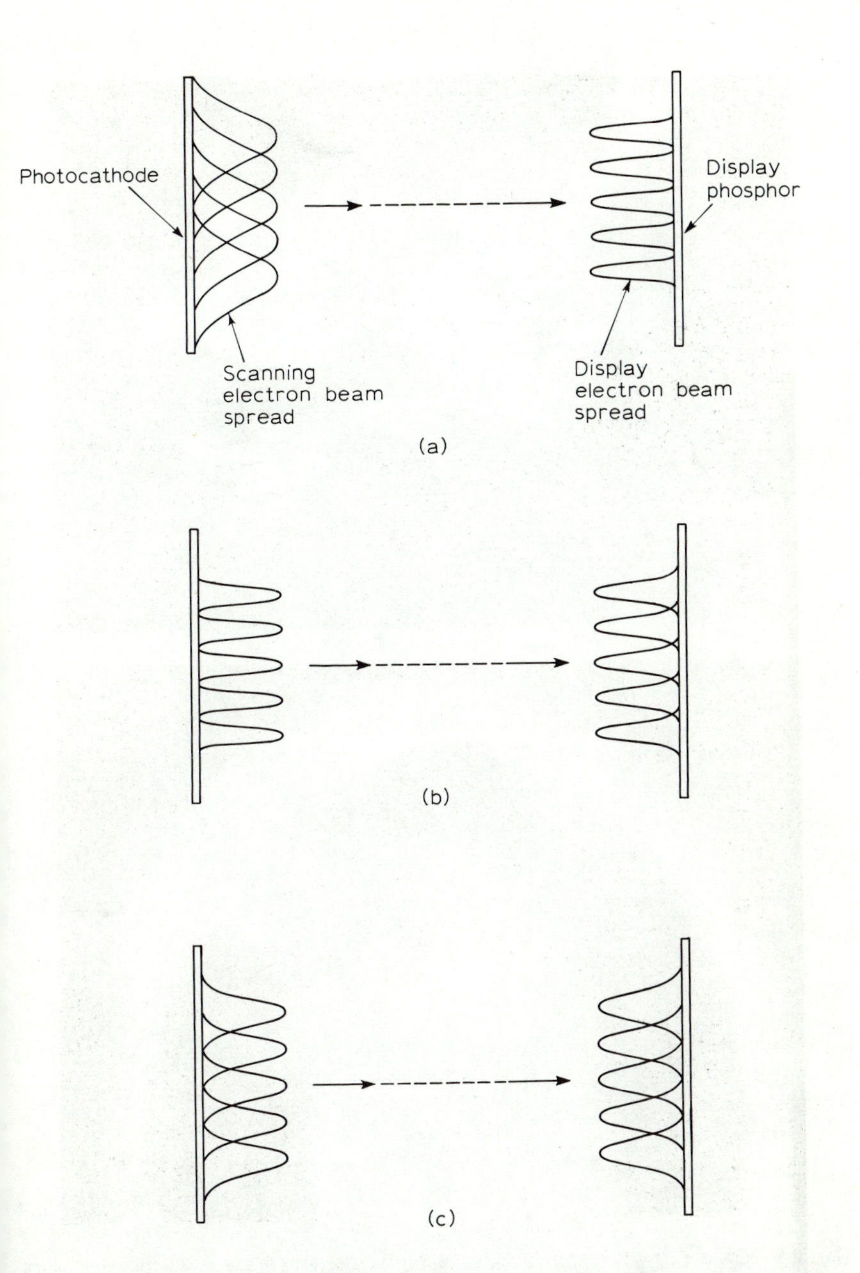

Fig. 9.4. Illustrating the effects of scanning and display electron beam spreads on imagery by a T.V. system. (a) Gross overscanning and excessively sharp display, (b) Under scanning and rather sharp display, (c) Ideal scanning and display.

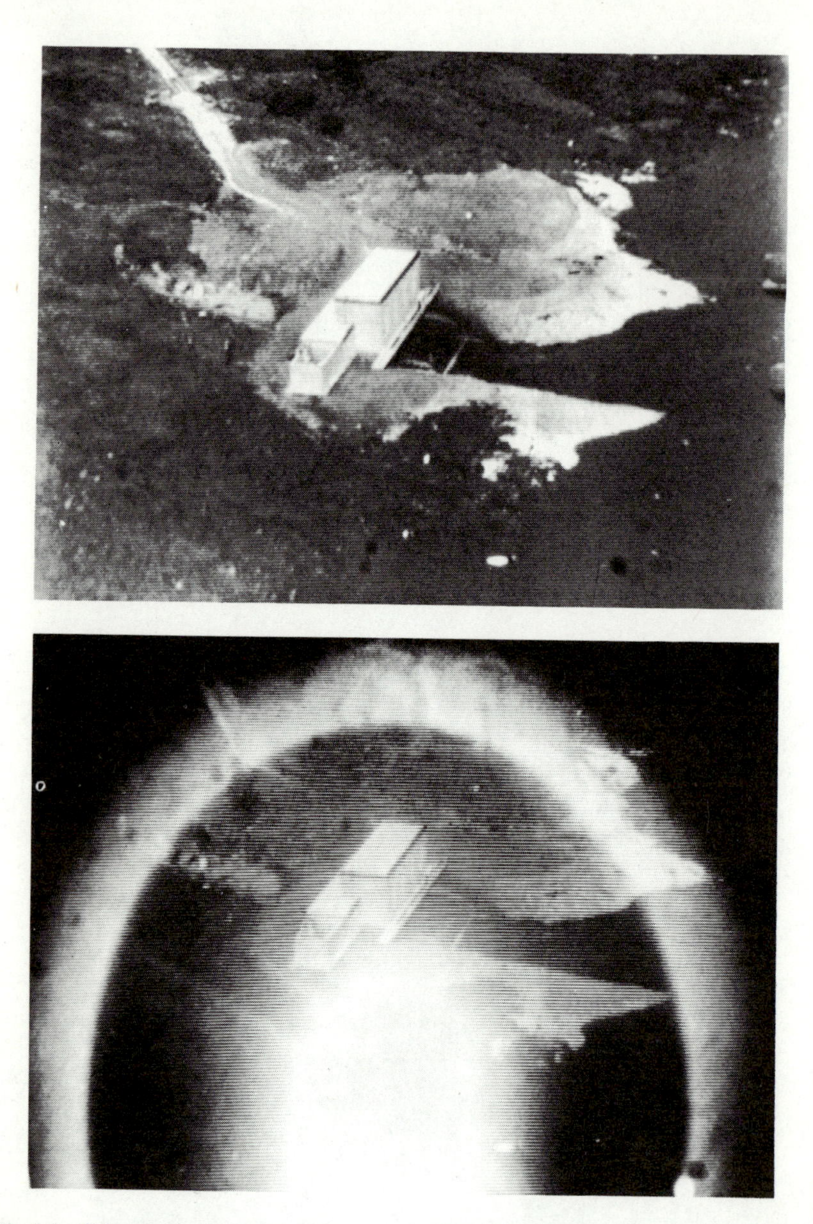

Fig. 9.5. Halation of a T.V. picture due to multiple reflections in the glass window in which the photocathode is mounted. Upper picture – typical scene – lower picture – same scene with bright light source in field of view.

television picture! It was in an attempt to overcome this problem, in part, that fibre face plates were first fitted to some television camera tubes[24] (see also Section 9.2.5).

9.2.4 Image tube quality factors

The prime quality factors associated with all image tubes, other than those associated with the necessary ancilliary optics, are diffusion effects in the photocathode and random photon noise. The diffusion effects are similar in extent and form to those in a television camera system. The photon noise is a particular characteristic of image tubes due to the construction and mode of usage. The usage of image tubes is frequently under low light conditions where energy is being received as individual quanta. Thus there is a basic noise in the image on the photocathode. Then the device amplifies the signal information – and with it the noise in the image[11]. The result is usually a very noisy final image.

Other forms of noise are introduced within the image tube, dependent on type. Earlier versions which relied on electrostatic or electromagnetic focussing of the imaging beam were prone to additional noise due to photon 'spillage' within the device[25]. Modern channel plate tubes are free from such lateral effects but exhibit a 'ghost' image of the channel plate structure and occasional blank spots where there are faulty channel plates[12]. The 'ghost' of the channel plate structure defines the limit of detail which can be recovered when viewing through such tubes. Glare effects in image tubes are similar to those in television systems. Again fibre face plates have been employed to overcome some of the glare problems[24].

9.2.5 Fibre bundles

A final form of secondary image which may be viewed is that produced at some stage on a fibre face plate. Such devices are widely used in television camera tubes and image tubes to cut down lateral halation problems in awkward illumination situations as already mentioned. Long, coherent fibre bundles ending in fibre face plates are also used as flexible fibrescopes to transmit optical image information from an inaccessible place (such as the interior of a furnace or the inside of a person's stomach) to a more accessible viewing position (e.g. Kapany[24] and Hirschowitz[26]). As with channel plate intensifiers fibre face plates of necessity superimpose a structure on the image being viewed, this structure usually being a regular hexagonal mesh[13]. Again, as with channel plate tubes, the mesh structure defines the limit of detail which can be recovered.

9.2.6 Interaction of quality and spatial noise

It is usually assumed that most information can be obtained from a noisy display system by making the image as sharp as possible, the information then being limited by spatial noise. It has been shown recently by Biberman that such an assumption can be, and often is, fallacious[23]. He has illustrated that it is often possible, both on television pictures and on displays exhibiting 2-dimensional noise, to increase substantially the detail information visualised by introducing a moderate degree of blurring of the display. This effect may be explained by considering the mathematics of sampling of image information[27] from which it can be shown that, for optimal performance, there should be a balance between image blur and grain structure. Such a balance is particularly appropriate when considering imagery exhibiting noise of a specific frequency, as with television, channel plate devices and fibre face plates.

9.3 CONCLUSIONS

It must be concluded that there are many distortions of luminance data from an object of interest which occur due to employment of indirect vision. These may be broken down into three basic categories – softening of profiles, modification of scene contrast and noise. Modern methods of image evaluation permit the distortions of data due to the first two categories to be analysed fairly rigorously. The methods available and their use will be described in Chapter 10. The noise factors are many – and often specific to a given type of equipment. They are not so readily handled – except as already referenced in this chapter – and will not be considered individually any further in this book. However, the general aspects of effects of noise on threshold performance, and some attempts at modelling random noise effects, will be discussed further in Chapters 11, 12 and 13.

REFERENCES

1. Kingslake, R. (1965). *Applied Optics and Optical Engineering*, Vol. 3, Academic Press
2. Born, M. and Wolf, E. (1964). *Principles of Optics*, Chap. 6.3, Pergamon Press
3. Emsley, H. H. (1963). *Visual Optics*, Vol. 2, Hatton Press
4. Emsley, H. H. (1963). *Visual Optics*, Vol 1, Chap. 5, Hatton Press
5. Born, M. and Wolf, E. (1964). *Principles of Optics*, Chapts. 8 & 9, Pergamon Press
6. De Velis, J. B. and Parrent, G. B. (Jr.) (1967). 'Transfer Function for Cascaded Optical Systems', *J. Opt. Soc. Am.*, 57, 1486
7. Overington, I. (1973). 'The Importance of Coherence of Coupling when viewing through Visual Aids', *Optics and Laser Technology*, 5, 216
8. Goldberg, E. (1922). *The Structure of Photographic Images*, (in German), Withelm Knapp, Halle

9. Martin, S. (1968). 'Review of Veiling Glare in Optical Instruments', *SIRA Report R. 433*
10. Amos, S. W., Birkinshaw, D. C. and Bliss, J. L. (1963). *Television Engineering*, Vol. 1, Pt. 1, Iliffe Books
11. Timm, G. W. and Ziel A. Van der, (1968). 'Noise in Various Electron Multiplication Methods used in Imaging Devices', *IEEE Trans. Electron Devices,* **ED–15**, 314, May
12. Sackinger, W. M. (1971). 'Noise Performance of the Channel Electron Multiplier,' *Photoelectronic Imaging Devices*, Vol. 1, Chap. 9, (Eds. L. M. Biberman & S. Nudelmann), Plenum
13. Ohzu, H. (1967). 'Image Transmission Characteristics of Fiber Bundles,' App. A of *Fibre Optics* by N. S. Kapany, Academic Press
14. James, T. H. and Higgins, G. C. (1960). *Fundamentals of Photographic Theory*, Chap. 16, Morgan & Morgan Inc., New York
15. Nelson, C. N. (1971). 'Prediction of Densities in Fine Detail in Photographic Images', *Phot. Sci. and Eng.*, **15**, 82
16. Selwyn, E. W. H. (1935). 'A Theory of Graininess', *Phot. J.*, **75**, 571
17. Fry, G. A. and Enoch, J. M. (1959). 'The Relation of Blur and Grain to the Visibility of Contrast Borders and Gratings', MCRL TP No. (696)–19–293, Ohio State University
18. Charman, W. N. and Olin, A. (1965). 'Image Quality Criteria for Aerial Camera Systems', *Phot. Sci. and Eng.*, 9, 385
19. Shaw, R. (1973). 'Some Detector Characteristics of the Photographic Process', *Optica Acta,* **20**, 749
20. Amos, S. W., Birkinshaw, D. C. and Bliss, J. L. (1963). *Television Engineering*, Vol. 1, Pt. 4, Iliffe Books
21. Cope, A. D., Gray, S. and Hutter, E. C. (1971). 'The Television Camera Tube as a System Component', Chap. 2 of *Photoelectronic Imaging Devices*, Vol.2, (Eds. L. M. Biberman & S. Nudelman), Plenum
22. Amos, S. W., Birkinshaw, D. C. and Bliss, J. L. (1963). *Television Engineering*, Vol. 1, App. 1, Iliffe Books
23. Biberman, L. M. (1973). 'The Mythology of Target Acquisition System Design and Performance', Rep. OA6201, Vol. 1,: *A Collection of Unclassified Technical Papers on Target Acquisition*, p. 39, Martin Marietta Aerospace, Orlando, Florida
24. Kapany, N. S. (1967). *Fiber Optics: Principles and Applications*, Chap. 5, Academic Press
25. Vine, J. (1971). 'Electron Optics', Chap. 10 of *Photoelectronic Imaging Devices*, Vol. 1, (Eds. L. M. Biberman & S. Nudelman), Plenum
26. Hirschowitz, B. I. (1964). 'Endoscopic Photography using Fibre Optics', *J. Soc. Mot. Pict. Tele. Eng.*, 73, 625
27. Lavin, H. and Quick, M. (1974). 'The OTF in Electro-optical Imaging Systems', in *Proceedings of the SPIE, Vol. 46, Image Assessment and Specification*, 279

10 Assessment of Image Quality

The need to be able to measure optical quality was recognised many centuries ago, but no really adequate concept was put forward until the 19th century, when several scientists progressively refined the concept of resolving power – the limit of ability to detect that two adjacent objects were present as the separation between them was reduced. This 19th century work culminated in the classical papers of Lord Rayleigh, which defined the limit of resolution for a perfect optical system to be the angular separation of two point objects when the first dark ring of the Airy disc produced by one object coincided with the peak of the Airy disc formed by the other. This concept was later extended by Rayleigh's followers to the case of two line objects, and later again became a general concept for specification of performance of most optical systems. Useful discussions of the concepts of resolution and resolving power are to be found in Ronchi,[1] Brock[2] and Perrin.[3]

Since the 1930's there has been an increasing awareness that this simple concept was other than satisfactory, because of the dependence of its value on numerous factors. As a result of this, other more adequate and universal measures have been sought. Some 35 years ago a concept already widely used in the fields of electronics and control systems was 'borrowed' by optical scientists – that of frequency of response.[4] From this grew the concept of Optical Transfer Function (OTF) which, in theory, was a complete measure of the degradation in sharpness due to an optical system.[5–7] Many methods have been developed for the instrumental measurement of the OTF of optical systems (see Section 10.2.2). If used in conjunction with some measure of gross change of scene contrast (e.g. veiling glare – see Section 9.1.6) such a measure should be capable of completely specifying an optical image. However, it soon became apparent that there was no obvious, direct relationship between OTF and visual performance in normal viewing, partly due to the unknown effects of coupling between the optical component and the eye, and partly due to an incomplete understanding of the relationships between visual performance when looking at periodic functions and when looking at isolated objects.

In recent years there has thus been a considerable effort made to find a figure of merit related to OTF which can define visual performance. Several possible empirically derived figures of merit have been investigated. These will be discussed later in this chapter (Section 10.3). Finally, from the modelling described in Chapter 7 it has been possible to propose a physically-based figure of merit (visual efficiency) which may be used as a starting point for the prediction of visual performance for many simple and definable viewing situations. This will also be discussed later in the chapter (Section 10.3.5).

Whilst progress was being made in the development of methods of measuring frequency response, attention was also being given to the possible forms of veiling glare. This has resulted in the definition of a Veiling Glare Index (VGI) as a standard for specifying general loss of contrast, and other glare functions applicable to ghost images and local glare. These will be discussed in Section 10.4.

10.1 LIMITATIONS OF RESOLUTION

As intimated earlier, the concept of resolution was originally applied to the minimum detectable separation between two point objects. This had some value

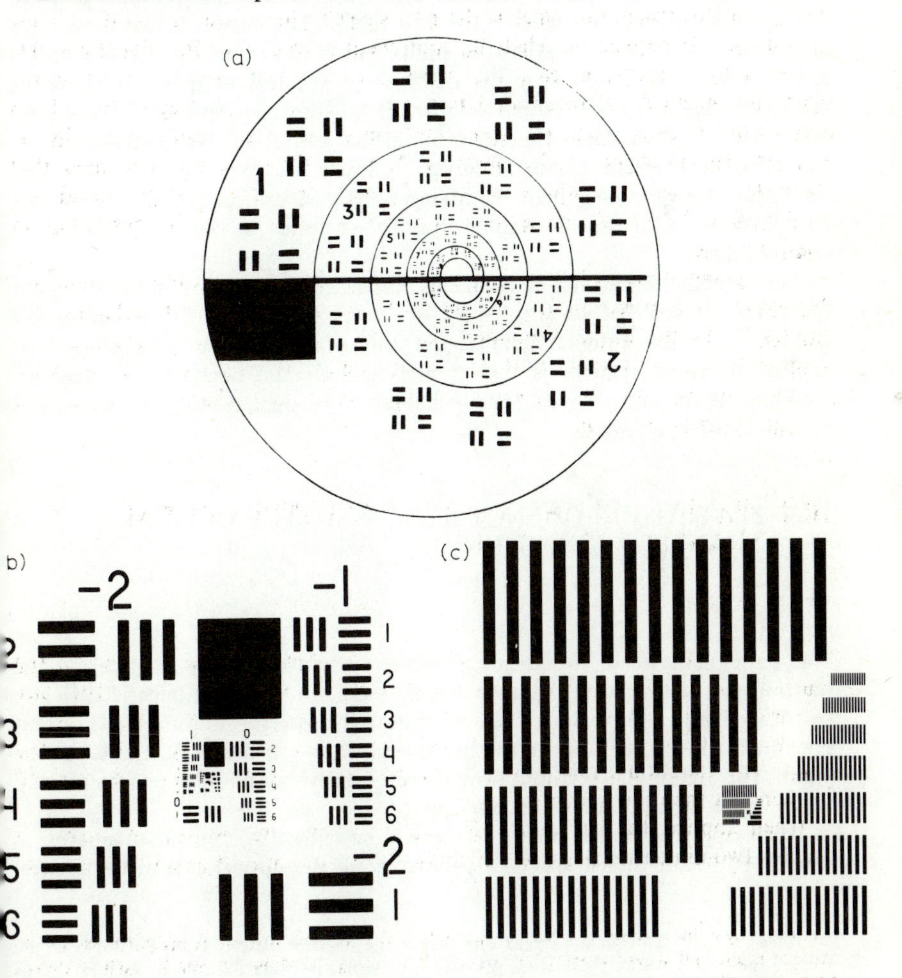

Fig. 10.1. Some of the many resolution test targets in current use. (a) British Cobb. (b) American 3-bar. (c) Sayce. Ealing High Resolution Test Target.

in astronomy, where the objects of interest were usually stars, but was of questionable use for any other application. In an attempt to overcome this for for practical measurement of the resolution of photographic equipment and visual aids, various resolution test charts were proposed. A number of these still retain considerable favour, e.g. the Cobb Chart,[8] and the American 3 Bar Chart.[9] Figure 10.1 shows some of these forms of chart. When one attempts to measure resolution using any of these charts, the first discovery is that no two charts give the same resolution figure. It appears from observation that the shape of the bars and the number of the bars comprising an *element* both affect the results. To someone conversant with all the threshold effects discussed in previous chapters this is not surprising, but what is the *best* shape? The answer is that there is no *best* shape – it depend on what the equipment is to be used for. Equally well it is found by observation that the *contrast* of the test pattern affects results markedly. Again from threshold data this is no surprise – and again there is no *best* contrast. Once again the scene luminance can affect results markedly, as can also the eyesight of the observer. It must therefore be concluded that resolution is a very inadequate method of specifying optical quality. Indeed, not only may it be grossly misleading in *absolute* terms, it *can* be misleading in relative terms.

For extended discussion on problems associated with resolution measurement the reader is referred to the papers of Vasco Ronchi[1] and Overington and Gullick.[10] In the author's opinion the only point in favour of resolution as applied to visual systems is that it does include the observer – a necessary requirement for any measurement technique to be used reliably for assessment of visual instrumentation.

10.2 SPATIAL FREQUENCY RESPONSE (THE OPTICAL TRANSFER FUNCTION)

10.2.1 Concepts

The basic concept of frequency response is arrived at by considering the transmission of sine waves through any system. If a system is linear,* then any sine wave input will produce a sine wave output. However, in general the output sine wave will be reduced in amplitude and have a phase-shift relative to the input. The functional relationship between relative amplitude, phase-shift and frequency is then the frequency response.

When applied to the quality of optical systems, the domain of interest is usually two-dimensional space perpendicular to the optical axis of the system.

*A *linear* system is defined here as one where the average output is proportional to the average input and where there is no distortion of signal information due to such factors as saturation effects.

Since the concept of two-dimensional frequency and phase-shift is very complex, it has become conventional to describe optical performance in terms of two orthogonal one-dimensional frequency responses. The test material is a one-dimensional bar pattern of effectively infinite extent, the brightness being modulated across the bars according to the law

$$B_L = B_m + (\Delta B)_{\max} \cos 2\,\pi f_s x \tag{10.1}$$

In the above B_L is the local luminance, B_m is the mean luminance, $(\Delta B)_{\max}$ is the maximum deviation from the mean luminance, f_s is the spatial frequency of the pattern and x is the distance from an arbitary zero.

Then, in general, the output will be

$$B'_L = B_m + \gamma(\Delta B)_{\max} \cos (2\pi f_s x + \phi) \tag{10.2}$$

where γ is the relative amplitude and ϕ is the phase angle.

The trend of γ as a function of f_s is then known as the Modulation Transfer Function (MTF), whilst the trend of ϕ as a function of f_s is the Phase Transfer Function (PTF). The combination of MTF and PTF is the Optical Transfer Function (OTF). The concept is illustrated in Fig. 10.2.

In order for the concept to be applicable it is necessary for the optical system to be *linear* in the space domain and also for the image field to be homogeneous

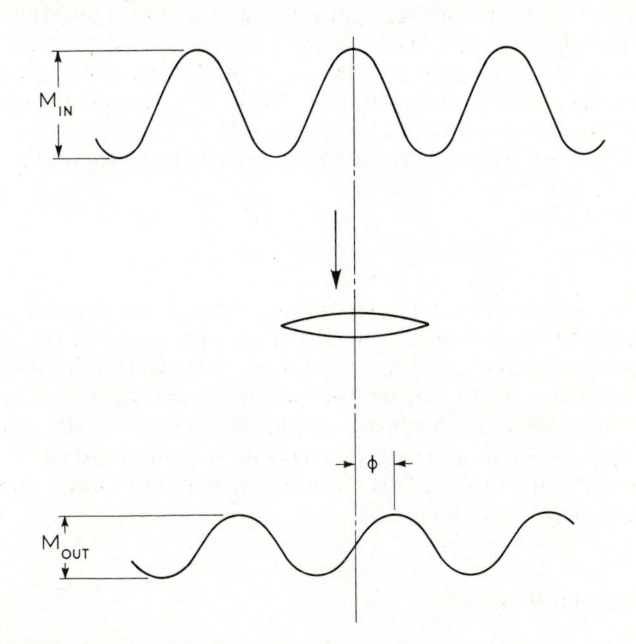

Fig. 10.2. Illustrating the transmission of spatial sinusoidal modulation through an optical system. $M_{OUT}/M_{IN} = \gamma = MTF$, ϕ = *phase displacement.*

over a large area compared to one cycle of the bar pattern. Most normal optical transmission systems are linear and many detector systems are adequately linear over restricted ranges of luminance for this concept to be applied. Homogeneity of the image field can usually be provided in a laboratory situation but care needs to be exercised.

It is usual to determine the OTF for bar patterns with the lines perpendicular to the radius from the optical axis (tangential OTF) and parallel to the radius from the optical axis (saggital or radial OTF) for other than axial imagery. In general the plane chosen for measurement of the radial and tangential OTF's must depend on the use the system is going to be put to, since most systems will exhibit some degree of astigmatism and curvature of field. For instance, in testing photographic lenses the OTF must be measured in some chosen optimum image plane (which will not usually be that yielding best on-axis performance). Conversely, for visual aids, the optimum imagery at any local field point may be considered, subject to the limitations of accommodation of the eye. However, in such cases the optimal radial and tangential imagery will usually be in different planes due to astigmatism, and due attention must be paid to the fact that lines and periodic patterns at orientations other than radial or tangential will not yield performance as high as either optimal radial or optimal tangential. Indeed, in such cases it is desirable to study the optimal OTF as a function of pattern orientation.[11,12] For axial imagery and a symmetrical system in optimum focus there is obviously only one OTF. However, in the case of practical systems exhibiting astigmatism or other defects such as mechanical centering errors, the OTF may *not* be symmetrical on axis. In such cases it is conventional to consider the *vertical* bars as tested to yield a tangential OTF.

The OTF is closely related to the line spread function of an optical system, the two being a Fourier transform pair, i.e.

$$F(j\omega) = \int_{-\infty}^{+\infty} G(x) \exp(-j\omega x) \, dx \tag{10.3}$$

where $\omega = Z\pi f_s$, $F(j\omega)$ is the OTF and $G(x)$ is the line spread function.

The integral of the line spread function is the edge profile of the image of a perfect edge, and the line spread function is itself the integral of the point spread function. The point spread function in turn, when convolved with any object profile, generates the image luminance distribution. Therefore, the OTF is seen to be directly related to all image functions of an optical system. The reader wishing to pursue the theory of OTF and its relationship to image formation is referred to Hopkins[13] and Brock.[14]

10.2.2 Measurement methods

Over the last two decades many methods have been devised for measurement of OTF. These include *direct* methods, where sinusoidally modulated patterns of

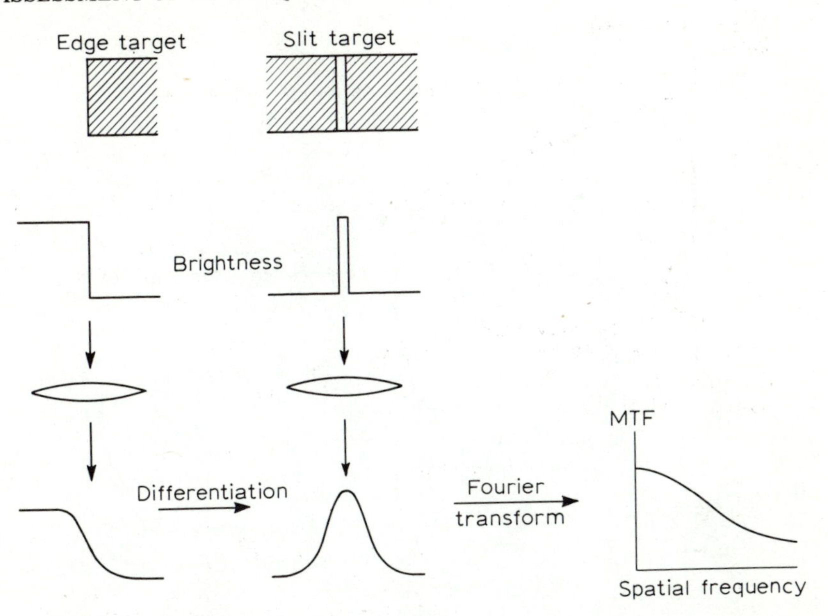

Fig. 10.3. Illustrating the determination of MTF from the images of slits and edges.

various frequencies are passed through a system under test, and various forms of *indirect* methods. These latter include methods where the image of a line or edge is scanned and the resultant output is mathematically processed to yield the OTF (simple Fourier transformation in the case of line images, differentiation and Fourier transformation in the case of edge images, Fig. 10.3) and also methods where the wavefront aberrations of a system are measured, the spread functions and OTF being computed from the wavefront aberrations and pupil size (Fig. 10.4). Below is a survey of some of the more popular methods. More general surveys are given by Rosenhauer and Rosenbruch.[15,16]

EROS.[17,18] This equipment developed by the SIRA Institute at Chislehurst, UK in conjunction with the Royal Aircraft Establishment, Farnborough, and marketed by the Ealing-Beck Corporation, produces a continuously varying frequency of sinusoidal pattern by rotating one slit pattern with respect to a second. The resultant moving Moiré fringe pattern is projected through the system under test and imaged on an analyser slit connected to a photo-multiplier tube. An immediate output is produced which can be displayed as MTF and PTF on an oscilloscope. Various versions of the equipment exist, ranging from a portable, modular system to a very expensive, rigidly mounted version for the most precise work.

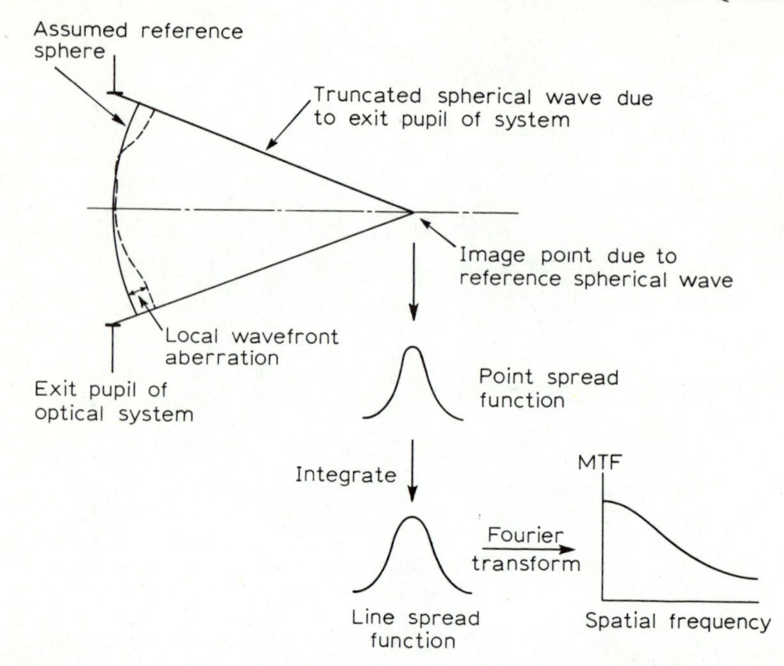

Fig. 10.4. Illustrating the derivation of MTF from analysis of wavefront aberrations. The point spread function is the resultant interface pattern from the truncated, aberrated wavefront.

ODETA.[19] This portable equipment is developed and marketed by the Oude Delft Company in Delft, Holland. It is based on a narrow object slit (2 μm) presented to the system under test and a Moiré scanner as an analyser (i.e. essentially the reverse of the EROS equipment). The output from the scanner is spectrum analysed and may be presented either as an oscilloscope trace of the MTF or as a graphical output on an $X - Y$ recorder. The equipment is claimed to be suitable for measurement of the MTF of lenses, image intensifier tubes, television systems, flourescent screens, etc.

ACOFAM.[20,21] This is a sturdily built, nonportable equipment originally developed by the AQMEL Company of Southern France. It is arranged to measure and plot out the profiles of the images of sine wave patterns of seven specific spatial frequencies. The input sine patterns are projected into the test lens via a collimator mounted on a robust pivoting arm, the output modulation being recorded via a microscope objective and slit assembly mounted on a traverse slide. A recent modification of the system[21] enables the equipment to be used for testing in the infra-red out to 20 μm.

Edge Trace Methods. Several edge trace methods are in use throughout the world. In the USA the Perkin-Elmer Corporation were one of the first to employ this technique (for photographic image assessment[22]). For such measurements the edge trace technique is ideally suited, of course, since the photographic emulsion is its own detector. Others in the USA employing similar techniques were R. A. Jones of Itek[23] and Hopkins and Dutton of Rochester University.[24] These latter used scanning photoelectric detectors in place of the photographic emulsion. More recently Tropel Inc. have developed a commercial equipment based on edge scanning.[16] Meanwhile, in the UK, Overington and Gullick were experimenting with photographic edge trace techniques as a *general* method of measuring OTF, using the photographic system in such a way that it contributed little to the overall performance.[25,26]

Slit Methods. A number of systems have been devised where a slit source is used in place of an edge, the image of the slit being analysed by a narrower slit. Such systems, of which the Rodenstock-Askania[16] and the Cannon Lens Analyser[27] are two, are limited to some extent by the available energy, but have the virtue of producing the line spread function directly (after due compensation for finite object slit width).

Interferometric Methods. An alternative approach to OTF measurement preferred by a number of workers is to study the wavefront aberrations of an optical system by interferometric methods and then to compute the OTF. For this method it is necessary first to compute the point spread function from the measured aberrations and the diffraction due to wavefront truncation.[28,29] Having derived the point spread function it is a relatively simple matter in theory to arrive at the line spread function (by integration) and subsequently the OTF (by Fourier Transformation) (Section 10.2.1). However, in practice significant errors can occur if great care is not taken in the computational stages.

The interferometric methods have one striking advantage over the more direct methods of OTF measurement, provided that the computation is reliable; they contain *all* the information about an imaging wavefront. Hence one measurement enables computation of 'quality' at any point in image space. Furthermore, effects of coupling of two systems such as a visual aid and the eye (Chapter 9) amount to no more than summation of wavefront aberrations. A typical facility for wavefront aberration measurement is the interferometer built by the National Bureau of Standards in the USA,[28] whilst other bodies use various other standard forms of interferometer for appropriate measurements (e.g. Haines and King,[30] Baker[31] and Herriot and Bruning.[32]

10.2.3 Limitations of OTF

As will be obvious from the preceding section, OTF as usually used is an *instrumental* technique. Thus, although one can derive the frequency response of the aerial image of an optical system by one of the many methods listed – or the frequency response of the final photographic image if desired – in all cases where the human being is involved the final image has subsequently to be viewed by the eye. Thus the *effect* of such a frequency response as may be measured depends, in the end, on how it is viewed – and what the eye does with the frequency information. Now we have already seen in Section 4.14 that the eye has a form of total frequency response known as the 'Contrast Sensitivity Function'. We have also observed that this function is dependent on the viewing conditions – in particular the state of eye focus and scene luminance. Now in general, for viewing incoherent images (i.e. on screens, etc.), the OTF of the system incorporated in that final image and the contrast sensitivity of the eye may be multiplied together. However, having done so one must decide on how the eye works in order to predict the effect of any instrumental system OTF. Alternatively, still for viewing incoherent images, one may multiply the instrumentally derived MTF by the measured MTF of the refraction optics of the eye (see Section 2.3) in order to arrive at the MTF of the retinal image.

A further difficulty arises in viewing through transmission optical systems – the implications of direct coupling via an aerial image as discussed in Chapter 9. In such circumstances one should ideally study retinal imagery in terms of wavefront aberrations as discussed earlier in the Section on Interferomic Methods. However, the limited data available on the aberrations of the eye appear to show them to be rather variable, both from person to person and according to viewing condition[33,34] (see also Section 2.3). Thus in practice such an approach is not entirely satisfactory. The alternative is the measurement of total system contrast sensitivity, with subsequent allowance for the transfer functions of the neural networks in order to arrive at the retinal MTF.[10,35]

Recent image evaluation studies at BAC(GW) are believed to have shown a possible way of deriving retinal MTF's from contrast sensitivity. As part of our modelling of the visual system it has been necessary to assume that the bipolar cells in the retina provide a first difference signal from the retinal receptors as a result of local inhibition (Chapter 7). Now, for the general case of spatial sinusoidal modulation, the maximum sampled first difference can be shown to be proportional to sin $\pi f_s \Delta x$ for

$$\pi f_s \Delta x \leqslant \frac{\pi}{2}$$

and proportional to $(1 - \cos \pi f_s \Delta x)$ for

$$\pi f_s \Delta x > \frac{\pi}{2}$$

where f_s is the spatial frequency at the retinal image and Δx is the local receptor spacing for like receptors. For foveal imagery Δx is approximately 0.2 mrad. Therefore for

$$\pi f_s \Delta x = \frac{\pi}{2}$$

$f_s \approx 2.5$ c/mrad. But the practical contrast sensitivity of the eye is very poor at spatial frequencies above 2.5 c/mrad (see Section 4.14). Therefore, for practical purposes one may assume the sampled first difference to be proportional to $\sin \pi f_s \Delta x$ for all spatial frequencies. It has been shown[36] that if one starts with a measured contrast sensitivity for the naked eye under known viewing conditions and operates on it by the sampling function one generates a predicted retinal MTF which is in very close agreement with that predicted by operating on the MTF of the refraction optics to allow for tremor, drift and retinal diffusion (Chapter 7). It would thus appear that a reasonable estimate of the retinal MTF for *any* visual aid/observer combination can be obtained by operating on a total system contrast sensitivity by the same neural sampling function, since this sampling function should be dependent on the observer alone.

10.3 OTHER QUALITY MEASURES

Because of the lack of generality of OTF data when applied to visual performance a number of attempts have been made to find figures of merit associated with the OTF which are adequate descriptors of the relative quality of systems. In the following sections various figures of merit are discussed. Others are discussed by Birch.[37] At this stage it is assumed that all presented images are effectively noise free. Some implications of noise are discussed in Chapter 13.

10.3.1 Modulation Transfer Function Area (MTFA)

The MTFA concept was proposed in the middle '60's', primarily for photographic systems, and has been largely studied by Snyder in particular.[38] Basically MTFA is defined as the area between the MTF curve of a lens system and the detection threshold curve for optimally-viewed photographic representation of the American 3-bar resolution test target (see Fig. 10.5). But the 3-bar resolution threshold is an approximate inverse measure of the retinal MTF for an observer and film combination. Thus the MTFA is approximately representative of the area under the total system MTF referred to the retina, since the lens and film MTF's may be multiplied together.

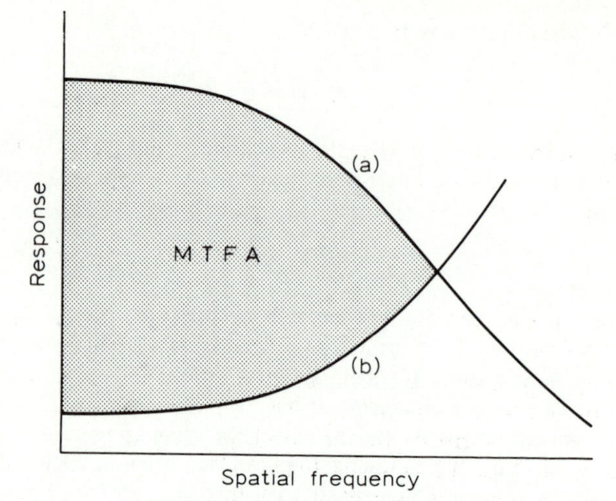

Fig. 10.5. Illustrating the concept of MTFA. (a) is the MTF of the transmission optics, (b) is the threshold detectivity curve for (typically) an American 3-bar target.

Several experimenters have attempted to relate subjective performance to MTFA and closely related factors with very considerable success. In addition to Snyder the most notable are Beidermann[39] and Higgins.[40] Snyder and Beidermann showed a strong correlation between MTFA for photographs of miscellaneous subjects and the subjective ranking of goodness. Higgins showed an even stronger correlation for a similar exercise when he compared subjective performance to the square of the area under the resultant curve from multiplying together the instrumental MTF and that due to the refraction optics of the eye.

10.3.2 Acutance measures

Whilst Snyder was investigating the range of validity of MTFA, large scale studies at Eastman Kodak were seeking to establish forms of quality factor which could predict subjective 'sharpness' of photographs well. Such factors had, of necessity, to take account of a chain of optical and photographic components – typically a camera lens, a camera film, a projector lens, a viewing screen and the observer's eye. These studies led to a proposal of system modulation transfer (SMT) acutance (Crane[41]), this being defined for a camera/projector system as

$$\text{SMT acutance} = 120 - 25\log\left[\left(\frac{200 \times \text{magnification}}{\text{MTC area}_{(\text{camera})}}\right)^2 + \right.$$
$$+ \left(\frac{200 \times \text{magnification}}{\text{MTC area}_{(\text{film})}}\right)^2 + \left(\frac{200 \times \text{magnification}}{\text{MTC area}_{(\text{projector})}}\right)^2 +$$
$$\left. + \left(\frac{200 \times \text{magnification}}{\text{MTC area}_{(\text{screen})}}\right)^2 + \left(\frac{200}{\text{MTC area}_{(\text{observer})}}\right)^2\right] \tag{10.4}$$

where magnification is taken to be the ratio of the image width *on the observer's retina* to the image width in the specified component (camera, film, etc.) and MTC area is the area, in units of mm^{-1}, under an experimentally determined MTF curve for the specified component.

The factor was then extended for general use to

$$\text{SMT acutance} = 120 - 25\log\left[\sum_{i=\text{camera}}^{\text{observer}}\left(\frac{200 \times \text{magnification}}{\text{MTC area}_i}\right)^2\right] \tag{10.5}$$

where the summation included a term for each component of the total system between camera and observer.

This factor was found to yield good correlation with sharpness judgements employing a scale of sharpness determined by pair comparison techniques,[42] so long as there were no significantly nonlinear components in the system.

Subsequently SMT acutance was found considerably lacking when using modern film materials exhibiting strong adjacency effects (see Section 9.2.2 for an explanation of adjacency effects). In consequence a modified acutance factor – contrast modulation transfer (CMT) acutance – was proposed (Gendron[43]). This factor essentially replaced the sum of component MTF functions by a product, yielding

$$\text{CMT acutance} = 125 - 20\log\left(\frac{200}{\text{MTC area}_{(\text{syst})}}\right)^2 \tag{10.6}$$

where

$$\text{MTC area}_{(\text{syst})} = \int_0^\infty \text{MTF}_c(\mu') \times \text{MTF}_f(\mu') \times \ldots \times \text{MTF}_0(\mu')\, d\mu',$$

MTF_c, MTF_f etc. being the MTF's at spatial frequency μ' as referred to the observer's retina for the camera, film, etc.

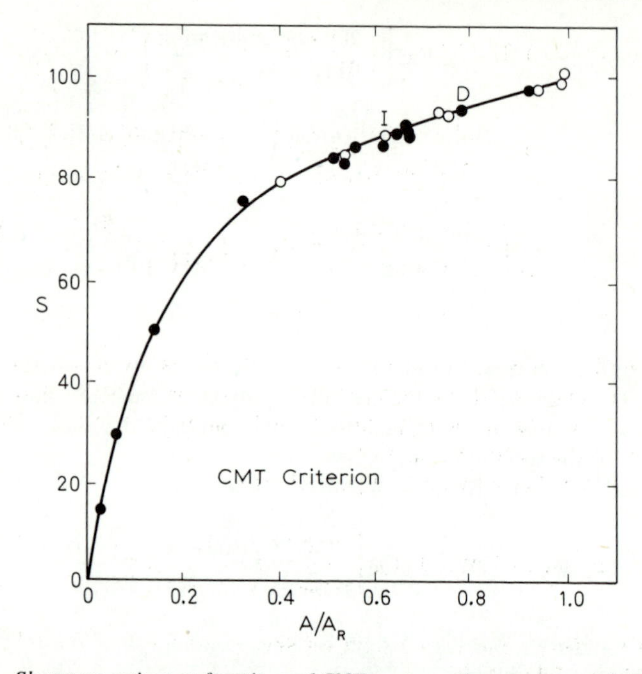

Fig. 10.6. Sharpness ratings as functions of CMT acutance for a variety of forms of image degradation. These include defocus, aberrations, photographic adjacency effects, halation and combinations of two or more forms of degradation. (Reproduced from Gendron[43] *by courtesy of the Society of Motion Picture and Television Engineers).*

It was found that this factor gave very high correlation with subjective sharpness judgements for a very wide range of forms of optical degradation, including asymmetric aberrations, gross adjacency effects and gross adjacency effects in the presence of halation. The goodness of fit for all these forms of degradation is illustrated in Fig. 10.6.

It has been shown by the present author that Equation 10.6 may be readily transformed to

$$\text{CMT acutance} = 40 \log (6.67 \text{ MTC area}_{(\text{syst})}) \qquad (10.7)$$

which immediately shows it to be closely related to MTFA.

10.3.3 MTFA Variants

Other workers have thought it desirable, for various reasons, to consider the most effective figure of merit to be the area under the instrumental MTF curve between certain frequencies. The frequency bands considered have ranged from

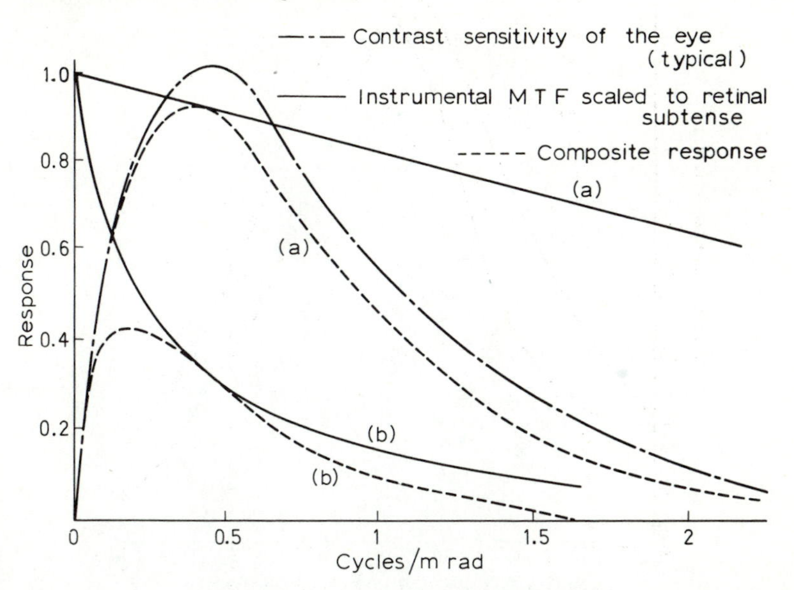

Fig. 10.7. The effect of interactions between instrumental MTF and the contrast sensitivity function of the eye. (a) High quality system, (b) low quality system.

the low frequencies, through the intermediate frequencies. Such ideas have usually been empirical, based on certain limited observations. Study of the implications of combining an instrumental MTF and a contrast sensitivity function for the eye will quickly show how such various selections can arise (Fig. 10.7). If the viewing conditions are such that the instrumental MTF *referred* to the angle subtended at the eye has good response out to 6 c/mrad or more, then the contrast sensitivity function will force the main weighting to be applied to the 'intermediate frequency' region. If, on the other hand, the referred response falls off at considerably less than 1 c/mrad, this will override the contrast sensitivity function and will force the main weighting to be applied to much lower frequencies.

The concept of a weighted or bandpass form of MTFA has recently been put on a firmer foundation by Granger,[44,45] who has considered to some extent the physiology of the eye, and has then determined a subjective quality factor (SQF) which, as with the acutance measures, considers all MTF data referred to the retinal image. He has then arbitrarily chosen a bandpass centred around the peak response of the visual system (as shown in Fig. 10.8). This function has again been shown to correlate very highly with subjective sharpness judgements over a considerable range of quality factors (including, in this case, colour contrast).[45] However, it is claimed by the team working with CMT acutance (private communication) that SQF does not work well for a combination of

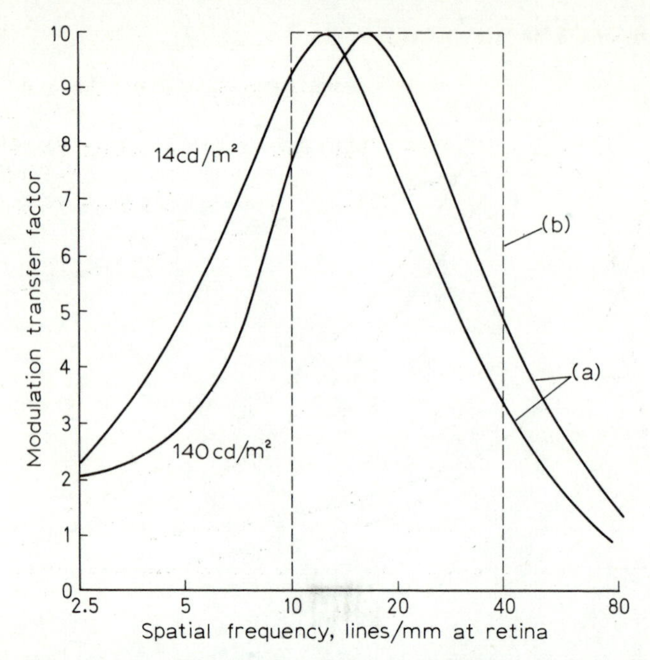

Fig. 10.8. Illustrating the concept of SQF. Curve (a): Typical contrast sensitivity functions for the eye as measured by O. Schade. Curve (b): SQF band as defined by Granger (Reproduced with permission of the Society of Photographic Scientists and Engineers Inc., as published in Granger and Cupery[44]. Copyright 1972).

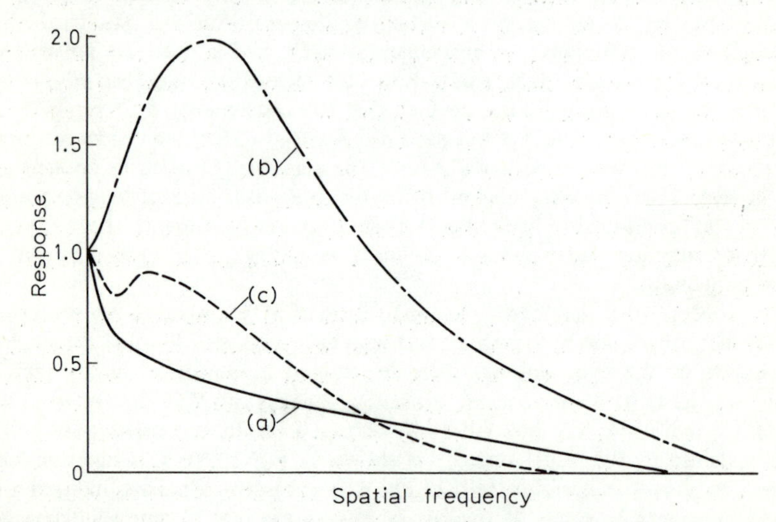

Fig. 10.9. The effect of a combination of halation and strong adjacency effects on low frequency response. Curve (a): halation 'MTF'. Curve (b): strong adjacency MTF. Curve (c): composite response.

adjacency effects and halation, this combination yielding a retinal system MTF with characteristic drop at very low frequencies (Fig. 10.9). There thus appears to be some conflict in terms of correlation with sharpness judgements as to whether it is only the intermediate frequencies in the retinal image, or both the low and intermediate frequencies, which contribute most.

10.3.4 Modulation Detectivity

An alternative approach to the variants on MTFA is that first developed for establishing performance limits for photographic systems by determining the intersection of the lens MTF curve and the threshold detection curve of the photographic emulsion for a given contrast of target,[46] (Fig. 10.10). This has been extended by Brown[47] for use in assessing visual systems. The present author considers such an approach of some use for incoherently coupled situations when viewing periodic object structure, but questions Brown's extension of the concept to visual aids involving direct coupling.

Since it seems inconceivable that any single frequency criterion can be applicable when viewing aperiodic objects (the normal form of visual task), it is not intended to pursue this form of quality factor further in this book.

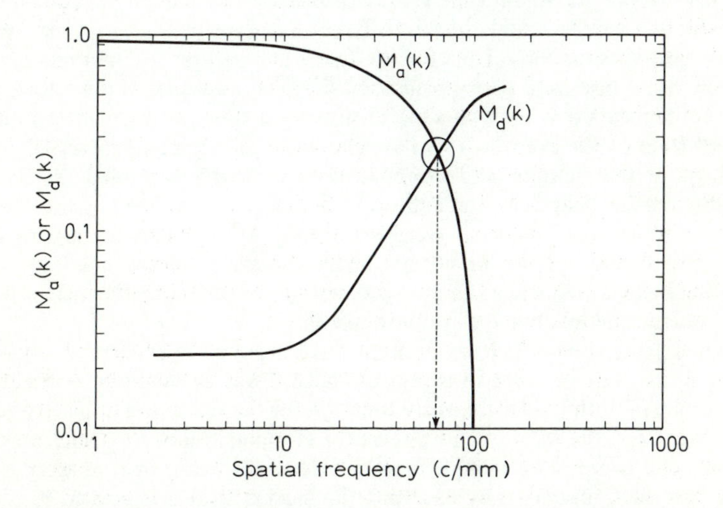

Fig. 10.10. Application of modulation detectability in determining the high contrast 3-bar resolution of a system. $M_a(k)$ = available image modulation. $M_d(k)$ = minimum modulation required for resolution. (Reproduced with permission of the Society of Photographic Scientists and Engineers Inc., as published in Scott[46]. Copyright 1966).

10.3.5 Visual Efficiency

Let us pause at this point to consider the implications of the relationship between OTF and line spread function as expressed in Equation 10.3, the high correlation between the MTFA, the square of the MTFA as applied to the retinal image, CMT acutance and subjective performance as discussed in Sections 10.3.1 and 10.3.2, and the visual modelling discussed in Chapter 7.

If, instead of defining the OTF in terms of the line spread function as in Equation 10.3, we define the spread function in terms of the OTF, then we get

$$G(x) = \frac{1}{\pi} \int_0^\infty F(j\omega) \cdot \exp(-j\omega x) \cdot \mathrm{d}\omega \tag{10.8}$$

Then, from Equation 10.8, the value of the spread function at $x = 0$ will be

$$G(x)_{(x=0)} = \frac{1}{\pi} \int_0^\infty F(j\omega) \cdot \mathrm{d}\omega \tag{10.9}$$

But this is the area under the MTF curve (i.e. approximately the MTFA) for symmetrical functions and is closely related to MTFA and CMT acutance for most situations. Furthermore, since most line spread functions have their central peak at $x = 0$, $G(x)_{(x=0)}$ as defined by Equation 10.9 is usually a very good estimate of the peak of the spread function. But for extended objects this peak value defines the maximum edge gradient, which is the basis of the visual model discussed in Chapter 7 and shown to be an accurate predictor of very many visual *threshold* situations. Thus for the first time we have a direct relationship between visual *threshold* performance and the OTF measured at the retina. With a few refinements it is then possible to propose a figure of merit based on the physical state of the eye which we have chosen to call *visual efficiency* (η_v). The derivation of this function and its application to viewing extended objects have been discussed at length by the author.[48] Basically it is defined as the ratio of the areas under the frequency-weighted retinal MTF curves for a composite visual system and for the naked eye under the same viewing conditions, the weighting being a frequency function accounting for the finite dimensions of the retinal mosaic and involuntary eye movements.

In theory the visual efficiency concept can be applied to imagery at any point on the retina, but in order to apply it in such a way it would be necessary to know in detail both the local quality function for the refraction optics (to use as a reference) and the local effective receptor grouping. Since for many practical purposes one is only interested in quality effects for near-foveal imagery – and in any case such imagery is by its nature the most critical – at present η_v is only considered for foveal imagery. For such imagery the quality factors for the refraction optics are fairly well established (see Section 2.3), whilst the frequency weighting functions for average tremor, drift and receptor spacing are considered to be roughly as shown in Fig. 10.11.

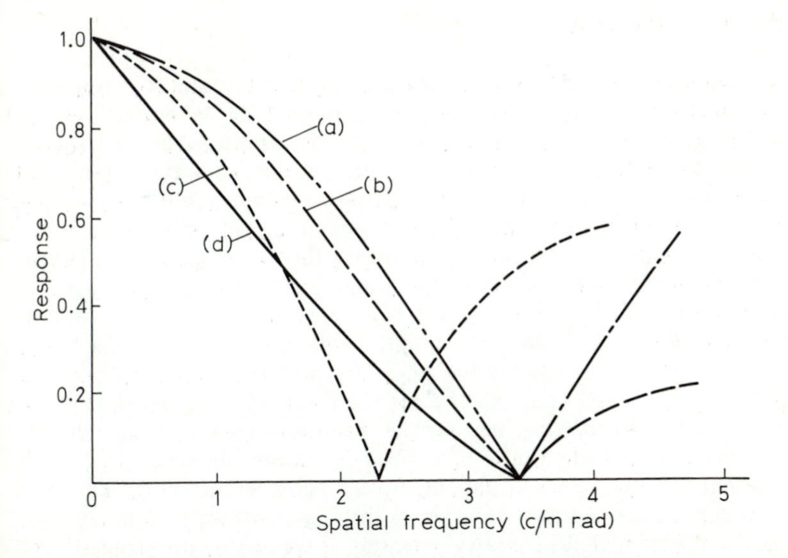

Fig. 10.11. The various frequency weighting functions to be applied in use of η_V. Curve (a): receptor sampling interval (central fovea). Curve (b): intersaccadic drift (0.3 mrads). Curve (c): tremor (±0.15 mrads). Curve (d): 2.5 mm eye pupil diffraction limited performance for comparison. (From Overington[48]).

Since the denominator of the ratio (the area under the weighted MTF for the naked eye) is a variable which is dependent on state of eye accommodation, pupil diameter, etc.,[48,49] it can be seen that the visual efficiency for a given instrumental quality of a visual system is not a constant. This, of course, is a necessary requirement for any visual quality measure if it is to be in any sense universal in application.

Unfortunately the ideal concept that it is only the peak regions of the retinal illuminance gradient which contribute to visual performance (which appears to be implied to be the case for 'sharpness' and 'goodness' judgements from MTFA and CMT acutance) does not apply universally, as illustrated by the results of blurred threshold experiments in which modelling was attempted in Section 7.7. Thus for general application of a quality factor such as visual efficiency to threshold situations it is necessary, in addition, to establish whether the form of blur broadens the central peak of the retinal line spread function sufficiently to cause additional probabilistic contributions to be made to the 'stimulus' from regions adjacent to the ribbon of maximum retinal illuminance gradient (see Section 7.7). In circumstances where such broadening occurs it is presently impractical to consider the idea of one simple single figure of merit, since complex interaction of various frequency responses contained within the MTF appears to be implied.

10.4 VEILING GLARE

All the measures of image quality discussed in Sections 10.2 and 10.3 assume tha all the energy from small object detail transmitted by an optical system i utilised in generating the image of that detail. Unfortunately this is frequentl not the truth. It is quite usual for some part of the tranmitted energy to b scattered and to fall on parts of the image plane widely removed from the imag detail being studied. This has already been discussed in Chapter 9. On established technique is available for measuring this veiling glare – the measure ment of 'Veiling Glare Index' (VGI). For several years it has become the practic to measure this VGI by viewing a small black area in the centre of a uniforml illuminated field of a few degrees subtense at the optical system under test (se Fig. 10.12). The exact size of the black area and the illuminated area hav depended under what discipline the measurement was undertaken.[50,51] Fo example, for binoculars it has become the practice to use a 17.5 mrad diamete circular black area in the centre of a 87 mrad diameter illuminated field. Now the amount of veiling glare in the image plane must depend on the extent an form of the illuminated field. Also, since many systems are used in condition where the illuminated field is very extensive, it seems that the standard VGI' above are not particularly useful in defining the extent of veiling glare i practical use. In an attempt to overcome this difficulty a new British Standar for veiling glare measurement has recently been compiled.[52] This Standar provides for more varied test conditions, and provides guidance for users i selection of the best test conditions for their particular requirements. In thi

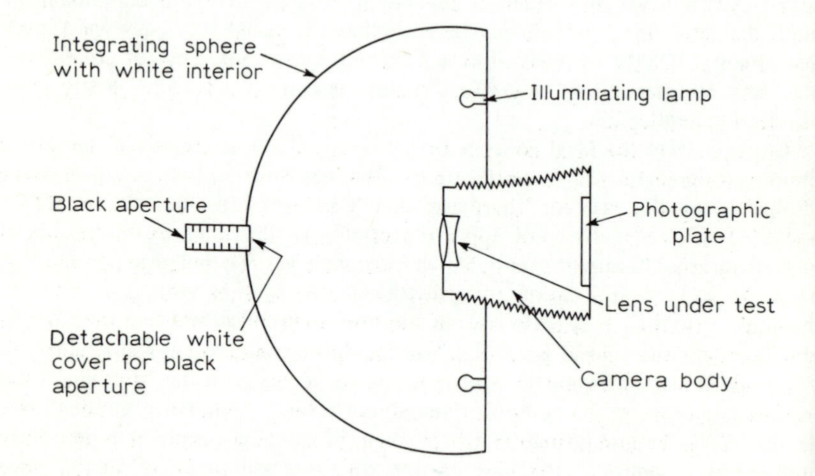

Fig. 10.12. Goldberg's apparatus for the measurement of veiling glare, on which mos commonly used systems are based.

Standard is also introduced the concept of a glare spread function, this being the distribution of light in the image plane from any one object point. The SIRA Institute have developed an equipment for measuring this latter function.[53]

Whilst veiling glare can be treated as external to the eye for incoherently coupled systems, with directly coupled systems care must be taken to consider the interaction of veiling glare due to the optical aid and that due to the eye itself, since these need not necessarily sum in a scalar fashion. Little information on his aspect has been published to the present author's knowledge.

REFERENCES

1. Ronchi, V. (1970). 'The Energetic Theory of Resolving Power', *Atti della Fondazione Giorgio Ronchi,* **25**, 565
2. Brock, G. C. (1967). 'Reflections on Thirty Years of Image Evaluation', *Phot. Sci. & Eng.,* **11**, 356
3. Perrin, F. H. (1960). 'Methods of Appraising Photographic Systems', *J. Soc. Motion Pict. Telev. Engrs.,* **69**, 151 and 239
4. Frieser H. (1938). *Z. wiss. Photogr.,* **37**, 261
5. Duffieux, P. M. (1946). *The Fourier Integral and its Application to Optics,* (in French), SA des Imprimeries Oberthus, Rennes
6. Luneberg, R. K. (1964). *Mathematical Theory of Optics,* University of California Press
7. Selwyn, E. W. H. (1948). 'The Photographic and Visual Resolving Power of Lenses', *Phot. J.B.,* **88**, 6 and 46
8. Cobb, P. W. and Moss, F. K. (1928). 'The Four Variables of Visual Threshold', *J. Frank. Inst.* **205**, 831
9. USAF Target – MIL Std. 150
10. Overington, I. and Gullick, S. A. (1973). 'Evaluation of a Total System – Optics plus Operator', *Optica Acta,* **20**, 49
11. Barton, N. P. (1972). 'Application of the Optical Transfer Function to Visual Instruments', in *Proceedings of the Electro-optics 1972 International Conference,* Brighton, England, 180
12. Barton N. P. (1974). 'MTF Testing and Specification for Afocal Systems', in *Proceedings of the SPIE, Vol. 46, Image Assessment and Specification,* 160
13. Hopkins, H. H. (1962). 'The Application of Frequency Response Techniques in Optics', *Proc. Phys. Soc.,* 79, 889
14. Brock, G. C. (1970). *Image Evaluation for Aerial Photography,* Focal Press
15. Rosenhauer, K. and Rosenbruch, K. J. (1967). 'The Measurement of the Optical Transfer Functions of Lenses', *Rep.Prog. Phys.,* **30**, 1
16. Rosenbruch, K. J. (1974). 'Trends in the Development of OTF Measuring Equipment' in *Proceedings of the SPIE, Vol. 46, Image Assessment and Specification,* 19
17. Bates, W. J. (1965). 'SIRA/BECK Equipment for measuring Optical Transfer Functions', *J. Sci. Inst.,* **42**, 538
18. Kenrick, P. (1974). 'Practical Aspects of the Use of Optical Test Methods in Production', in *Proceedings of the SPIE, Vol. 46, Image Assessment and Specification,* 220
19. Veenenga, K. R. (1971). 'Test Equipment for Electro-optical Systems and Components', in *Proceedings of Electro-optics 71 International,* Brighton, 473
20. Pouleau, J. (1967). 'Apparatus for the Control of Objectives by the Measurement of MTF' (in French), *Revue d'Optique,* 46, 202
21. Pouleau, J. (1974). 'MTF Measurements of Optical Systems in the Infra-red Range

(5–20 μm)', in *Proceedings of the SPIE, Vol. 46, Image Assessment and Specification* 299
22. Scott, F., Scott, R. M. and Shack, R. V. (1963). 'The Use of Edge Gradients in determining Modulation-transfer Functions', *Phot. Sci. & Eng.*, 7, 345
23. Jones, R. A. (1967). 'An Automated Technique for Deriving MTF's from Edge traces' *Phot. Sci. & Eng.*, **11**, 102.
24. Hopkins, R. E. and Dutton, D. (1970). Lens Standardisation Study Technical Report AFALTR 70–93. University of Rochester Institute of Optics Contract No AF33(657)9158, July
25. Gullick, S. A. (1970). 'The MTF of an Optical System by a Photographic Technique' *Optics Technology*, **2**, 88
26. Gullick, S. A. (1970). 'Photographic Technique of MTF Measurement for comparing Sighting System with its Simulator', *Optics and Laser Technology*, **2** 139
27. Asaeda, T., Watanabe, A. and Sayanagi, K. (1974). 'A Computerised Real-time MTF Measuring System', in *Proceedings of the SPIE, Vol. 46, Image Assessment and Specification*, 148
28. Swing, R. E. (1974). 'The Case for the Pupil Function', in *Proceedings of the SPIE Vol. 46, Image Assessment and Specification*, 104
29. Overington, I. (1973). 'The Importance of Coherence of Coupling when viewing through Visual Aids', *Optics and Laser Technology*, **5**, 216
30. Haines, J. H. and King, J. P. (1969). 'Approximate Computations of Physical MTF's' in *Proceedings of the SPIE, Vol. 13.*
31. Baker, L. R. (1955). 'An Interferometer for Measuring the Spatial Frequency Response of a Lens System', *Proc. Phys. Soc. B.*, **68**, 871
32. Herriot, D. R. and Bruning, J. H. (1974). 'Modulation Transfer Function by Measurement of the Pupil Function', in *Proceedings of the SPIE, Vol. 46, Image Assessment and Specification*, 114
33. Ivanoff, A. (1950). 'On the Subject of the Asymmetry of the Eye', (in French) *Comptes Rendus, Academie des Sciences*, **231**, 373
34. Westheimer, G. (1955). 'Spherical Aberration of the Eye', *Optica Acta*, **2**, 151
35. Overington, I. (1975). 'Some Considerations of the Eye as a Component of an Imaging System', *Optica Acta*, **22**, 365
36. Overington, I. (1975). 'Practical Application of Contrast Sensitivity for the Determination of the Performance of Visual Aids plus Observer', BAC(GW) Rep. Ref ST13071
37. Birch, K. G. (1969). 'A Survey of OTF Based Criteria used in the Specification of Image Quality', NPL Report, Op.Met. 5
38. Snyder, H. L. (1973). 'Image Quality and Observer Performance' in *Perception of Displayed Information*, L. M. Bieberman (Ed.) Plenum
39. Biedermann, K. (1967). 'Investigation of the Correlation between the Subjective Goodness and the Physical Quality of Photographic Images', (in German), *Phot. Korr* **103**, 5, 25, 41
40. Higgins, G. G. (1971). 'Methods for Analysing the Photographic System, including the Effects of Nonlinearity and Spatial Frequency Response', *Phot. Sci. & Eng.* **15**, 106.
41. Crane, E. M. (1964). 'An Objective Method of Rating Picture Sharpness: SM Acutance', *J. Soc. Motion Pict. Telev. Engrs.*, **73**, 643
42. Jackson, J. E. (1973). 'Scaling Techniques for Subjective Judgements of Picture Quality', paper presented at the *114th Technical Conference of the Society of Motion Picture and Television Engineers*, New York
43. Gendron, R. G. (1973). 'An Improved Objective Method for rating Picture Sharpness CMT Acutance', *J. Soc Motion Pict. Telev. Engrs.*, **82**, 1009
44. Granger, E. M. and Cupery, K. N. (1972). 'An Optical Merit Function (SQF) which correlates with Subjective Image Judgements', *Phot. Sci. & Eng.*, **16**, 221

5. Granger, E. M. (1974). 'Subjective Assessment and Specification of Color Image Quality', in *Proceedings of the SPIE, Vol. 46, Image Assessment and Specification*, 86
6. Scott, F. (1966). 'Three-Bar Target Modulation Detectability', *Phot. Sci. & Eng.*, **10**, 49
7. Brown, E. B. (1972). 'Detectability Curves and Visual System Optimisation', presented at the Topical Meeting on *Design and Visual Interface of Biocular Systems*, Annapolis, Maryland, May
8. Overington, I. (1974). 'Visual Efficiency: A means of Bridging the Gap between Subjective and Objective Quality', in *Proceedings of the SPIE, Vol. 46, Image Assessment and Specification*, 93
9. Overington, I. (1973). 'Visual Efficiency: A New Figure of Merit for Optical Quality', Study Note No. 5 of MOD Contract K64A/77/CB64A, BAC(GW) Ref. ST9430
0. Rosenhauer, K. and Rosenbruch K. J. (1968). Flare and Optical Transfer Function', *Applied Optics,* 7, 283
1. Anon (1956). 'Photometry of Telescopes and Binoculars', NPL Notes on Applied Science, No. 14, HMSO
2. Anon (1973). 'Recommendations for Measurement of the Veiling Glare Index of Lenses and Optical Systems', *British Standard BS4995*
3. Martin, S. (1972). 'Glare Characteristics of Lenses and Optical Instruments', *Optica Acta,* 19, 499

11 Varieties of Structure and Texture

It was stated in Chapter 7 that in real life one rarely meets acquisition problem involving simple, structureless objects viewed against plain fields. However, a that point we were not ready to consider more complex situations in depth. I the next four chapters we shall consider the forms of complex situations whic may be met with in real life, and the extent to which laboratory exper mentation enables us to understand and model them. The present chapter i intended to act as an introduction to the range of complexity which may b encountered and to refer the reader on to the relevant following chapters wher the various complexities will be studied in detail.

11.1 TARGET LUMINANCE STRUCTURE

The first factor which usually differs from classical experiments in practical task is the luminance structure of the target. Rarely is an object of interest c

Fig. 11.1. Air to ground photograph illustrating one of the simplest features encountered real life. The Pump House (circled), although a simple structure, still exhibits multip contrasts against its surroundings.

uniform luminance over its entire surface. Thus the ideas of both area and contour detection become difficult to handle. Figures 11.1 and 11.2 show two air to ground viewing situations which are of the simplest categories met with in practice. In Fig. 11.1 the object of interest is the pump house on the edge of the lake. Although this object is conceptually of simple block form it contains at least three separate areas of contrasting luminance. In Fig. 11.2 we see another conceptually simple object of interest – the gasholder – which is of cylindrical shape. Here, owing largely to polar reflection effects which will be discussed in Chapter 14, the surface luminance is gradually varying over the entire surface, again adding complication to considerations of modelling in terms of area or profile. The realisation that this sort of object is the rule rather than the exception has led the author to pose the question – already stated in Section 4.1 – 'What is contrast?'. This question has been considered in some depth by the author[1].

If distribution of luminance within the main body of the target was the only problem, possibly matters would not be too bad. However, when one considers the retinal image not only do gross distributions of object luminance matter but fine detail *shape* structure of the object, when convolved with the eye's spread function, gives rise to considerable modifications of the *illuminance* distribution in the retinal image. For instance, Fig. 11.3 shows silhouette pictures of a tree

Fig. 11.2. Air to ground photograph illustrating the all too common distributed illumination within an object of interest (the gasholders). Such distributed illumination makes nonsense of the concept of contrast.

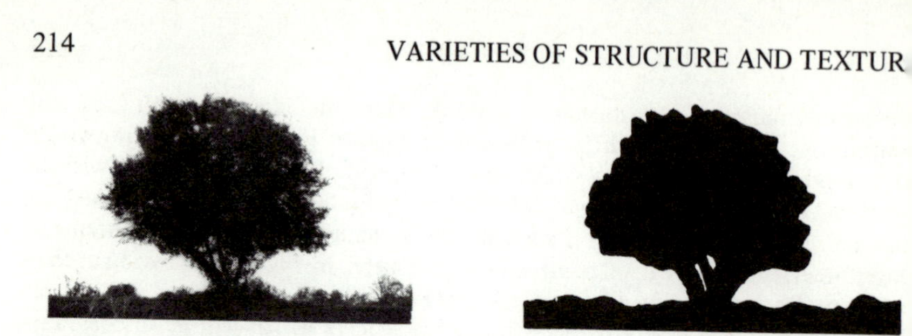

Fig. 11.3. Silhouettes of a tree and a sharp edge profile of a tree-like shape, illustrating th extreme importance of edge detail structure in a recognition task.

and a sharp profile of the same shape as the tree. It should be obvious to al readers that they are *very* different – there being no difficulty in recognising the real tree from the silhouette. This must be entirely due to the structure in the profile (since there is no structure *within* the object) and if the page of the book is viewed from a long way away one might expect that the two silhouettes would look the same. In the limit this is true, but it is necessary to view from a much greater distance than that necessary to remove the profile structure before the hidden cues are removed. It is suggested that it is the unnatural softening of the simple profile in the retinal image which allows differentiation between the two silhouettes at ranges where there is no *obvious* profile structure.

Experimental studies aimed at providing clues as to the importance of luminance structure within an object are considered in Chapter 12, together with attempts at modelling simple recognition thresholds based on the findings from the experimental studies.

11.2 BACKGROUND STRUCTURE

In many viewing situations, structure in the background surrounding the object of interest can have a strong influence on detectability. For instance, it is not unusual to find that the scene surrounding an object of interest interacts with the latter's profile, thus producing a complex set of local contrasts at different parts of the object profile. Figure 11.4, taken from Brown[2], illustrates this

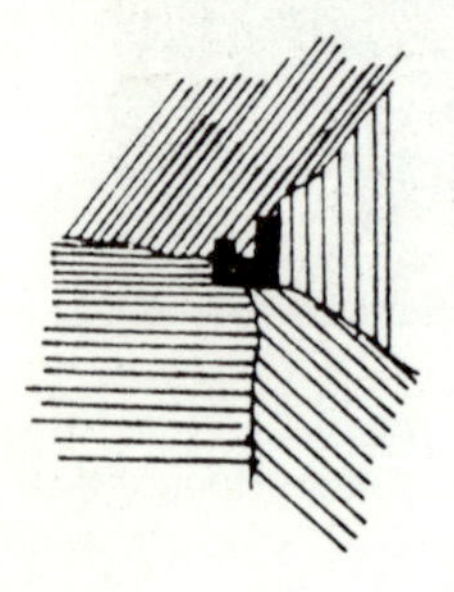

Fig. 11.4. Illustrating the concept of background structure interacting with an object of interest. (Reproduced from Brown[2] by courtesy of the Advisory Group for Aerospace Research and Development of NATO).

situation conceptually, whilst a further look at Fig. 11.1 will show how this situation occurs in real life. Brown chooses to call this object/background interaction 'complicacy'.

Other effects of background structure are those due to close proximity of unwanted strong stimuli, which tend to modify threshold performance in unexpected ways, and general background structure, which introduces a form of spatial noise into the presentation. Finally, both sub-threshold local background structure and preadaption to particular structures can have marked effects on visual performance. All the foregoing are discussed in depth in Chapter 13.

11.3 TEXTURE

In addition to all the gross luminance variations and profile effects discussed in Sections 11.1 and 11.2, there will often be small scale luminance variations within both objects of interest and their backgrounds. Such variations may be random – for instance, photographic or screen grain in indirect imagery or stone and grass textures in direct imagery. Alternatively they may be of a characteristic structure – as the 'chicken wire' effect in viewing through fibre optics (Section 9.2.5) or the organised street patterns of many towns and cultivated field patterns as seen from satellites (e.g. Fig. 11.5)[3,4]. Such textural

Fig. 11.5. Typical satellite photograph illustrating the intelligence structure in a ground scene viewed from a satellite due to field patterns.

effects may aid or hinder acquisition, dependent on whether they can be related to the object of interest or not.

Some experimental work on the effects of texture is reported in Section 12.7.

11.4 STRUCTURED SEARCH

A further way in which background structure can influence acquisition is in the field of search. In Chapter 8 the problems of search in an *empty* field were discussed. If, however, there is other *intelligence*, as opposed to random noise, in the search field in addition to the object of interest, then the search situation may be severely modified. Attention will tend to be drawn to objects other than that of current interest, according to their nature and similarity to the object of interest. Considerable studies have been carried out in recent years into this subject. These will be reported in Section 13.3.

11.5 DYNAMIC STRUCTURE

Finally, under the heading of 'structure and texture', we must include what may be called 'dynamic' structure – that is, structure which is changing with time. Under this heading must be included dynamic noise (as for instance occurring in cine projection where the grain noise changes from frame to frame), perspective effects (which modify background structure in, for instance, oblique viewing of the ground from the air) and variations of surface reflectivity due to changes of viewing direction. Of the foregoing, dynamic noise may be considered to be similar to the signal/noise situations for multiple glimpses already discussed in Section 7.3 – that is, variations of noise within an acceptable 'integration' time may be considered to suppress the basic instantaneous noise to a greater or a lesser degree. Also perspective effects are definable absolutely in terms of the geometry of the viewing situation (although it is likely that severe complexities of modelling are introduced because of them). On the other hand, changes in surface reflectivity are a subject in their own right and, as such, will be discussed at length in Chapter 14.

REFERENCES

1. Overington, I. (1972). 'Some Aspects of the Variation and Measurement of Contrast of Targets in Fields Trials', in *Proceedings of the NATO/APOR Conference on Field Trials & Acquisition of Tactical Operational Data*, Vol. 1, p. 137
2. Brown, M. B. (1972). 'The Effect of Complex Backgrounds on Acquisition Performance', in *AGARD Conference Proceedings No. 100*, (Ed. H. F. Huddleston), London, p. B5–1

3. Corbett, F. J. (1974). 'Sensor Performance Evaluation of the Skylab Multispectral Photographic Facility', in *Proceedings of the SPIE, Vol. 46, Image Assessment and Specification*, 239
4. Welch, R. (1974). 'MTF Analysis Techniques applied to ERTS – 1 and Skylab – 2 Imagery', in *Proceedings of the SPIE, Vol. 46, Image Assessment and Specification*, 258

12 Structured Targets

As intimated in the previous chapter, there are various forms of structure existing in many of the objects of interest met with in real life. These range from gross trends in luminance (e.g. Fig. 11.2) to multiple levels of luminance (Fig. 11.1), fine profile detail (Fig. 11.3) and surface texture. The aim of this chapter is to present a survey of laboratory data which shed some light on this problem of target structure. These data will be considered in terms of the contour modelling of threshold of Chapter 7 where appropriate. From the findings the possibilities of extending modelling to cover certain forms of simple recognition will be considered.

12.1 THE ANNULUS/DISC EXPERIMENT

One experiment which is considered to shed some light on the influence of target structure on thresholds is that carried out at BAC (GW) and known as the 'annulus/disc experiment'[1]. In this experiment, originally aimed at distinguishing between area and contour theories of visual processing, a series of three disc stimuli and three annular stimuli of equal outside diameter were presented to a group of observers in random order. The presentations were made with positive contrast and at high (approaching visually perfect) quality on a back projection screen at an adaptation luminance of 55 cd/m^2. Presentation times were fixed at 1/3 second and all presentations were made at a known, central position on the screen. The stimuli subtended angles of 3, 6 and 9 mrad when viewed from a distance of 1 m. The annuli were of such form that a central disc of diameter one third the outer diameter was removed, the central region thus having the same luminance as the background. The stimuli were presented in random order to each of 16 observers, the thresholds being determined by the threshold tracking technique[2].

The results of the experiment are shown in Fig. 12.1, where it will be seen that the relative detectability of the annulus and disc depend strongly on the angular subtense. At the largest size (9 mrad) the annulus is the more easily detected, despite the reduced area relative to the disc, whilst for the smallest size (3 mrad) the converse is true. In order to study these unexpected results further the original object shapes were convolved with the eye's spread function as measured by Westheimer and Campbell[3], with the results shown in Fig. 12.2. It will be seen from this figure that the retinal images of the three *disc* stimuli are all fully resolved – that is, there is an illuminance plateau in the centre of the image. On the other hand, of the retinal images of the *annuli*, only the largest of the three contains a plateau region. Of the other two the 6 mrad annulus has the

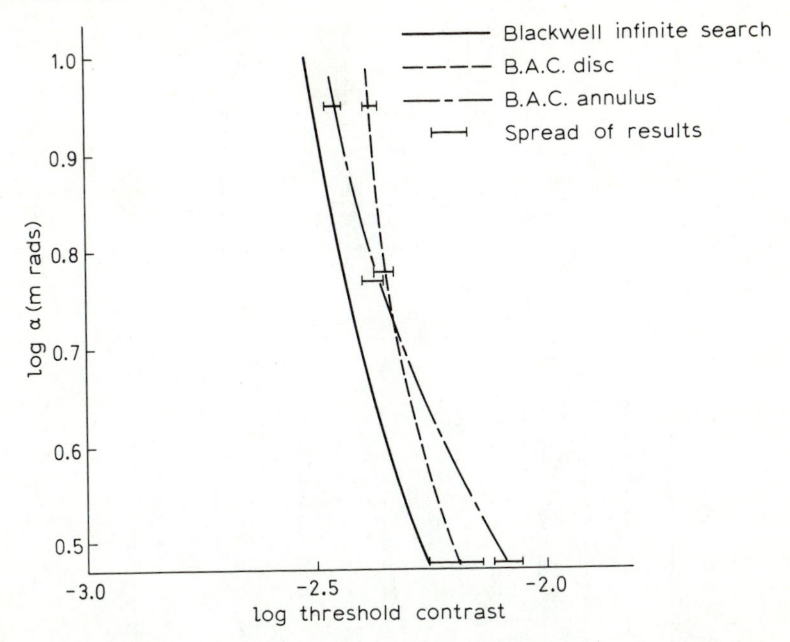

Fig. 12.1. Comparative detection thresholds for annular and disc stimuli of various outside diameters.

outer profile just resolved, whilst the inner profile is softened by some 15–20%. The 3 mrad annulus, on the other hand, has both outer and inner profiles softened, the inner some 15–20% *more* than the outer. If it is assumed that the differential softening of the inner contour by 15–20% is sufficient to suppress its influence effectively, as was discussed in Section 7.7 for parts of a blurred profile, then it is found that the remaining contour, relative to that of the equivalent disc stimulus, is just sufficient to account for the observed results. For the 9 mrad annulus there is a threshold enhancement equivalent to that for a 30% increase in contour length relative to the 9 mrad disc – as provided by the inner contour. For the 6 mrad annulus the threshold is as for the disc, i.e. the outer contour only. Finally, for the 3 mrad annulus the degradation is as expected for the softening of the outer contour, again assuming this to be the only one contributing to the threshold.

The annulus/disc experiment was, of course, only able to provide any information for one high photopic luminance level. Other data of a similar nature exist for lower light levels, particularly for the scotopic region (e.g. Hills[4]), which appears to show similar behaviour to the photopic region but at a much larger scale. Similar comparative behaviour has been observed between photopic and scotopic light levels in forms of *receptive field* studies (see Section

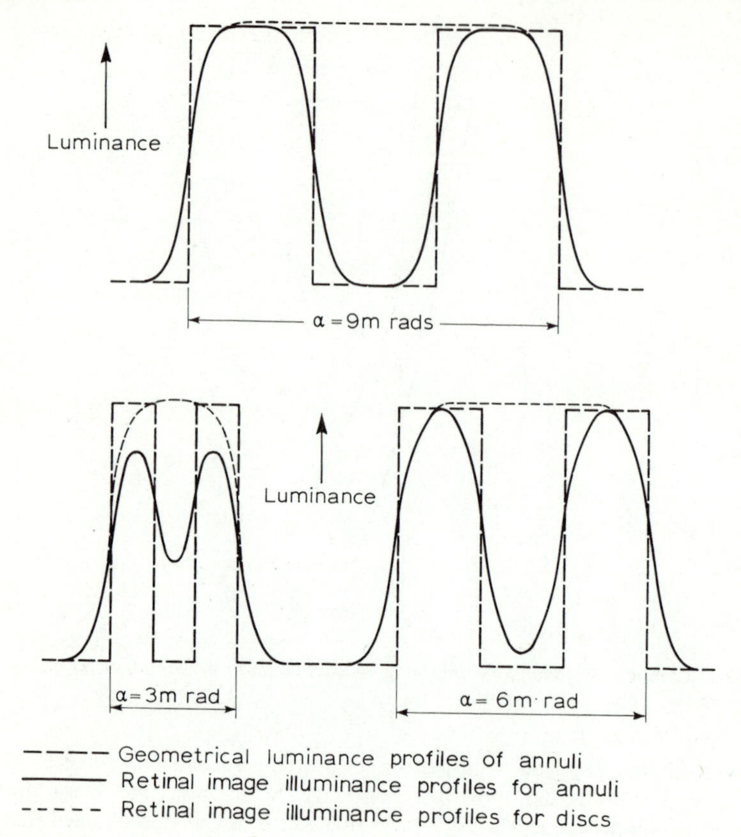

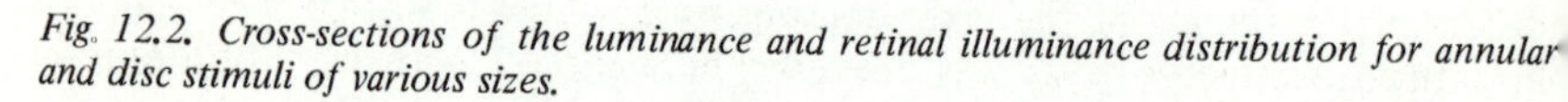

Fig. 12.2. Cross-sections of the luminance and retinal illuminance distribution for annular and disc stimuli of various sizes.

13.1) which suggests that, although not presently modelled, there may be similar effective illuminance gradient sensitivity for scotopic (or rod) vision as we appear to have found for photopic (or cone) vision.

12.2 THE MULTICONTRAST EXPERIMENT

In an attempt to provide guidelines for modelling of the detection and recognition thresholds for the simplest form of multicontrast stimulus an experiment was carried out by BAC (GW) in which the size thresholds associated with a high contrast square and a set of low contrast appendages were studied[5]. The basic material for this experiment is illustrated in Fig. 12.3; it consisted of

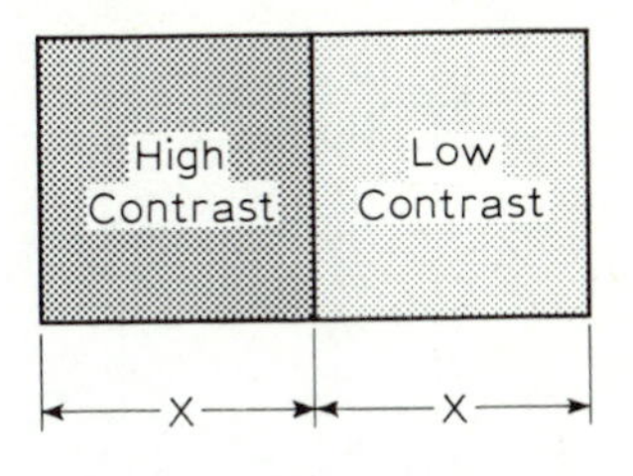

Fig. 12.3. The form of stimulus used for the multicontrast experiment.

transparencies of two high contrast squares (nominal contrasts of – 0.89 and – 0.66), each permuted with a series of medium to low contrast squares. The material was viewed, in random presentation order, by eight observers against a high luminance (3200 cd/m^2) back illuminated screen, the observers being able to move towards the stimuli from a distance of some 38 m by means of a very slowly moving, motorised chair. The only limit on viewing time was that associated with the slow, but finite, growth rate of the target.

The essence of the experiment was that observers were required to report firstly when they detected the presence of the total stimulus, and secondly when they could detect the presence of the appendage. As a supplementary exercise stimuli comprising the appendage only were also provided, and a similar exercise was carried out to determine the thresholds for the various contrast appendages in isolation. This was considered desirable to check how closely Blackwell's Tiffany findings[6] could be reproduced with the particular experimental arrangement used for this study. The results are shown in Fig. 12.4. It will be seen that the mean detection sizes for the appendages in isolation are in very satisfying agreement with the Tiffany results, except for the lowest contrast, where the size required is considerably higher than found by Blackwell. The agreement is particularly satisfying, since no serious attempt was made to provide uniform illumination over the entire visual field, as attempted by Blackwell. It is suspected that the degraded performance for the lowest contrast appendage is as a result of decisions being made at free choice level, as opposed to Blackwell's forced choice decisions (see Section 3.3). This would be expected to increase δ in Equation 7.4 without affecting K_1, the result being a degradation of large size (low contrast) thresholds whilst leaving small size (high contrast) thresholds relatively unaffected (see Fig. 4.1).

Having established that the thresholds for the appendages in isolation are in agreement with those by other workers, we are in a position to consider the thresholds for the complex targets. If we firstly consider the thresholds for the *detection* of the complex targets we find them to be hardly different from those for the high contrast squares in isolation, until such time as the appendage contrast is approaching that of the basic square. Then, progressively, the threshold size falls to that which would be expected from a simple rectangular stimulus comprising the main square and the appendage at equal luminance. More spectacular are the results for detection of the existance of an appendage.

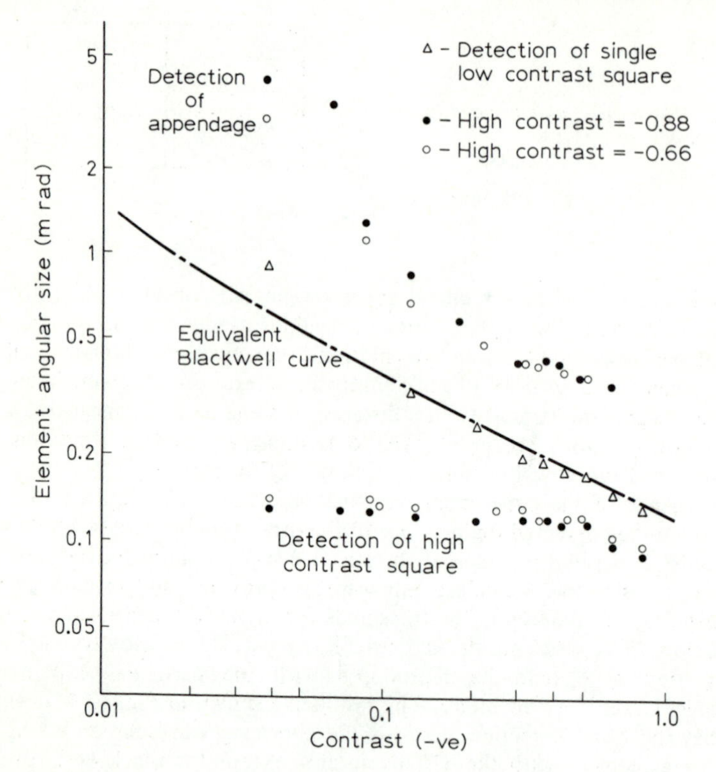

Fig. 12.4. Detection thresholds for presence of an appendage attached to a high contrast square compared to the thresholds for simple square stimuli.

For the majority of contrasts the thresholds for presence of an appendage are displaced by a constant factor of approximately 0.3 in logarithmic size (linear) or 0.6 in logarithmic contrast, suggesting that the effective 'size' contributing in the Ricco's Law region is constant at 50% of the appendage size! This seems to apply *whatever* the contrast between the main square and the appendage, so long as all detections are made in the Ricco's Law region.

In an attempt to shed light on the form of this detection task, a stimulus of basic dimension 0.44 mrad, with an appropriate appendage threshold contrast, was convolved with the Westheimer and Campbell eye spread function. The resulting retinal image isophot diagram, together with a cross-section illuminance profile, are shown in Fig. 12.5. It will be seen that there is no semblance of the two separate illuminance plateaux left in the cross-section, and that there is only a small tendency to an ovoidal shape in the isophot diagram to indicate the existence of an appendage. Detailed modelling of this form of stimulus has not been pursued at this date, but once again there appears to be evidence that

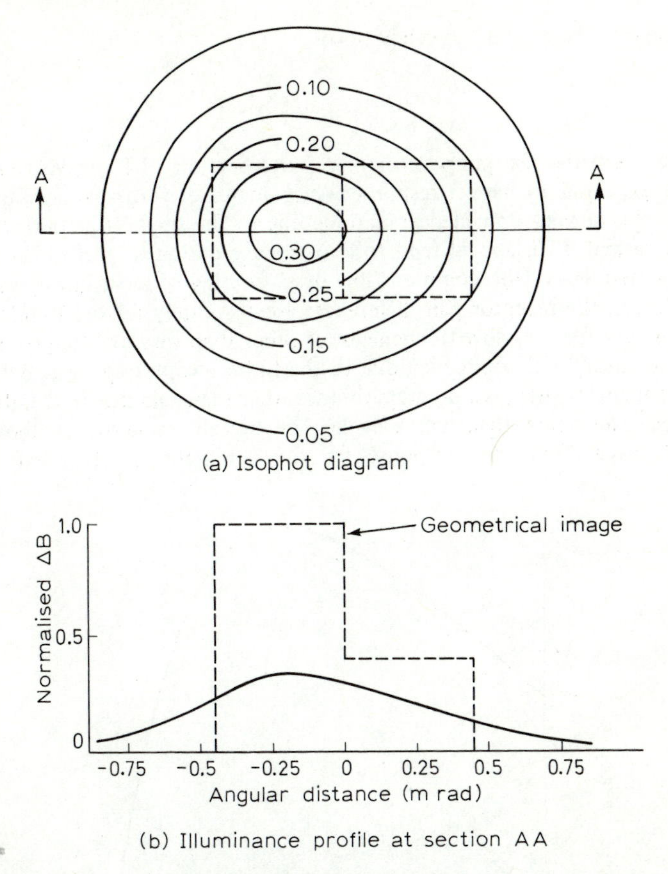

Fig. 12.5. Retinal image illuminance profile and isophot diagram for a multicontrast stimulus consisting of high and low contrast squares of 0.44 mrad side length.

differences of retinal image illuminance gradient of some 20% are controlling factors in producing differential threshold performance.

12.3 CONTRAST SENSITIVITY AND THE MACH EFFECT

With the information gained from the study of the complex targets of Sections 12.1 and 12.2 it is now time to look in depth at the contrast sensitivity functions described in Section 4.14. The form of unsharp object used, which has received considerable attention in recent years is, at its simplest, a one dimensional, modulated sine bar pattern as described in Section 10.2. Such a pattern exhibits a series of long borders of maximum gradient which is proportional to the spatial frequency since

$$\frac{dB}{dx} = (\Delta B')_{max}\ \omega \cos(\omega x).$$

Therefore, from the reasoning in Chapter 7 and Section 12.1, the visual system might be expected to have a response which increases with spatial frequency. However, the normal optical degradations due to the spread function, tremor, drift and retinal diffusion all tend to lead to degradation of performance with increasing frequency (the common 'low pass' filtering of most linear systems). Also, although the receptor pair differences are absolutely related to maximum gradient dB/dx for very low frequencies, at other than very low frequencies the gain is not linear with frequency due to the finite receptor spacing. With high spatial frequencies there is a progressive loss rather than gain due to the effective integration over more than half a cycle. The overall result of the above is a predicted foveal frequency response for the complete visual system which

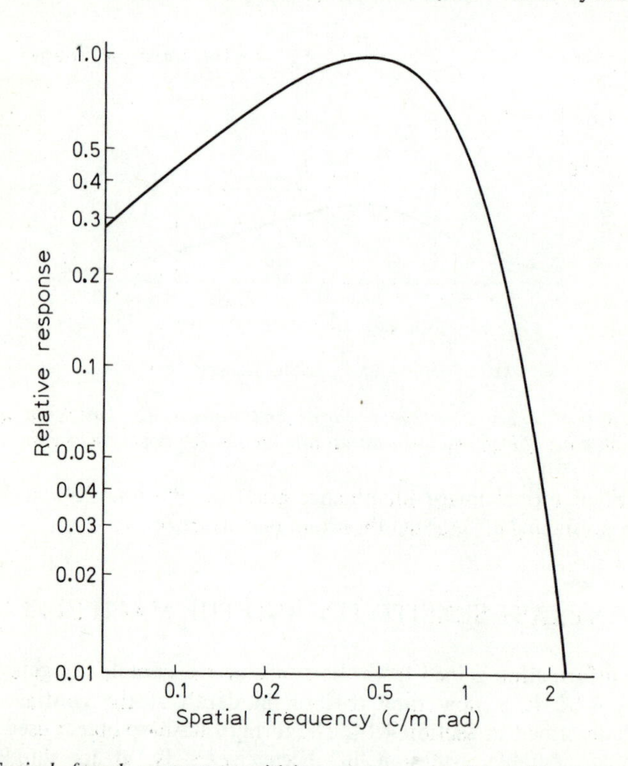

Fig. 12.6. Typical foveal contrast sensitivity curve as predicted from the combined frequency response functions due to optical degradations, eye movements, retinal receptor spacing and finite receptor size.

increases from zero at zero spatial frequency to a maximum around 0.35 to 0.6 c/mrad and then progressively reduces again[7]. A typical predicted response is shown in Fig. 12.6. Reference back to Figs. 4.21 and 4.22 will show a striking similarity except for a slight shift of the position of peak response. It is believed that this difference in frequency for peak response is in keeping with field size effects discussed later in this section, the typical extent of patterns used by Campbell and co-workers being around 2 degrees diameter.

It was found[8,9] that the predicted contrast sensitivity curve was relatively insensitive in its shape to retinal receptor spacing within the range 0.15 to 0.3 mrad and to change of pupil diameter for small pupils. It was, however, found to be considerably sensitive to the amounts of tremor and drift assumed and to the actual shape of the eye's spread function. It was concluded that the insensitivity to receptor spacing was a desirable feature, since retinal receptors could not all be regularly distributed within close tolerances – in fact Østerberg's work[10] suggests they are not. The lack of sensitivity to pupil diameter, on the other hand, is probably a result of the well balanced spread functions of the refraction optics for pupil diameters from 1.5 mm to at least 3 mm[11].

Having shown that the basic contrast sensitivity functions experimentally measured by Campbell and co-workers could be predicted in form by the simple maximum edge gradient concept, it was of interest to consider what the result of carrying out an inverse Fourier transform and integration on such a function would be. It will be remembered from Section 10.2 that the MTF is the Fourier transform of the line spread function. The line spread function is, in turn, the 1st differential of the edge response function of a system. Thus, if we inverse Fourier transform and integrate the contrast sensitivity function (which is essentially a frequency response function similar to the MTF), we should obtain the edge response function of the complete visual system if the system may be considered in any sense linear. Figure 12.7 shows the result of such a computation. This will be seen to bear a strong similarity to the enhancement of borders known as the Mach effect (Section 2.9). It would thus appear that the Mach effect is possibly a necessary side effect of the mechanisms of retinal processing which can be predicted readily for a given viewing situation.

To this point contrast sensitivity has been considered as a general form of frequency response function, without giving any serious consideration to its constancy of form. In fact it is found that the form of the function, whilst always retaining, to some extent, its band-pass properties, does change shape and absolute frequency dependence as a function of size of test field, retinal position of test field and luminance level.

Hoekstra *et al*[12] have studied the variation of contrast sensitivity as a function of test field size for foveally-centred fields and have found that, as the test field is increased, so the low frequency attenuation is reduced, whilst the high frequency attenuation remains largely unaltered (Fig. 12.8). They suggest that this is due to insufficient cycles of the test pattern for small test fields and

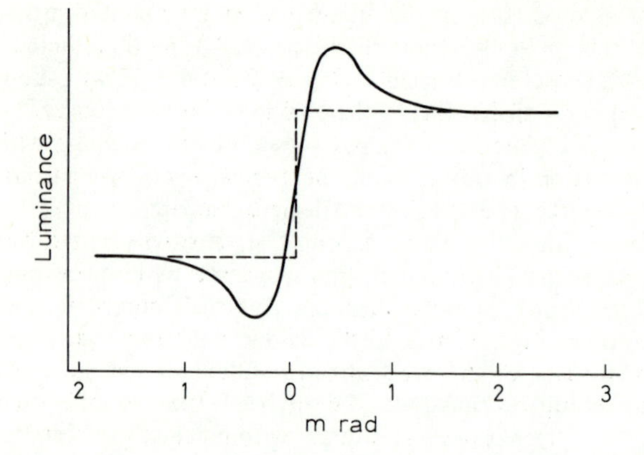

Fig. 12.7. The predicted Mach effect obtained by inverse Fourier transformation of a typical foveal contrast sensitivity function.

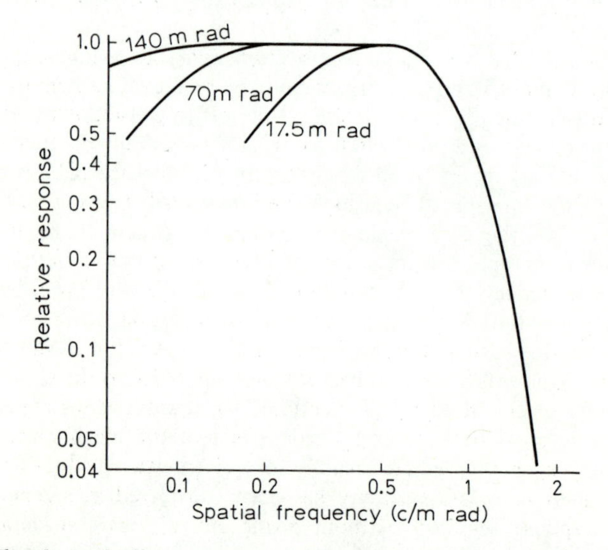

Fig. 12.8. The observed effect on contrast sensitivity of increasing the size of the test field. (Reproduced from Hoekstra et al.[12] *by courtesy of Pergamon Press).*

low frequencies. The author does not like this as a complete explanation, since Hoekstra *et al* find a significant low frequency attenuation even with several cycles of test pattern present. However, it must be true that, in the limit, if frequency is reduced to a *very* low value, such a low frequency attenuation will occur. In fact, according to McCann *et al*[13], at the *lowest* of frequencies detection is controlled by contrast (or modulation depth) and number of cycles present, not being dependent on gradient at all. An alternative explanation, of all but the very lowest frequency region, which is in keeping with observed evidence of 'parallel processing' (see Section 12.4) is that, for each local region of the retina, there is a contrast sensitivity function of the form shown in Fig. 12.6, this comprising the composite MTF of the refraction optics, involuntary eye movements and retinal diffusion, together with the sampling function due to the retinal mosaic. However, due to the varying quality factors and retinal spacing, this function has a frequency scale which is dependent on retinal position. If only a small foveal field is presented, then the central foveal contrast sensitivity is all that applies. If, however, the field is of several degrees diameter, then there are a whole series of annular zones exhibiting different pass-band contrast sensitivities, and the overall response will be governed by the envelope containing all possible contrast sensitivities. The series of such envelopes will obviously have a common high frequency response and a gradually widening low frequency response much as found by Hoekstra *et al.* The foregoing is discussed

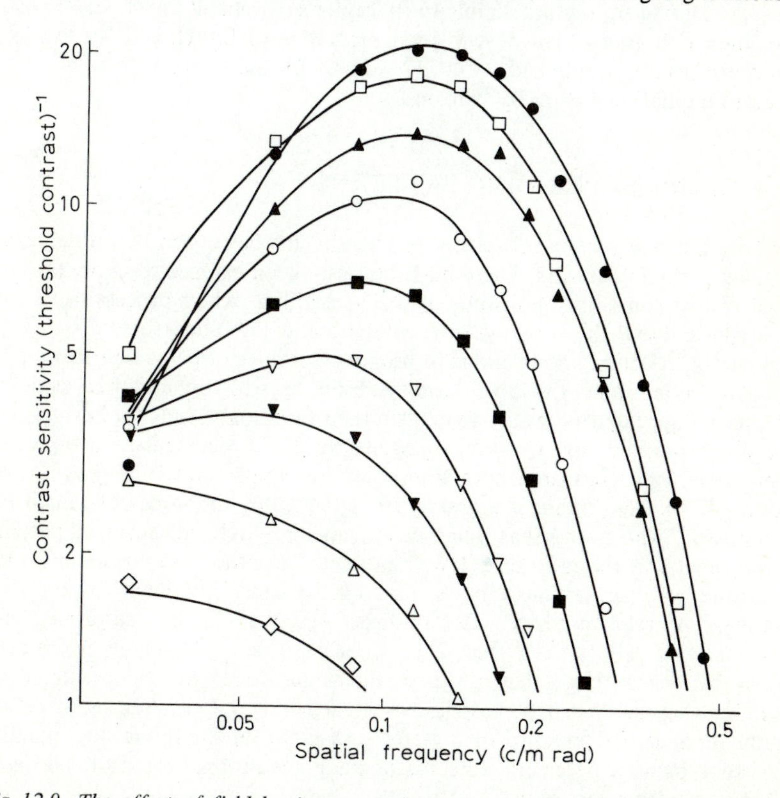

Fig. 12.9. The effect of field luminance on the contrast sensitivity measured at 0.21 rad from the fovea. (Reproduced from Daitch and Green[14] by courtesy of Pergamon Press).

in detail by the author[7], and predictive modelling of Hoekstra's results carried out.

Some indication of the effect of retinal position and luminance on contrast sensitivity is to be found in the work of Daitch and Green[14], who carried out experimentation with a small test field centred 0.21 rad from the fovea at various field luminances. Their collected findings are shown in Fig. 12.9. As can be seen, the contrast sensitivity remains fairly stable at low frequencies for all luminances, whilst falling markedly at high frequencies as the luminance is reduced. Similar trends as a function of luminance for foveal vision were found by Valois *et al.*[15] These results are believed by the present author to be as expected due to the composite effects of quality factors, neural noise and quantum noise. It will also be seen, if the high luminance function in Fig. 12.9 is compared to the foveal contrast sensitivity as shown in Figs. 4.21, 12.6 and 12.8, that at 0.21 rad from the fovea the whole function has shifted to lower frequencies. If the overall quality is assumed to fall off inversely as receptor spacing, as has been found desirable to fit Taylor's peripheral thresholds (Section 7.5), then it is found that a very good prediction of Daitch and Green's high luminance results is obtained[7]. Further support for the foregoing is to be found in recent unpublished work by Williams[16].

12.4 COMPLEX PERIODIC PATTERNS

Over the last few years a miscellany of experiments have been performed using complex periodic patterns. These have variously been claimed to imply that the visual system contains a multitude of band-pass filters which process visual data in parallel and that the visual system contains marked non-linearities.

Some of the first experimenters to propose existence of parallel processing by tuned circuits were Campbell and Robson[17], who found that such an assumption appeared necessary to explain their threshold trends for detection of periodic patterns of square wave, triangular wave and rectangular wave profiles when these were compared with thresholds for simple sinusoidal gratings (see Section 4.14). The finding was greatly strengthened by the work of Graham and Nachmias,[18] who found that, when two incoherently related sinusoidal patterns of frequencies in the ratio of 3:1 were presented together, the threshold for the composite was nearly the same as that of the more prominent of the two, regardless of relative phase. The findings were yet further confirmed, for frequencies of ratio greater than 2:1, by Sachs *et al*[19]. However, it has been shown by the author[7] that all these threshold trends can be predicted by assuming that the visual process is one of detecting 1st differences of retinal illumination in the presence of noise at *each point* on the retina, the 'parallel' processing being a necessary outcome of the graded retinal receptor spacing in regions away from the fovea.

Whilst the foregoing studies have been going on, Burton[20] has been investigating the visual performance when complex periodic patterns are presented which consist of two sinusoidal patterns of nearly the same frequency, or nearly in the ratio of 2:1. He has found that, under certain conditions, it is possible to see the 'beat' frequency when neither of the two presented patterns is above threshold. This, he argues, shows the existence of a non-linearity at an early stage of processing, since only the two frequencies can be present in the analysis networks. The present author has pointed out[7] that in such a situation, as in electronic systems, it is quite possible to redefine the composite waveform in terms of the sum and difference frequencies and that, having done this, there is a range of frequencies (almost exactly that range used by Burton) over which the difference signal will be more detectable than the two component signals due to the band-pass characteristic of the contrast sensitivity function.

12.5 BRIGHTNESS ILLUSIONS

As was intimated in Section 2.9, a range of visual phenomena given the general title of brightness illusions appear to be closely related to the Mach phenomenon. It is found that certain forms of sudden luminance discontinuity in the visual field result in the perceived brightness being grossly distorted. Cornsweet[21] has studied this subject and reported at some length, whilst Mach himself, as reported in Ratliff's recent collected translation[22], experimented with many forms of local field discontinuity. Now brightness illusions are usually assumed to be subjective, but it is the present author's belief that they are largely explained in terms of the combination of optical degradation of the retinal image and the signal/noise situation in which the visual response is obtained.

As an illustration of the mechanisms by which the author believes the general range of brightness illusions are generated, let us consider two of the more striking, yet simple, ones described by Cornsweet – firstly the case of a circular, complex border as shown in Fig 2.12 applied to an otherwise uniform field, and secondly a similar, but stronger, border applied to a field whose centre is lighter than the surrounding annulus, as shown in profile in Fig. 12.10. Both of these patterns are *observed* as having a centre *darker* than the surround. If we now consider one edge of each of these borders, and derive the forms of the retinal images by convolution, we find that the resultant illuminance profiles each exhibit relatively gentle positive gradients and much stronger inverse gradients. Now it was suggested in Chapter 7 that one of the characteristics of the visual system required to explain threshold performance, and one in keeping with the known responses of bipolar and ganglion cells, was that there should be a threshold of response followed by a relatively linear increase in response to subsequent increases in retinal illuminance gradient. With this in mind, and looking again at Fig. 12.11, it can be seen that certain of the bipolar responses on the shallow portions of the illuminance profile will be suppressed due to

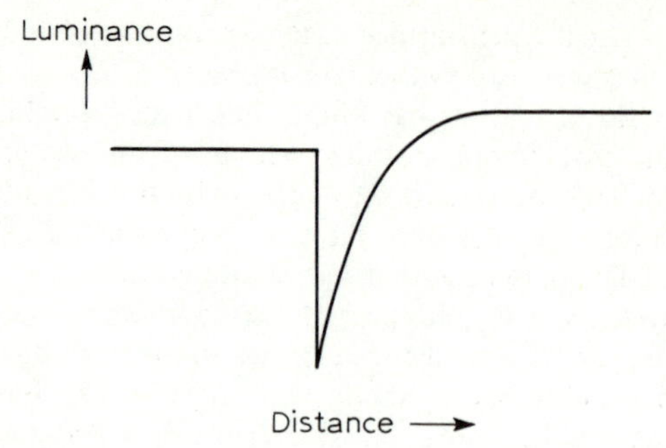

Fig. 12.10. A form of luminance profile used to yield an illusion of a brightness reversal of the centre and surround.

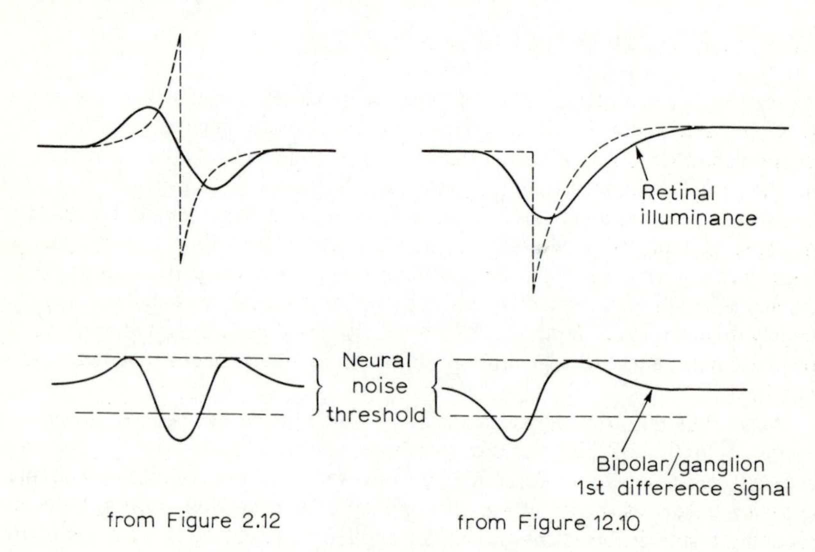

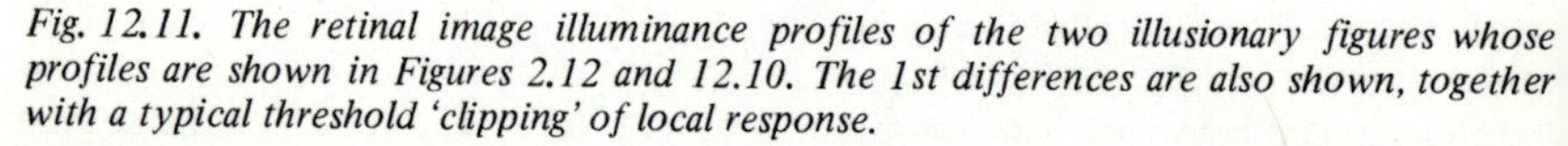

Fig. 12.11. The retinal image illuminance profiles of the two illusionary figures whose profiles are shown in Figures 2.12 and 12.10. The 1st differences are also shown, together with a typical threshold 'clipping' of local response.

insufficient signal. Conversely the inverse gradient region will produce a strong response. In the limit *only* the inverse gradient portion might be expected to yield a response. In such circumstances the brightness illusions observed are a necessary and predictable outcome of the response functions of bipolar and ganglion cells. To the author's knowledge no controlled studies allied to this explanation of brightness illusions have been carried out at this time but it is expected that most, if not all, brightness illusions should be predictable from this approach.

12.6 MODELLING OF SIMPLE RECOGNITION

Thus far all our modelling has been concerned with 50% thresholds for *detection.* In practical visual situations it is much more frequent that we are concerned with recognition rather than detection. As an introduction to the attempted modelling of recognition let us consider what possible contributory factors allow the recognition decision. Firstly we have an object which has certain characteristics which differentiate it from its surroundings. Let us consider this statement for a moment. 'Differentiate' suggests, as with our detection modelling a differencing technique, but this time it is not a simple difference of luminance on adjacent receptors but a difference of local luminance structure from that which would be expected for an alternative, confuseable, object. Again consider the wording 'would be expected'. Expectation implies a required input from the memory – it is only by experience that we can learn what visually discriminates one object from another. Thus immediately, with recognition, we have a problem – how do we define state of learning and ability to recall from memory? These are variables associated with the human observer and have been discussed to some extent in Chapter 3. A deeper study than discussed there is beyond the scope of this book.

Returning to the more straight-forward problem of differentiation, it must again be remembered, possibly more importantly than with detection, that the differential effects for which one is looking must be differences *in the retinal image*, not in the original object. This is an important difference, as can be illustrated by convolution of the profiles of various complex objects with the eye's spread function at 50% recognition size threshold. For instance Fig. 12.5 shows the retinal image isophot for 50% recognition threshold of the presence of the appendage in Lavin's multicontrast experiment[5] for a 0.44 mrad square. It will be seen that very little of the basic structure of the original object remains. Gone are the corners of both squares and gone also is the luminance step between the high and low contrast squares. What remains is an egg-shaped spread function which provides enough data from the right-hand portion to allow the recognition decision to be made. Similarly, reference back to Figs. 5.10 and 5.11 will illustrate the gross degradation of the retinal images of certain complex shapes at recognition threshold.

An attempt has been made to predict the recognition thresholds of certain relatively simple shapes where the recognition task was believed not to be seriously confounded with learning effects – in particular the prediction of the recognition sizes of the four individual high contrast objects used in the experiment on recognition of simple shapes described in Section 5.1[2,3]. In order to do this it was first necessary to attempt to assess what parts of the contour of the object probably constituted a 'recognition' profile. The shapes are shown in Fig. 5.3, together with the names ascribed to them. They were all constructed from six unit blocks and were presented at high contrast. The recognition decision, when they were presented singly, was the choice of one of the four, all

TABLE 12.1. Estimated and Predicted Recognition Threshold Data for 'Castle', 'Church', 'Line' and 'Block' Targets (from Lavin and Overington[23]).

Target	Mean steepest slope over interrogation region (cd/m^2/receptor spacing)	Predicted equivalent circle circumference (mrad)	Estimated contour for interrogation at mean recognition range (mrad)
Line	29.5	1.17	1.17
Church	26.0	1.25	1.15 to 1.3
Castle	39.5	1.02	0.9 to 1.05
Block	62.0	0.82	0.73

of them being familiar to the observers. It was considered that one could fairly confidently define the regions of contour which it would be necessary to 'interrogate' in order to decide which one was being viewed. These were essentially the 're-entrant' portions of the 'castle' and 'church', and the upper edges of the 'block' and 'line'. It was found that the retinal images of the castle and church had significantly attenuated illuminance gradients along a well-defined region associated with the re-entrant portions of the objects. When these gradients were taken, together with the gradients associated with the upper regions of the block and line, the lengths of contour for simple detection in terms of Blackwell's disc stimuli agreed very closely with the predicted portions of contour required for recognition (see Table 12.1). This was considered very significant, and was taken as tentative verification of the concept that recognition is achieved when important portions of differential contour associated with a particular object reach their own simple detection threshold. Much more work remains to be done in this area of modelling in order to determine, if possible, just what parts of a contour are important in a given recognition task, and whether it is the sum of the parts, or each individually, which must reach detection threshold in a complex situation. It is the author's current belief that it is only necessary for the contour represented by the sum of the individual differential parts to rise above simple detection threshold.

12.7 THE EFFECT OF TEXTURE

As mentioned in Section 11.3, the effects of texture in a target are dependent on whether that texture is random noise or 'intelligence' structure. Furthermore, whether random texture is classified as noise or intelligence depends on the characteristics of the background. If an object of interest and its background are both imaged on a grainy film, then the random graininess overlaying both the object and its background serves to break up the outline of the object, and hence degrades its detection threshold. The effects of 'granularity'* have been studied theoretically by Selwyn[24], Fry[25] and Jones[26], amongst others. Equally the practical effects of graininess** have received attention in experimentation by Fry and Enoch[27], Jones and Higgins[28], Selwyn[29], Charman and Olin[30] and Freiser and Biedermann[31]. Unfortunately only limited agreement between models of granularity and observed graininess have been found, which is suspected by the author to be partly due to the effects of the eye's own noise and the effective differentiation believed to take place in the retina.

If, instead of object and background having common imposed texture due to film grain, the object alone has its own discrete texture, there is a chance that the threshold may be enhanced. This chance of enhancement would seem to depend on whether the total 'strength' of stimulus due to the multitude of small local gradients within the object of interest is greater than the total 'strength' of the stimulus due to the outline. To the author's knowledge there is only a small amount of literature on this subject, and from what there is it is difficult to draw common conclusions. However, certain experimental studies recently reported provide at least qualitative information on the effects of texture. For instance Greenwood[32] has studied the contrast sensitivity for grain patterns and compared it to that for sine bars. He finds the same broad form of function, but with the contrast sensitivity for the grain patterns studied being only about 1/25th of that for sine bars. Other studies on the contrast sensitivity of grain patterns are those of Koenderink and Doorn[33,34]. Meanwhile Spillman and Coderre[35] have studied the thresholds for striped and plain flash-presented test fields of 0.5 rad diameter. They have found major interactions between the spatial frequency of the stripes and the retinal illuminance, there being a major decrease in thresholds for striped fields compared with uniform fields at high photopic levels, whilst the reverse is true at scotopic levels. It is believed by the author that this reversal may possibly be explained by considering that the perimeter of the 0.5 rad test field, which is broken up by the striped pattern, occurs at an optimum part of the visual field for scotopic response, whilst the stripes should provide strong foveal stimulation at photopic levels.

A further study worthy of mention concerned with texture is that of Rietveld *et al*[36] who have studied the visual evoked response (EEG) for a variety of checkerboard, striped and blank test fields. In particular they find that there is a stronger response from striped fields than from blank fields and a yet stronger response from checkerboard fields. Also they have studied checkerboard patterns of a fixed size (approximately 0.2 rad square with 6 mrad square pattern elements). For these they have studied the variation of EEG as a central circular portion of the pattern is progressively obscured. They find that the response reduces markedly with the central 35 mrad obscured and that is is almost non-existent with the central 70 mrad obscured.

**Granularity* is the objective microstructure of a grainy or textured image.
***Graininess* is the subjective visual appearance resulting from the microstructure

A possible but largely untried approach to the predictive modelling of the effects of graininess and texture on thresholds has been suggested by the author[37,38]. It is suggested that the vision model developed in Chapter 7 contains, in its constant K_1, a term which is largely dependent on the spatial noise in the input to the eye. However, whilst it is easy to see, in principle, how Gaussian or Poissonian noise functions can be combined in this term K_1, it is by no means so clear how one is to combine other frequency dependent noise functions due to texture with the Poissonian noise due to the quantum nature of light.

REFERENCES

1. Lavin, E. P. and Overington, I. (1972). 'Visual Modelling', Annex E of *Final Report on the Third Visual Studies Contract*,(BAC (GW) Ref. L50/196/1535), Sect. 2
2. Clare J. N. (1970). 'Threshold Tracking: a Shorter Method of determining Visual Detection Thresholds', *BAC (GW) Human Factors Study Note, Series 7, No. 2*, BAC (GW) Ref. L50/20/HF/26
3. Westheimer, G. and Campbell, F. W. (1962). 'Light Distribution in the Image Formed by the Living Human Eye', *J. Opt. Soc. Am.*, **52**, 1040
4. Hills, B. L. (1970). 'Visual Perception with Electronic Imaging Systems', PhD. Thesis, University of Nottingham
5. Lavin, E. P. and Overington, I. (1972). 'Visual Modelling', Annex E of *Final Report on the Third Visual Studies Contract*, May, (BAC (GW) Ref. L50/196/1535). Sect. 4
6. Blackwell, H. R. (1946). 'Contrast Thresholds of the Human Eye', *J. Opt. Soc. Am.*, **36**, 624
7. Overington, I. (1974). 'An Exploratory Study into the Various Observed Complex Functional Characteristics of Vision and their Compatability with a Unified Simple Modelling', BAC (GW) Ref. ST12386
8. Overington, I. and Lavin, E. P. (1970). 'A Theory of Foveal Vision', App. 6 of *Final Report on Visual Studies II Contract*, BAC(GW) Ref. L50/20/PHY/196/1214
9. Overington, I. and Lavin, E. P. (1971). 'A Model of Threshold Detection Performance for the Central Fovea', *Optica Acta*, **18**, 341
10. Østerberg, G. (1935). 'Topography of the Layer of Rods and Cones', *Acta Ophthal.*, **13**, Suppl. 6
11. Campbell, F. W. and Gubisch, R. W. (1966). 'Optical Quality of the Human Eye', *J. Physiol.*, **186**, 558
12. Hoekstra, J. J., van der Goot, D. P. J., van den Brink, G. and Bilsen, F. A. (1974). 'The Influence of the Number of Cycles upon the Visual Contrast Threshold for Spatial Sine Wave Patterns', *Vision Research*, **14**, 365
13. McCann, J. J., Savoy, R. L., Hall, J. A. (Jr.), and Scarpetti, J. J. (1974). 'Visibility of Continuous Luminance Gradients', *Vision Research*, **14**, 917
14. Daitch, J. M. and Green, D. G. (1969). 'Contrast Sensitivity of the Human Peripheral Retina', *Vision Research*, **9**, 947
15. Valois, R. L. de, Morgan, H. and Snodderly, D. M. (1974). 'Psychophysical Studies of Monkey Vision, III. Spatial Luminance Contrast Sensitivity Tests of Macaque and Human Observers', *Vision Research*, **14**, 75
16. Williams, T. L. (1975). Unpublished studies on contrast sensitivity as a function of retinal image position carried out by SIRA Institute and reported to a working party on image evaluation

17. Campbell, F. W. and Robson, J. G. (1968). 'Application of Fourier Analysis to the Visibility of Gratings', *J. Physiol.*, **197**, 551
18. Graham, N. and Nachmias, J. (1971). 'Detection of Grating Patterns containing Two Spatial Frequencies: A Comparison of Single and Multiple Channel Models', *Vision Research,* **11**, 251
19. Sachs, M. B., Nachmias, J. and Robson, J. G. (1971). 'Spatial Frequency Channels in Human Vision', *J. Opt. Soc. Am.*, **61**, 1176
20. Burton, G. J. (1973). 'Evidence for Non-linear Response Processes in the Human Visual System from Measurements on the Thresholds of Spatial Beat Frequencies', *Vision Research,* **13**, 1211
21. Cornsweet, T. N. (1970). *Visual Perception*, Academic Press
22. Ratliff, F. (1965). *Mach Bands*, Holden-Day Inc.
23. Lavin, E. P. and Overington, I. (1972). 'Visual Modelling', Annex E. of *Final Report on the Third Visual Studies Contract*, BAC (GW) Ref. L50/196/1535. Sect. 5.4.5
24. Selwyn, E. W. H. (1935). 'A Theory of Graininess', *Phot. J.*, **75**, 571
25. Fry, G. A. (1963). 'Coarseness of Photographic Grain', *J. Opt. Soc. Am.*, **53**, 361
26. Jones, R. C. (1955). 'New Method of Describing and Measuring the Granularity of Photographic Materials', *J. Opt. Soc. Am.*, **45**, 799
27. Fry, G. A. and Enoch, J. M. (1959). 'The Relation of Blur and Grain to the Visibility of Contrast Borders and Gratings', MCRL T.P. No. 696–19–293, Ohio State University
28. Jones, L. A. and Higgins, G. C. (1951). 'Photographic Granularity & Graininess, VI; Performance Characteristics of the Variable-magnification Graininess Instrument', *J. Opt. Soc. Am.*, **41**, 64
29. Selwyn, E. W. H. (1948). 'The Photographic and Visual Resolving Power of Lenses', *Phot. J. B.*, **88**, 6 and 46
30. Charman, W. N. and Olin, A. (1965). 'Image Quality Criteria for Aerial Camera Systems', *Phot. Sci. and Eng.*, **9**, 385
31. Frieser, H. and Biedermann, K. (1963). 'Experiments on Image Quality in Relation to Modulation-Transfer Function and Graininess of Photographs', *Phot. Sci. and Eng.*, **7**, 28
32. Greenwood, R. E. (1973). 'Visibility of Structured and Unstructured Images', *J. Opt. Soc. Am.*, **63**, 226
33. Koenderink, J. J. and Doorn, A. J. van (1974). 'Detectability of Two-dimensional Band Limited Noise', *Vision Research,* **14**, 515
34. Koenderink, J. J. and Doorn, A. J. van (1974). 'Spatial Noise for Visual Research', *Vision Research,* **14**, 721
35. Spillman, L. and Coderre, J. (1973). 'Increment Thresholds for Striped and Uniform Test Fields as a Function of Background Level', *J. Opt. Soc. Am.*, **63**, 601
36. Rietveld, W. J., Tordoir, W. E. M., Hagenouw, J. R. B., Lubbers, J. A. and Spoor, Th. A. C. (1967). 'Visual Evoked Responses to Blank and to Checkerboard Patterned Flashes', *Acta Physiol, Pharmacol, Neerl.*, **14**, 259
37. Overington, I. (1974). 'An Investigation into the Reasons for the Degraded Thresholds obtained in the Size Probability Experiment', BAC (GW) Rep. No. ST10961
38. Overington, I. (1974). 'Visual Efficiency: A Means of Bridging the Gap between Subjective and Objective Quality', in *Proceedings of the SPIE, Vol. 46, Image Assessment and Specification*, 93

13 Background Structure

As stated in Chapter 11, there are many forms of background structure and several ways in which they can influence detectability of a target. The structure may be adjacent high contrast objects which modify thresholds due to their energy content; local background structure which interacts with the target destroying the concept of one contrast or one illuminance gradient around a contour; a local region of dissimilar luminance to the rest of the field against which to view a target; a sub-threshold structure which nevertheless interacts with threshold; a temporal structuring of the background which interacts with target interpretation; random noise or 'intelligence' structure which may change the confidence level at which a decision is made. Considerable study has been given to several of these areas. In this chapter what the author considers to be some of the more important parts of the large body of data will be summarised. It will be shown that certain of the phenomena observed may be tentatively explained by considering the retinal image gradients associated with the complex situations.

13.1 LOCAL BACKGROUND STRUCTURE

13.1.1 'Receptive Field' Studies

A class of study which has received a considerable amount of attention in recent years is concerned with the influence of symmetrical local surrounds on thresholds of simple stimuli. These studies are mostly classed 'receptive field' studies because of the belief that they illustrate the extent to which a signal in a given optic nerve fibre is influenced by incoming data from an extended portion of the retinal image. These studies involve a number of situations.

Some of the most important studies are those of Enoch[1,2] and Westheimer[3,4], where a small flash stimulus is viewed against various sizes of circular plateaux of strong local luminance superimposed on a general lower background luminance. In these experiments there is usually a differential luminance between the local surround and the general background of the order of 1 to 2 orders of magnitude. At high photopic levels the findings are that with small surrounds (of diameter less that one mrad) the threshold is, if anything enhanced. As the plateau diameter is increased to around 1.5 mrad the thresholds degrade markedly before recovering a great deal as the plateau size is increased to 6 mrad diameter or more. For stimuli presented extra-foveally the form of threshold trend is similar, but with the point of maximum degradation moving to larger sizes. At low light levels it is found that the recovery is

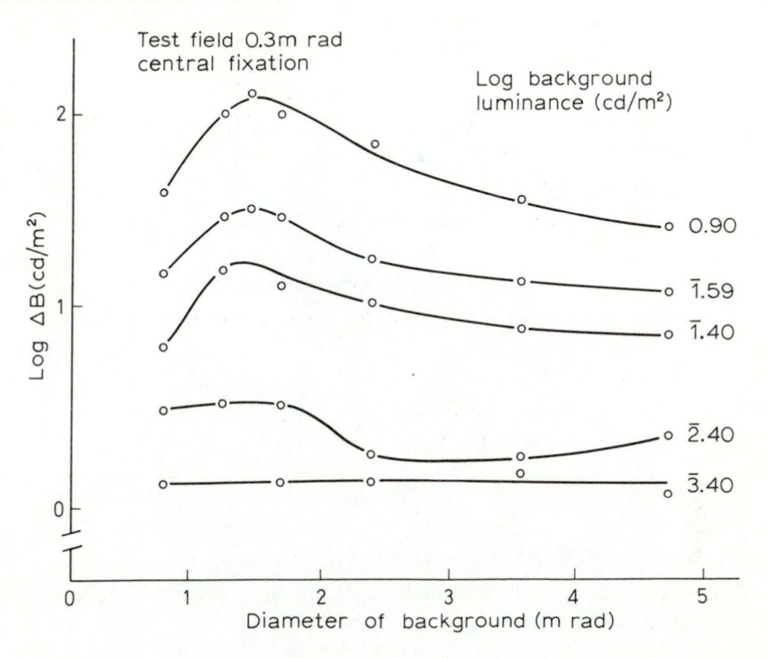

Fig. 13.1. Threshold trends for a small, flashing stimulus presented foveally at the centre of local backgrounds of various sizes and luminances. (Reproduced from Westheimer[3] by courtesy of the Journal of Physiology).

inhibited, whilst for the lowest light levels there is no measurable degradation. Sample results are shown in Fig. 13.1. The general behaviour is usually explained at high luminance as due to a local summation field of a few minutes diameter and a surrounding antagonistic or inhibitory field stretching out some 10 mrad diameter or more. It is then further argued that away from the fovea the summation and inhibitory fields expand in size, whilst at the very low luminances the inhibitory field is suppressed.

A modified explanation is believed by the author to be possible, based on the modelling of Chapters 7 and 12. Let us study the retinal images of a selection of Westheimer's stimulus presentations at high photopic luminance. In Fig. 13.2 are shown the presentation situation in sketch form for plateaux of the same size as the flash stimulus, about 1.5 mrad diameter and about 6 mrad diameter. Below is shown the equivalent retinal illuminance distribution. It will be seen that the smallest plateau (a) is unresolved on the retina, with the result that the flash stimulus merges with it, being enhanced in the process. Conversely the 1.5 mrad and 6 mrad diameter plateaux are fully resolved, so that the flash stimulus is effectively presented against an illuminance 1 to 2 log units above the general adapation level. Now the range of response of cone receptors at a given

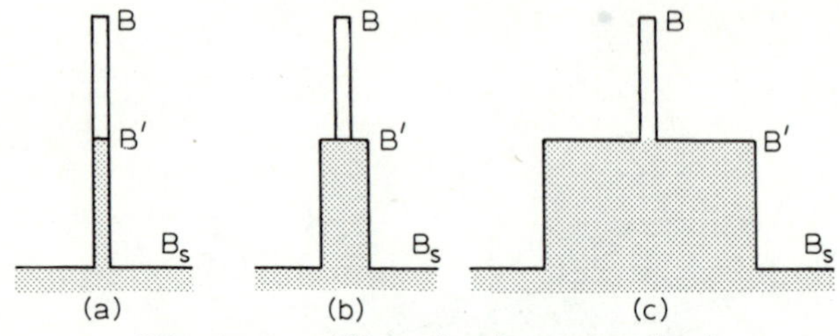

Stimulus conditions (luminance profiles)

B = Flashing stimulus luminance, B' = Local background luminance, B_s = Surround luminance

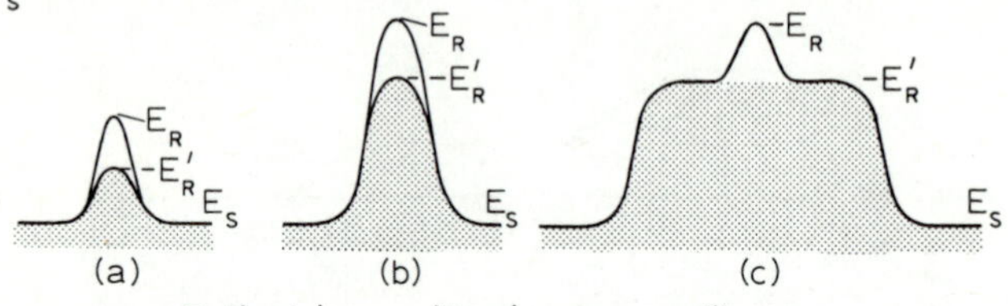

Retinal image illuminance profiles

E_R = Flashing stimulus illuminance, E'_R = Local background illuminance, E_s = Surround illuminance

Fig. 13.2. Illustrating the retinal images for various background/stimulus combinations as studied by Westheimer[3].

adaptation level is only of the order of ± 1½ log units of luminance[5,6], the typical response function being as shown in Fig. 13.3. Thus, if the stimulus is presented on a plateau some 1 to 2 log units above adaptation level, its effective 'strength' will be reduced due to saturation effects (in much the same way as happens to grossly overexposed parts of a photograph). Hence the threshold will be degraded. This is in keeping with the results obtained by Westheimer for the 1.5 mrad plateau. As the diameter of the plateau is still further increased, it is believed by the author that there will come a time when the plateau itself effectively controls the local adaptation level due to inhibitory influences at the ganglion cell level (see Section 2.4). The threshold for stimuli presented at the centre of the plateau will then return to an unsaturated level, but will be controlled by the luminance of the plateau rather than that of the surround. This concept of control of threshold by the plateau rather than the surround is supported by recent studies reported by Esen and Novak[7]. This is believed to be an explanation for the threshold behaviour found by Westheimer for large plateaux, such experimental data as Westheimer's containing a calibration of the extent of local adaptation in the human retina – very valuable data for the study of complex fields as will be seen later. The *reason* for this local adaptation, be it photochemical (e.g. Rushton[8,9]), electrostatic[10], or due to lateral neural

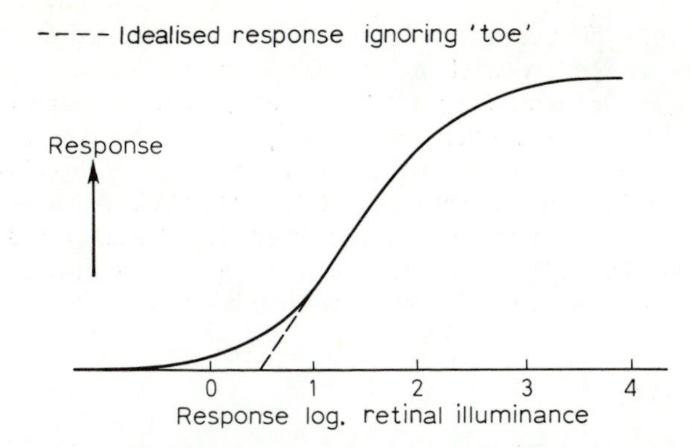

Fig. 13.3. Typical response function of a cone receptor (after Werblin[5]).

interconnections (e.g. Brindley[11]), seems unimportant to an appreciation of the behaviour of the visual system to complex, high contrast fields. The fact that there is no rise of threshold at very low luminances is believed to be due to the fact that at such low luminances the eye is fully dark adapted and the 'normal' operating point therefore moves from the centre of a response curve such as Fig. 13.3 towards the toe. Under such circumstances it would require a local difference of 3 or 4 log units of luminance to drive the local response into saturation, a condition not studied to the knowledge of the author.

If we now consider the work of Enoch[1,2], we find that his initial reported work is very much on the same lines as Westheimer's – and with the same findings. Both authors present threshold trends for other than foveal viewing, showing a maximum degradation moving to larger plateau diameters with increasing distance from the fovea, but with more sudden recovery, so that all experimental curves have fully recovered with plateaux of 12 mrad diameter and greater. This is in keeping with the known increasing separation between receptors plus a widening of the optical spread function for such off-axis situations, whilst suggesting that, over the region tested, the local adaption scaling remains substantially constant.

Other studies of Westheimer's using a 'double border' are of particular interest because it was here that he hoped, by use of a pair of adjacent 0.3 mrad wide annuli, one white and one black on a mid grey background, to show that presence of a border as such had no effect on threshold. What he did, in fact, find was that there was a very minor enhancement of threshold for 1.5 mrad and 3 mrad border diameters and a return to normal threshold for border diameters of 10 mrad and more. It is not stated in Westheimer's paper whether his borders were black inside white or white inside black, but it is believed by the author that it might make a significant difference – and that the observed results suggest he used black inside white. Figure 13.4 shows the convolutions of a 3 mrad twin border and a 0.3 mrad diameter target (as used by

Westheimer) with the spread function for a naturally focussed eye. It will be seen that the two borders largely cancel each other in amplitude but provide a very powerful border sensation since the majority of the combined transition from one to the other occurs between two receptors (> 0.2 mrad). The effect of convoluting a target with a 1.5 mrad diameter twin border is also illustrated for the two cases of black inside white and white inside black. It will be seen that the only major effect on the maximum gradient is to change the diameter at which it occurs. This might be expected to produce possible differences due to an effectively increased size in the case of white inside black.

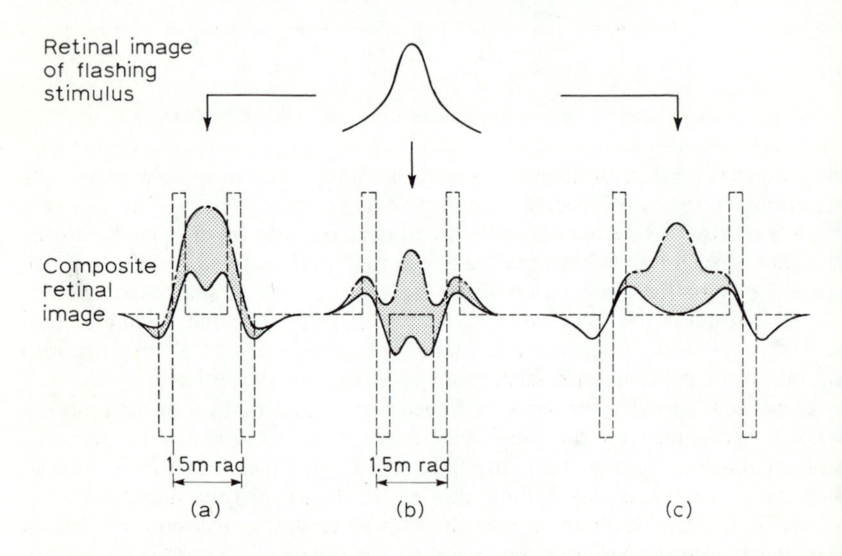

Fig. 13.4. Illustrating the combined retinal image distribution due to various Westheimer 'twin borders' and a disc stimulus, (a) 1.5 mrad diameter border, white inside black (b) 1.5 mrad diameter border, black inside white, (c) 3 mrad diameter border, white inside black. Shaded areas indicate addition of stimulus.

Other studies of interest to receptive field considerations are those of Rentschler and Arden[12], who studied the effect of the slope of a bipartite field on the detection of an edge presented within it, and of several workers[13,16] on the threshold trends for small stimuli presented in the vicinity of a strong luminance discontinuity. These latter have regularly shown evidence of marked suppression of threshold for stimuli presented close to the edge at the high luminance side, whilst Vassilev[15] has also shown larger effects on thresholds for rectangular stimuli than for disc stimuli.

Finally Alexander[17] has recently presented some data on the effect of stimulus *size* on the form of receptive fields at scotopic levels.

13.1.2 Sub-threshold local structure

Further studies closely allied to the 'receptive field' studies are typified by those of Fiorentini and Maffei[18], who studied the effect of sub-threshold annular stimuli of varying diameter but constant area on a flash stimulus presented within the annuli. The annuli studied were all of 0.77 mrad^2 area and were all presented at 0.1 log units below their own mean thresholds. It was found that, as with Westheimer's work, for a small annulus the threshold of the flash stimulus was improved, for a moderate annulus (again 3 mrad diameter) the threshold was degraded and by the time the annulus had been increased to 6 mrad diameter the threshold had recovered to a level equal to that with no annulus present (Fig. 13.5). This time the reason for the results is believed by the author to be completely different to that for Westheimer's results. Fig. 13.6 shows the effective retinal images for flash stimuli and three annuli in the range used by Fiorentini and Maffei. The images of the flash stimulus and the respective annuli are then combined, with surprising results. It is found that the maximum

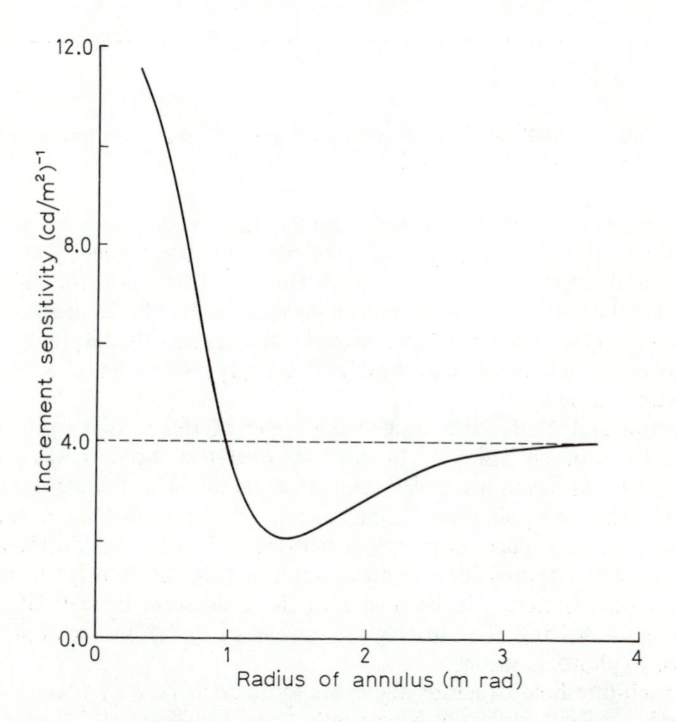

Fig. 13.5. Photopic threshold trends for a small, flashing disc stimulus in the presence of sub-threshold annuli. (Reproduced from Fiorentini and Maffei[19] by courtesy of Pergamon Press).

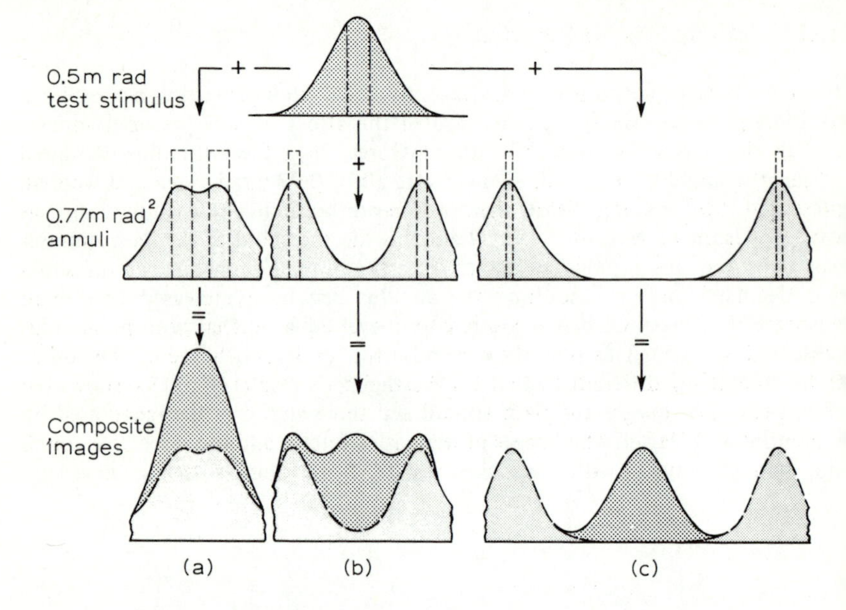

Fig. 13.6. Illustrating the superposition of retinal images of a small test stimulus and variou annuli as studied by Fiorentini and Maffei[19].

illuminance *gradient* for the presentation on the smallest annulus is greatl enhanced and also that the maximum gradient region is extended. This will lea to a two-fold improvement in threshold. On the other hand, for the 3 mra annulus it will be seen that the stimulus sinks into the annulus in such a way tha the maximum gradients are reduced severely, thus raising the threshold. Finall for the 6 mrad annulus the flash stimulus fits snugly *into* the annulus, remainin virtually unmodified.

Fiorentini and Maffei also studied the effect of this second form of loca structure for scotopic vision[19]. In this case their test stimulus was a 6 mra diameter disc with various larger subliminal annuli. The findings were ver similar to those for photopic vision, except that the maximum threshol degradation was no longer with annuli between 1.5 and 3 mrad diameter, bu rather with annuli approaching 18 mrad diameter (Fig. 13.7). It is felt that suc a set of results is closely in keeping with the differences in scale of optima *receptor unit* spacing (i.e. spacing of groups of rods) for scotopic visio compared to photopic vision.

Other sub-threshold structure studies are those carried out by Bagrash *et al*[20] Kulikowski[21], Kulikowski and King-Smith[22] and Gelade *et al*[23]. In the first these Bagrash *et al* studied the trend of the composite threshold of a 6 mra diameter disc and a surrounding annulus of variable luminance. They found tha

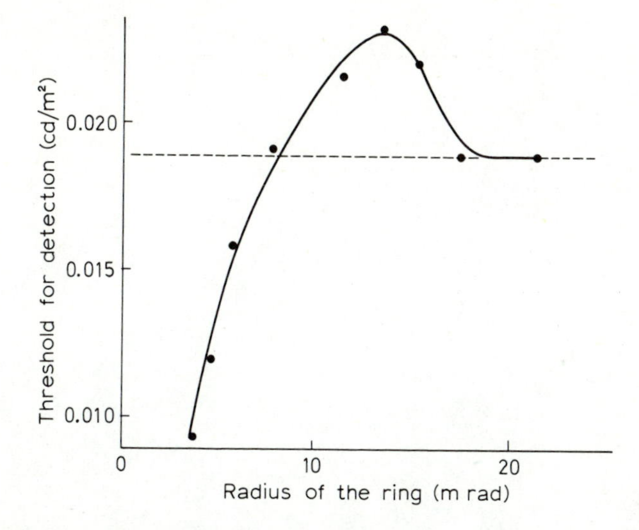

Fig. 13.7. Scotopic threshold trends for a small disc stimulus in the presence of sub-threshold annuli. (Reproduced from Fiorentini and Maffei[18] *by courtesy of the Journal of Neurophysiology).*

there was a worst threshold with the annulus luminance 50% of that of the disc, the general background being essentially dark. On the other hand, Kulikowski presented narrow dark line stimuli against sub-threshold sinusoidal patterns of 0.29 c/mrad and 1.15 c/mrad at photopic luminance levels. He found that the threshold for the line stimulus against the 0.29 c/mrad pattern was significantly altered, being degraded when presented against the bright striations of the pattern and enhanced when presented against the dark striations. Conversely he found no effect of the 1.15 c/mrad pattern on the threshold for the line. Kulikowki and King-Smith reason that these results are evidence of parallel processing in the visual system (see Section 12.4). On the contrary, the present author finds no reason to assume parallel processing as an explanation of these results. If the retinal images are generated for the various cases as shown in Fig. 13.8, it will be seen that a similar situation exists for the 0.29 c/mrad pattern as has been discussed for the Fiorentini and Maffei patterns. Conversely, for the 1.15 c/mrad pattern, the retinal image of the line spreads over more than one cycle of the pattern, there thus being little difference in the maximum gradients for the 'peak' and 'trough' superpositions. Following from this work Kulikowski and King-Smith studied thresholds for thin lines in the presence of sub-threshold lines and edges[22]. For pairs of sub-threshold lines either side of the test stimulus a characteristic summative/inhibitive response function was generated, there being a maximum threshold suppression with sub-threshold lines separated by approximately 1.75 mrad from the test stimulus (see

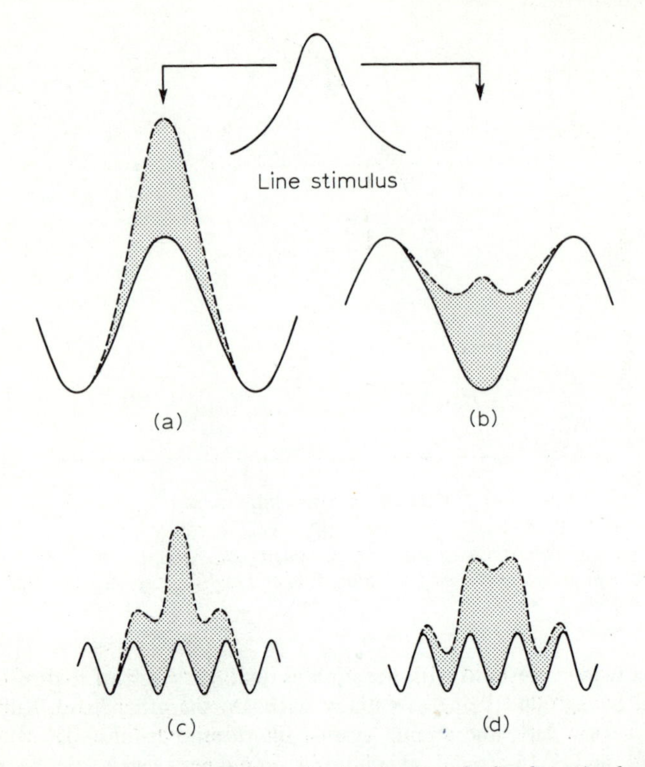

Fig. 13.8. Illustrating the superposition of retinal images of a fine line stimulus and various sub-threshold sine bar patterns, (a) superposition on peak of 0.29 c/mrad pattern, (b) superposition on trough of 0.29 c/mrad pattern, (c) superposition on peak of 1.15 c/mrad pattern, (d) superposition on trough of 1.15 c/mrad pattern.

Fig. 13.9). For edges, on the other hand, a simple suppression occured for edges close to the stimulus, the threshold recovering as the edge was moved further away. In this case the maximum degradation was with the edges approximately 0.9 mrad from the stimulus (see Fig. 13.10). Again the present author considers that the entire body of data may be potentially explained in terms of the retinal images of the complex objects and the edge gradient detection model desribed in Chapter 7.

Finally Gelade *et al*[23] have studied the effect on line thresholds of sub-threshold lines and edges at scotopic luminance levels. Their findings complement those of Fiorentini and Maffei for annuli, again showing trends very similar to those of Kulikowski and King-Smith, but on a much larger scale.

The above interpretations of receptive field and local sub-threshold structure data appear to leave little need for any *complex* neural summation and inhibition effect as supposed to exist according to Enoch *et al*[1], Westheimer

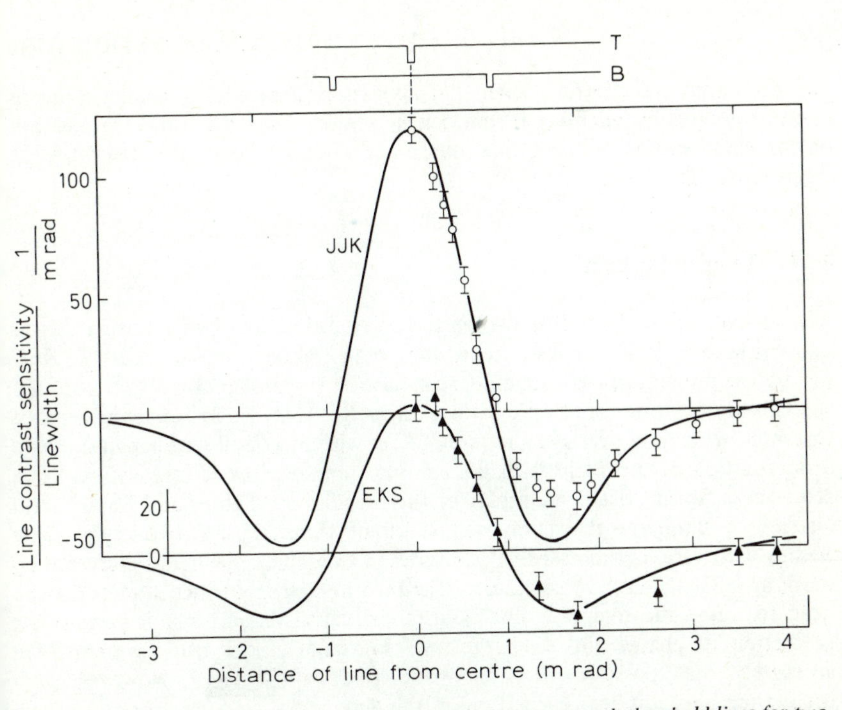

Fig. 13.9. Threshold trends for a fine line imaged between two sub-threshold lines for two observers (JJK and EKS). (Reproduced from Kulikowski and King-Smith[22] by courtesy of Pergamon Press).

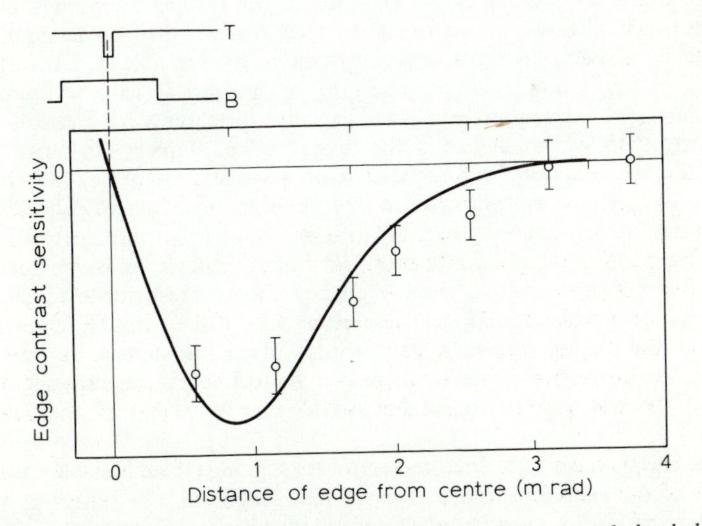

Fig. 13.10. Threshold trend for a fine line imaged adjacent to a sub-threshold edge. (Reproduced from Kulikowski and King-Smith[22] by courtesy of Pergamon Press).

and Fiorentini and Maffei[18]. All that appears necessary is to assume a simple form of progressive adaptation function in conjunction with initial examination of the complexities of the retinal image and effective neural differentiation as discussed in Chapter 7.

13.1.3 Temporal Effects

In addition to the effects described in the previous section there are other effects of local background. For instance, a short pre-exposure to a sub-threshold flash just before presentation of a stimulus can have a very similar effect to that of the spatially interacting sub-threhold line stimuli – that is, an enhancement of threshold with near overlay, a reduced effect with increased time separation and a degradation of threshold with the stimulus presented at certain critical times after the adapting flash. Examples of such studies are Fehrer and Raab[24] and Herrick[25]. Similarly the composite resultant threshold for two equal short flashes with various time separations varies in a complex manner as discussed by Clark and Blackwell[26] in particular. The information from such studies may be used to refine the temporal threshold modelling of vision, since it permits the prediction of charge and discharge time constants of the visual system more accurately than possible from the work cited in Chapter 7. However, such predictions have not been carried out at this time.

Another aspect of temporal effects is that of pre-adaptation or 'tuning'. It has been found by several workers that the threshold performance against spatial bar patterns is affected by a previous time history of viewing supra-threshold bar patterns, such changes in performance occurring whether pre-adaptation is achieved by fixation on the adapting pattern or by scanning it. For instance, Gilinsky[27] has found that pre-exposure to horizontal bars will enhance thresholds for horizontal bars and degrade thresholds for vertical bars. Pre-exposure to vertical bars has the reverse effect. Equally Tolhurst[28] has found that pre-exposure to bar-patterns of a specific frequency will inhibit subsequent threshold performance for that frequency to an extent dependent on the contrast of the adapting bars. To a lesser extent the third harmonic of a square wave has an adapting influence. Yet again Campbell and co-workers have shown that pre-exposure to a given frequency affects thresholds to a significant extent for frequencies within an octave either side of the adapting frequency[29] and that the tuning due to a pattern of a given orientation, as found by Gilinsky, is progressive – that is, there is a gradual tuning or detuning as the angle of the test pattern approaches or deviates from that of the adapting pattern[30].

Since the tuning effects discussed above are suprathreshold and due to several seconds of pre-exposure, it is considered that they cannot be retinal in origin. Instead, such effects *must* be attributed to disturbance of central processes, in contrast to the main body of 'local structure' effects discussed previously. Also, and perhaps more importantly, such temporary adaptation is not, in general, randomly variable from glimpse to glimpse and hence must be considered to be similar to learning processes or a temporary variation of a threshold signal/noise level.

13.2 GENERAL BACKGROUND STRUCTURE

13.2.1 Random Noise

Random noise can occur in a background due to many causes. Some of these, such as the noise due to photographic grain, projection screen structure and characteristic graininess of image intensifier tubes have already been mentioned (Section 9.2). Where such structure interacts with the object of interest, its main effect is to soften the boundary between object and background. Such an interaction, one might expect, would degrade thresholds for seeing the object, on the basis of the findings of Chapter 7. Van Meeteren[31] reported on several experiments using image intensifiers where such a degradation of performance was indeed observed. Fry and Enoch[32] similarly report on experiments involving controlled amounts of photographic grain and attempt to relate the effects of grain to the effects of blurring of a contour. Yet again Stromeyer and Julesz[33] and Pollehn and Roehrig[34] have studied the effects of random one-dimensional noise on detectability of a bar pattern. Stromeyer and Julesz have found that noise frequencies within an octave of the bar pattern frequency have some effect, whilst Pollehn and Roehrig find that the effects of noise are strongly dependent on bar pattern frequency distribution. Finally Rasmussen[35] has studied the effects of number of quantisation levels and contrast available in a noisy display, and has found performance to be largely insensitive to these parameters.

However, it is important to realise that the effects of random noise must be dependent on both the type of object structure and the task involved. If the task is simple detection and the object is a long border then one would expect the noise to play a minimal part, since the brain can effectively compare various parts of the border and the available 'signal' is large. Conversely, if the task is still detection but the object is small and compact, one might expect the noise to be very important. Yet again, if the task is one of recognition, the noise may be expected to hide some of the detail required for recognition, the importance of the noise then being critically dependent on the form of the detail required for recognition.

Other forms of essentially random noise are of much lower spatial frequency – such as unimportant components of a natural scene, which break up the background and possibly produce strong, large-scale interactions with the object of interest. The effect of such structure is hard to generalise upon – indeed it is likely that it must be handled for each specific case. Present knowledge only allows a very limited speculation as to the typical effect but the 'receptive field' data referred to in Section 13.1.1 should provide a lead.

13.2.2 'Intelligence' Structure

The other form of background structure which one comes across is what may be termed 'intelligence structure' – that is, other detail within the scene which is visually resolved as discrete, recognisable objects. Such structure can have a variety of effects, dependent on its relationship to the object of interest in space and in association. For instance, if one is flying over strange terrain and looking for a specific object, it is to be expected that other ground features, where they can be related to the object of interest, will be of considerable help in confirming that an otherwise meaningless patch of luminance is the object of interest. On the other hand, again in viewing the ground from the air, parts of the background may from time to time partially obscure the object of interest (screening). Yet again the presence of many objects in close proximity to the object of interest may not only upset the retinal image luminance distribution, with consequent degradation of threshold performance, but may so confuse the observer with excessive information that he is unable to isolate the object of interest from within the complex pattern. Brown[36] has chosen to call this form of complexity 'confusibility', and has reported on limited experiments carried out to investigate the effect of presence of other target-like objects on ability to recognise.

In the experiments the stimuli were scenes consisting of an array of relatively simple high contrast shapes such as that shown in Fig. 13.11. All the shapes were constructed of an assemblage of six unit squares, with a constraint that each one

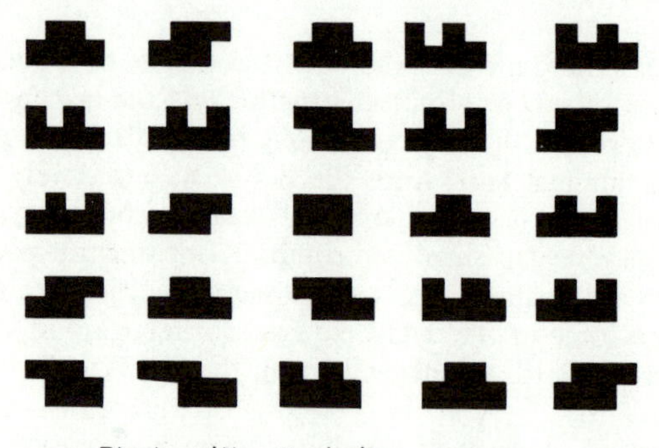

Fig. 13.11. A typical matrix of confusible shapes as used in the study of recognition in complex scenes by Brown. (Reproduced from Brown[36] by courtesy of the Advisory Group for Aerospace Research and Development of NATO).

must have a base of at least three units in a horizontal line. The shapes were arranged in groups of 25 in 5 x 5 matrices, each matrix containing one of four 'target' shapes as shown in Fig. 5.3 and as previously used in a simple recognition experiment. Two main effects were studied – packing density and the degree of similarity to the target of the other shapes used. It was found that the recognition range was significantly affected by packing density, being lower for the higher packing density. Equally the similarity of shapes to the target affected the recognition range significantly. However, when both similar and dissimilar shapes were used, their *position* in the matrix relative to the target had no effect on performance!

13.3 STRUCTURED SEARCH

The last mentioned experimental work by Brown leads us naturally into the field of structured search – that is, wide scale search for an object of interest where the background contains many objects similar to the one being searched for. The major work on this subject know to the author is that of Williams[37,38] and Howarth and Bloomfield[39,40].

Williams has investigated at considerable length the recognition of specific shapes in the presence of considerable numbers of similar shapes. In his experiments he has varied several parameters of the similar shapes – the actual shape, the contrast, the chromaticity, the size and the orientation (for irregular shapes). In his experiments he monitored his subjects' eye fixation points relative to the presented scene. He found that some 95% of all fixations fell on specific objects, the probability of next glimpse being cued to a particular object depending on its proximity to the present fixation and its similarity to the object of interest. He has produced distribution functions of probability of next fixation against difference from target in the domains previously listed. Typical distribution functions are shown in Fig. 13.12.

Howarth and Bloomfield, on the other hand, have carried out several experiments involving varying numbers of objects similar to the object of interest and have studied the *time* taken for detection. They have found that the time taken to detect is a roughly exponential function of the number of objects present and the degree of similarity. Their main experiments have involved black ball-bearings of different sizes and multisided polygons within which a single ball-bearing of equal area is to be found. Enoch[41], in work some years ago, studied the distribution of glimpses in a variety of real-life structured scenes. He found this distribution, which must be associated with performance, to vary dramatically from scene to scene (as perhaps might be expected). In many cases the glimpse distributions were rather obvious – such as a concentration along a railway track when looking for a station – but, in scenes where one might have expected there to be a symmetry in the glimpse distribution, the glimpses were found generally to be concentrated below the centre of the field being searched.

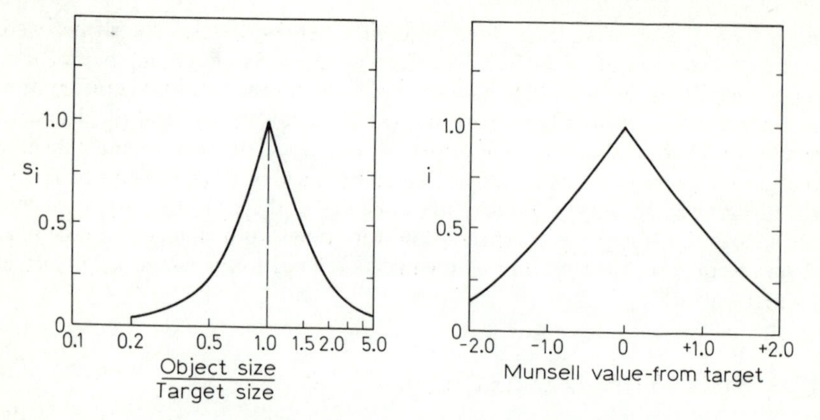

Fig. 13.12. Typical discrimination probability functions as determined by Williams for search in complex scenes. (a) as a function of similarity of size, (b) as a function of similarity of colour measured on the Munsell scale of chromaticity differences. (Reproduced from Greening[42] by courtesy of the Advisory Group for Aerospace Research and Development of NATO).

There was also a frequent observer bias to left or to right of the centre line. This finding, which to the present author's knowledge has been neither confirmed nor questioned by other experimentation, would appear to have some considerable importance in real-world search tasks, being presumably in some way associated with subconscious expectancy.

REFERENCES

1. Enoch, J. M., Sunga, R. N. and Bachman, E. (1970). 'Static Perimetric Technique believed to test Receptive Field Properties', *Am. J. Ophthalmol.*, **70**, 113
2. Enoch, J. M., Berger, R. and Birns, R. (1970). 'A Static Perimetric Technique believed to test Receptive Field Properties: Extension and Verification of the Analysis', *Documenta Ophthal.*, **29**, 127
3. Westheimer, G. (1967). 'Spatial Interaction in Human Cone Vision', *J. Physiol.*, **190**, 139
4. Westheimer, G. (1965). 'Spatial Interaction in the Human Retina during Scotopic Vision', *J. Physiol.*, **181**, 88
5. Werblin, F. S. (1973). 'The Control of Sensitivity in the Retina', *Scientific American*, January, 71
6. Werblin, F. S. (1971). 'Adaptation in a Vertebrate Retina: Intracellular Recording in *Necturus*', *J. Neurophysiol.*, **34**, 228
7. Esen, J. S. Van and Novak, S. (1974). 'Detection Thresholds within a Display that Manifests Contour Enhancement and Brightness Contrast', *J. Opt. Soc. Am.*, **64**, 726
8. Rushton, W. A. H. (1965). 'Visual Adaptation', *The Ferrier Lecture, Proc. R. Soc. B.*, **162**, 20
9. Rushton, W. A. H. and Powell, D. S. (1972). 'The Rhodopsin Content and the Visual Threshold of Human Rods', *Vision Research*, **12**, 1073

10. Fry, G. A. (1965). 'Physiological Irradiation across the Retina', *J. Opt. Soc. Am.*, **55**, 108
11. Brindley, G. S. (1970). *Physiology of the Retina and Visual Pathway*, 2nd edn. Edward Arnold
12. Rentschler, I. and Arden, W. (1974). 'Edge Detection in Luminance and Colour Discrimination', *Vision Research,* **14**, 1043
13. Hills, B. L. (1970). 'Visual Perception with Electronic Imaging Systems', Ph.D. Thesis, University of Nottingham
14. Hawkins, K. and Church, N. T. (1969). 'Contrast Sign Dependence', Study Note No. 2 of *Research into Factors affecting the Detection of Aircraft through Optical Sights*, BAC (GW) Ref. L50/20/PHY/186/1059
15. Vassilev, A. (1973). 'Contrast Sensitivity near Borders: Significance of Test Stimulus Form, Size and Duration', *Vision Research*, **13**, 719
16. Wildman, K. N. (1974). 'Visual Sensitivity at an Edge', *Vision Research',* **14**, 749
17. Alexander, K. R. (1974). 'Sensitization by Annular Surrounds: the Effect of Test Stimulus Size', *Vision Research,* **14**, 1107
18. Fiorentini, A. and Maffei, L. (1970). 'Transfer Characteristics of Excitation and Inhibition in the Human Visual System', *J. Neurophysiol,* **33**, 285
19. Fiorentini, A. and Maffei, L. (1968). 'Perceptual Correlates of Inhibitory and Facilitatory Spatial Interactions in the Visual System', *Vision Research,* **8**, 1195
20. Bagrash, F. M., Thomas, J. P. and Shimamura, K. K. (1974). 'Size-Tuned Mechanisms: Correlation of Data on Detection and Apparent Size', *Vision Research,* **14**, 937
21. Kulikowski, J. J. (1969). 'Limiting Conditions of Visual Perception', *Prace Instytutu, Automatkyi PAN., Warsaw,* **77**, 1
22. Kulikowski, J. J. and King-Smith, P. E. (1973). 'Spatial Arrangement of Line, Edge and Grating Detectors revealed by Subthreshold Summation', *Vision Research,* **13**, 1455
23. Gelade, G. A., Poole, C. L. and Beurle, R. L. (1974). 'The Pooling of excitation in Threshold Bar Stimuli', *Vision Research,* **14**, 317
24. Fehrer, E. and Raab, D. (1962). 'Reaction Time to Stimuli masked by Meta-contrasts', *J. Exptl. Psychol.*, **63**, 143
25. Herrick, R. M. (1973). 'Foveal Increment Thresholds for Multiple Flashes', *J. Opt. Soc. Am.,* **63**, 870
26. Clark, W. C. and Blackwell, H. R. (1960). 'Relations Between Visibility Thresholds for Single and Double Pulses', University of Michigan Rep. No. 2144–343–T
27. Gilinsky, A. S. (1968). 'Orientation-specific Effects of Patterns of Adapting Light on Visual Acuity', *J. Opt. Soc. Am.*, **58**, 13
28. Tolhurst, D. J. (1972). 'Adaptation to Square-wave Gratings: Inhibition between Spatial Frequency Channels in the Human Visual System', *J. Physiol.,* **226**, 231
29. Blakemore, C. and Campbell, F. W. (1969). 'On the Existence of Neurones in the Human Visual System selectively sensitive to the Orientation and Size of Retinal Images', *J. Physiol,* **203**, 237
30. Campbell, F. W. and Kulikowski, J. J. (1966). 'Orientation Selectivity of the Human Visual System', *J. Physiol.,* **187**, 437
31. Meeteran, A. Van. (1973). 'Visual Aspects of Image Intensification', Institute for Perception TNO, Soesterberg
32. Fry, G. A. and Enoch, J. M. (1959). 'The Relation of Blur and Grain to the Visibility of Contrast Borders and Gratings', *MCRL TP No. (696)–19–293,* Ohio State University
33. Stromeyer, C. F. III and Julesz, B. (1972). 'Spatial-Frequency Masking in Vision: Critical Bands and Spread of Masking', *J. Opt. Soc. Am.,* **62**, 1221
34. Pollehn, H. and Rohgrig, H. (1970). 'Effect of Noise on the Modulation Transfer Function of the Visual Channel., *J. Opt. Soc. Am.,* **60**, 842
35. Rasmussen, R. A. (1972). 'Effects of Data Quantisation and Display Contrast on Detection in Intensity Modulated Displays', California University, Scripps Institution of Oceanography, Rep. Ref. 72–73

36. Brown, M. B. (1972). 'The Effect of Complex Backgrounds on Acquisition Performance', in *AGARD Conference Proceedings No. 100*, (Ed. H. F. Huddleston), London p. B5–1
37. Williams, L. G. (1966). 'Target Conspicuity and Visual Search', *Human Factors*, 80
38. Williams, L. G. (1967). 'A Study of Visual Search using Eye-Movement Recordings', Honeywell Inc. Rep. No. 12009–IR2
39. Howarth, C. I. and Bloomfield, J. R. (1969). 'A Rational Equation for Predicting Search Times in Single Inspection Tasks', *Psychonomic Science*, 17, 225
40. Howarth, C. I., Bloomfield, J. R. and Dewey, M. E. (1972). 'The Effect of Multiple Targets or of Objects confusible with the Target on Detection Times', in *AGARD Conference Proceedings No. 100*, (Ed. H. F. Huddleston), London, p. B4–1
41. Enoch, J. M. (1959). 'Natural Tendencies in Visual Search of a Complex Display', in *Visual Search Techniques, Publ. No. 172. National Academy of Sciences, National Research Council*, pp. 92, 187, 251
42. Greening, C. P. (1972). 'The Likelihood of looking at a Target', in *AGARD Conference Proceedings, No. 100.* (Ed. H. F. Huddleston), London, p. B1–1

14 Surface Reflectivity

To this point, in all considerations of object/background presentations, it has been assumed, by implication, that whatever the surface luminance structure is, it is invariant with time during presentation. Now, if one considers the laws of reflection, this assumption must imply one of two things – either the viewing and illumination conditions are invariant or all surfaces in the viewed scene are diffuse reflectors*. In a practical situation it is relatively rare for one to be looking at something from an invariant angle and with invariant illumination. On the other hand it is convenient to assume that most natural objects and many man-made objects, seen from a distance, will approximate to diffuse reflectors for modelling purposes. Obvious exceptions are glass and water, which are partially reflecting mirrors and therefore have luminance which is very dependent on incident light distribution. It is the purpose of this chapter to show how real surfaces behave as reflectors and to review the methods available for specifying and measuring surface reflectance.

14.1 GENERAL REFLECTION CHARACTERISTICS

Before considering forms of reflectance and methods of measurement let us look at the reflection characteristics of some natural and man-made surfaces as a function of viewing angle and form of illumination. We shall restrict ourselves for the present to the two simplest forms of illumination – uniform distribution over a hemisphere and a collimated beam. Considerable data are available illustrating the reflection characteristics of many forms of surface for the latter of these conditions (e.g. Coulson *et al*[2,3], Brennan and Bandeen[4], Hodgson and Overton[5] and Crowther[6,7]) and the reflectance under uniform hemispherical illumination may be simply derived by integration. In addition considerable data also exist of reflection from various surfaces under natural hemispherical illumination as measured directly (e.g. Duntley *et al*[8], Boileau and Gordon[9] and Gordon and Church[10,11]).

14.1.1 Reflection from grass and similar vegetation

One very common natural background against which objects are viewed is grass and similar vegetation. It is easy to consider such surfaces to be entirely specified

*A diffuse surface is one which appears of equal luminance from whatever direction it is viewed. In other words it obeys Lambert's cosine law[1]

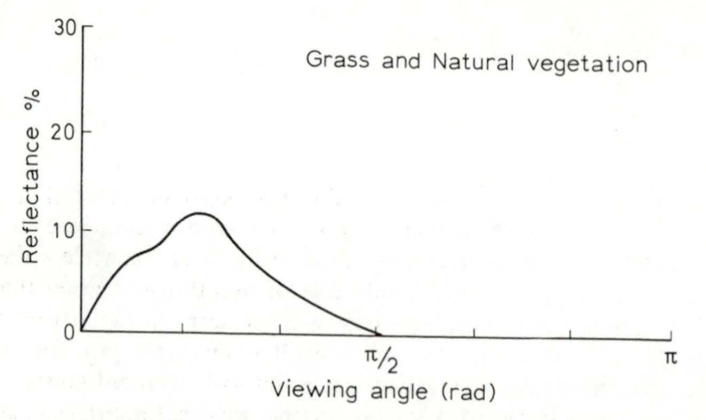

Fig. 14.1. Typical reflectance characteristic for grass. (Reproduced from Hodgson and Overton[5] by courtesy of the Directorate of Stores and Clothing Development).

in terms of mean diffuse reflectance, texture and colour. In practice there are severe changes in reflectance, and hence luminance, dependent on viewing direction. A little thought illustrates that this must be so, for how else can one explain the well known stripes produced on a lawn when it is mown by a roller-mower. The effect is, of course, produced by a layering of the individual grass blades such that their mean angle is other than vertical. This is turn must imply a directional reflectance effect, which is exactly what is recorded when reflectance is studied in detail (see Section 14.3.1). Fig. 14.1 shows a typical reflection characteristic for naturally growing grass as a function of direction of viewing. There is a region of peak reflectance at 0.5 rad viewing elevation, with lesser reflectance at both grazing angles and high elevations. This characteristic peak is due to statistical summation of a multitude of specular components of reflected light at various angles from the shiny surfaces of individual grass blades. Multiple reflections and diffuse reflectance of part of the incident light account for the regions of lower reflectance. It is the statistical summation of components of specular reflectance which is distorted in mean elevation angle by the action of a roller-mower.

14.1.2 Reflection from Foliage

Another major natural background component (or object of interest in some circumstances) is tree foliage. Again, as with grass, it is easy to think that the characteristics are specified by mean luminance, texture and colour, with the addition of broken outline in general. Once again, however, the sum total of reflectance is defined by a mixture of components due to shiny top surfaces and relatively diffuse under surfaces of individual leaves. Since there is a tendency for leaves to align themselves with shiny surfaces upwards, once again there is a predominent direction of peak reflectance defined by the statistical alignment of the leaves. Typical reflectance characteristics for foliage thus have strong similarities to those for grass in Fig. 14.1.

14.1.3 Reflection from soils and rocks

The third major natural background component is the group of surfaces comprising soils and rocks. For this class of surface it would seem, on the face of it, very reasonable to assume an approximately diffuse reflectance characteristic. However, although many soils and rocks do approximate to diffuse surfaces when dry and illuminated predominantly with light at normal incidence, there are marked deviations from the ideal diffuse characteristics for incident light other than near normal incidence. In addition the characteristics of reflectance change markedly when such surfaces are wet. This is due to the water acting as a partially reflecting mirror, with the result that strong lobes of local reflection are introduced, the strength and broadness of the lobes depending largely on the roughness and absorptivity of the surface. In general, the rougher and the more absorbent is the dry surface, the less severe and broader will be the additional reflectance due to the surface wetness.

14.1.4 Reflection from concrete and brickwork

We have just seen how most of the common natural surfaces deviate from being diffuse surfaces. What about man-made surfaces?

One of the commonest modern man-made surfaces encountered in outdoor

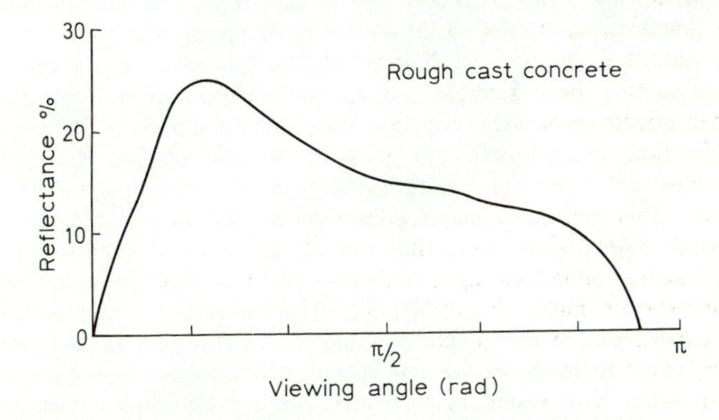

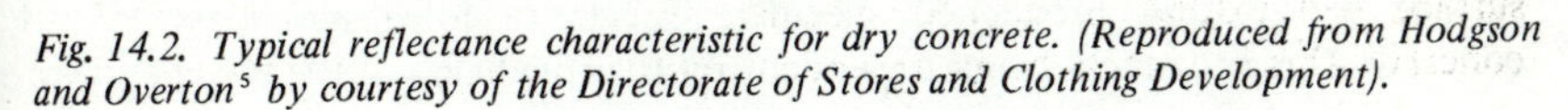
Fig. 14.2. Typical reflectance characteristic for dry concrete. (Reproduced from Hodgson and Overton[5] by courtesy of the Directorate of Stores and Clothing Development).

viewing is concrete. By its appearance one would think that concrete would behave reasonably like a diffuse surface (when dry). However, tests have shown that even here there is a significant departure from the ideal diffuse surface. A typical reflectance characteristic for dry concrete is shown in Fig. 14.2. Brickwork behaves in a similar manner when dry.

If the surface being viewed is wet then the water again acts as a partial reflecting mirror and introduces much stronger lobes of local reflection in the same manner as discussed for soils and rocks.

14.1.5 Reflection from painted and similar surfaces

Other common man-made surfaces are painted surfaces and various plastics. These two types of surfaces have in common a smooth surface with a limited opacity. The very smooth surface will always behave as a partial mirror, the difference between a matt and a gloss surface being largely the difference in degree of surface roughness. Underlying this will be the basic, largely diffuse, reflectance of the underlying opaque surface. Thus once again we have a total reflectance characteristic which contains potentially prominent lobes.

14.1.6 Reflection from glass and water

Both glass and water are characteristically surfaces whose reflectance is specular. Hence, with collimated light falling on a smooth glass or water surface there is a narrow reflection lobe obeying the basic laws of specular reflection from dielectrics (e.g. Born and Wolf[12]). Equally, for smooth surfaces and an absolutely uniform hemispherical illumination, there will be a uniform luminance at all viewing angles. This is fine as far as it goes, but one must remember that, in practice, conditions are often not as simple. For instance, in a typical outdoor situation the natural illumination is frequently far from uniform. Whilst, neglecting direct sunlight, natural non-uniformities in illumination have only small effects on surface reflection from many materials – due to the broad lobe reflection characteristics of partially diffuse random surfaces – with smooth glass and water the non-uniformities in illumination will *totally* define luminance. This can have major effects when glass or water are part of a background. Similarly, in other than natural illumination, unless great care is taken to exclude unwanted light, contributions of incident light from windows, etc., can produce major directional reflectance effects. Such effects are frequently experienced in viewing pictures, meters, etc. through a glass cover.

So far, under this heading, we have specifically considered *smooth* surfaces of glass and water. Now water, in particular, often has a rippled surface. Such a surface will exhibit broadened specular reflectance lobes much as the wet concrete, this broadness being dependent on the roughness of the water surface.

It is interesting to note that, whilst such effects are usually troublesome, they are sometimes useful. An example of this is the recent use of specular lobe broadness to estimate ocean roughness from satellite pictures[13] and other aerial photography (e.g. Cox and Munk[14]).

14.1.7 Polarisation Effects

When considering surface reflection, it is wise to remember that light reflected specularly is, to a greater or lesser extent, polarised[12]. This is particularly true for reflections from dielectrics such as glass and water, where the reflected light may be very highly polarised. Thus the diffuse and specular (or gloss) components of surface reflection can often be at least partially separated out. Where it is essential to minimise angular variations in reflectance the use of a viewing polariser can be very effective. This principle is, of course, used to great effect in polarised sunglasses. It must, however, be realised that, even with dielectric surfaces, total blocking of specular reflection is only achieved at a certain angle (the Brewster angle), some components of the specular reflection remaining unpolarised at other angles[12].

14.2 FORMS OF REFLECTANCE

Having discussed the practical variations of reflectance as a function of viewing angle, type of surface and illumination conditions it is time to consider into what forms reflectance can be broken down. We have already mentioned the two classical forms – specular and diffuse – but these are obviously inadequate as complete descriptors. We require descriptors which can be related to surfaces, illumination and viewing angles.

A common practice is to refer to the overall reflectance characteristics in terms of two components, the basic diffuse reflectance and a *gloss* component, this latter being a broad descriptor of the strength and broadness of any specular reflectance lobe additional to the diffuse reflection characteristic. However, whilst such a pair of components provide a *general* indication of the surface properties, they are still inadequate for rigorous description of surface luminance properties under complex illumination conditions.

A thorough treatment of the problem of specifying reflectance is that due to Nicodemus[15]. He derives an expression defining a fundamental property which he calls *bidirectional reflectance* and also illustrates its meaning.

Consider a small source of intensity I_i in the direction of a surface element Δa. The source subtends a solid angle

$$\Delta\Omega_i = \sin\psi_i \Delta\psi_i \Delta\theta_i$$

at Δa, where ψ is the angle from the surface normal and θ is the azimuth angle in

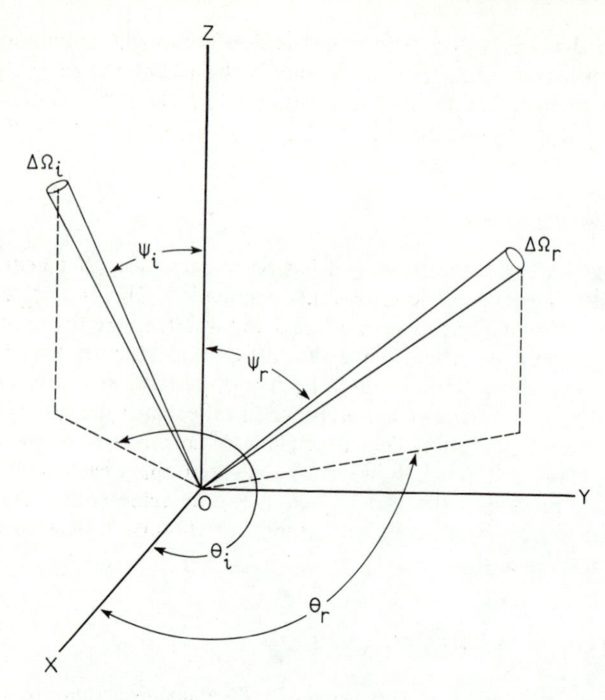

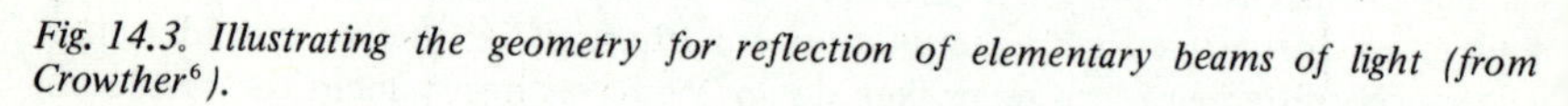
Fig. 14.3. Illustrating the geometry for reflection of elementary beams of light (from Crowther[6]).

the plane of the surface (see Fig. 14.3). The luminous flux incident on Δa is then

$$\Delta P_i = I_i \cos \theta_i \Delta a \Delta \Omega_i$$

or

$$\Delta P_i = I_i \Delta a \Delta \Omega_i' \tag{14.1}$$

where

$$\Delta \Omega_i' (\equiv \cos \psi_i \Delta \Omega_i \equiv \sin \psi_i \cos \psi_i \Delta \psi_i \Delta \theta_i)$$

is the 'projected' solid angle of the elementary incident beam.

Next consider the reflected light in a direction ψ_r, θ_r within a small element

$$\Delta \Omega_r = \sin \psi_r \Delta \psi_r \Delta \theta_r$$

The luminous power is then

$$\Delta P_r = \Delta I_r \Delta A \Delta \Omega_r' \tag{14.2}$$

where ΔI_r is the reflected light (due to the elementary source) in the direction ψ_r, θ_r and $\Delta\Omega_r' \equiv \cos\psi_r \Delta\Omega_r$.

The reflectance is then given as

$$\rho = \frac{\Delta P_r}{\Delta P_i} = \frac{\Delta I_r \Delta\Omega_r'}{I_i \Delta\Omega_i'} \tag{14.3}$$

If $\Delta\Omega_i$ is small, then ΔI_r is proportional to I_i and $\Delta\Omega_i'$, so ρ is invariant with respect to I_i and $\Delta\Omega_i'$. However, ρ is proportional to the solid angle of the receiver system, and a more appropriate measure would be

$$\rho' = \frac{\rho}{\Delta\Omega_r'} = \frac{\Delta I_r}{I_i \Delta\Omega_i'}$$

or

$$\rho'(\theta_i \psi_i \theta_r \psi_r) = \frac{\Delta I_r}{\Delta E_i} \tag{14.4}$$

where ΔE_i is the illuminance at Δa produced by the source. Strictly, the bidirectional reflectance should be written $\rho'(\theta_i \psi_i \theta_r \psi_r \lambda)$ to include the wavelength dependence. If the incident light is well-collimated within an element of solid angle $\Delta\Omega_i$, then the *total* reflectance from a surface, the directional-hemispherical reflectance ρ_{dh}, is given by

$$\rho_{dh}(\theta_i \psi_i) = \int_h \rho'(\theta_i \psi_i \theta_r \psi_r) d\Omega_r' \tag{14.5}$$

where the integration is carried out over the upward (or reflected) hemisphere.

In the special case of a perfectly diffuse reflector,

$$\rho_{dh} = \rho' \int_h d\Omega_r' = \pi\rho'$$

Therefore, because $\rho_{dh} = 100\%$, $\rho' = 31\%$. (Some authors in fact plot graphs of ρ' on a scale multiplied by a factor π).

Albedo, on the other hand, which should be termed the hemispheric-hemispheric reflectance since it is the proportion of incident illumination which is reflected, is the integration of ρ' over the downward (incident) and upward hemispheres.

Finally the reflectance in a given direction from a generally distributed incident light, the hemispherical-directional reflectance ρ_{hd} is the integration of ρ' over the downward hemisphere.

It can thus be seen that one may obtain any reflectance data from a knowledge of the bidirectional reflectance and the hemispherical distribution of incident illumination.

14.3 PRACTICAL REFLECTANCE DATA

14.3.1 Bidirectional Reflectance

The only reliable data available for bidirectional reflectance are a limited number derived from laboratory measurements, where incident and reflected directions are closely controlled. Only a few materials have been thoroughly studied in this way, these mainly comprising samples of soil, sand, grass and a few artificial surfaces.

Soils and rocks. The angular distribution of bidirectional reflectance for light of wavelength 0.643 μm incident on samples of sand and soil has been measured by Coulson[2,3]. None of the samples is perfectly diffuse, although for small

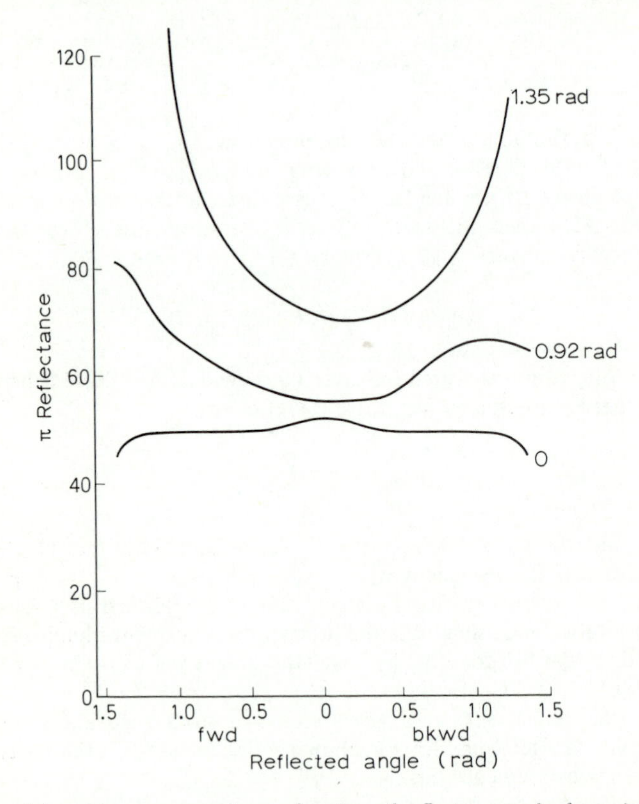

Fig. 14.4. ρ' for white quartz sand, as a function of reflectance angle, for various angles of incidence at a wavelength of 0.643 μm. (Reproduced from Coulson et al[3] *by permission. Copyright by the American Geophysical Union).*

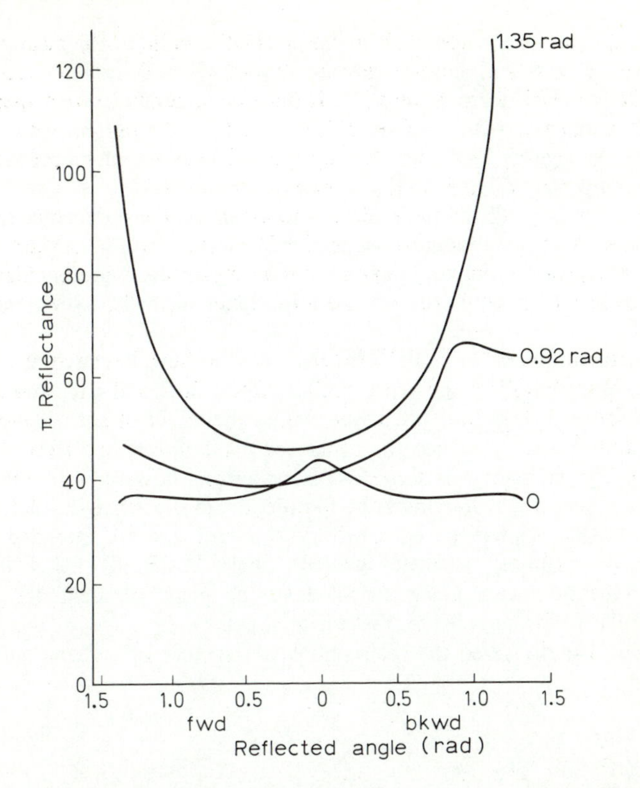

Fig. 14.5. ρ' for red clay soil, as a function of reflectance angle, for various angles of incidence at a wavelength of 0.643 μm. (Reproduced from Coulson et al[3] *by permission. Copyright by the American Geophysical Union).*

angles of incidence (relative to the surface normal) all samples approximate this ideal case. Each sample exhibits backscatter (also called back-gloss) of an amount which increases with incident angle, (see Figs. 14.4 and 14.5 for example). Forward scattering also develops with increasing incident angle, leaving a minimum reflectance in the normal direction. Generally, at any given viewing direction, the amount of reflected light increases with the incident angle. At the larger incidence angles (over about 0.7 rad) the characteristics of the samples begin to differ. Forward scattering becomes greater than backscatter for both sand samples (desert sand and white quartz sand), whereas the soils (red clay and black loam) exhibit the opposite trend. In fact for black loam the forward scattering is almost non-existent.

The relative magnitudes of the bidirectional reflectance of these four samples (for radiation 0.643 μm) can be illustrated by comparing the values found under normal incidence. The bidirectional reflectance is almost constant with viewing direction, except for a slight peak in the normal direction. The mean values for quartz sand, desert sand and red clay are about 16%, 10% and 13% respectively, whilst that for black loam is only 2%. It must be remembered that these figures should be compared with a value of $100/\pi = 31\%$ for a perfect white diffuser. The graphs in Figures 14.4 and 14.5 are plotted in terms of πx reflectance, on this scale the perfect diffuser having a reflectance of 100%.

Coulson has investigated the effects of moisture on the reflectance of soil[3]. A considerable change of angular dependence occurs as the surface becomes wetter. The forward reflection maximum of a clay surface is greatly increased by the surface moisture, while the relative importance of the backscattering peak is decreased.

Measurements made at SCRDE (Stores and Clothing Research and Development Establishment)[5,16] showed a similar effect. Both soil and sand exhibited increased forward reflection when wet, whilst the effect of surface moisture on rough concrete was to reduce the backward maximum to the level of forward scattering. The geometry of the SCRDE system is, however, not very closely defined and does not strictly measure bidirectional reflectance. In fact, measurements are made relative to an arbitrary standard and are intended only for comparative purposes, a single incident angle (1.05 rad) being used, and reflection measurements being confined to the plane of incidence, which is perpendicular to the surface (i.e. the principal plane).

Coulson[2] has measured the bidirectional reflectance of soil and sand outside

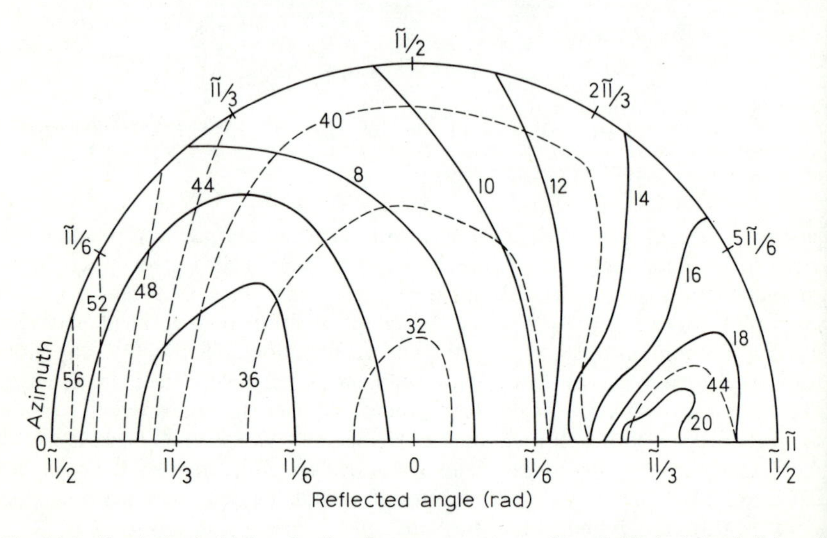

Fig. 14.6. Hemispheric bidirectional reflectance patterns of sand (dashed isopleths) and black loam soil (solid isopleths). Values shown are π x reflectance. Incidence angle 0.92 rads; $\lambda = 0.643$ μm. (Reproduced from Coulson[2] by courtesy of the Optical Society of America).

the plane of incidence. Radiation of wavelength 0.643 μm was incident at 0.92 rad. For this incidence condition, complete contour maps of bidirectional reflectance (isopleths) are available (see Fig. 14.6). The patterns of isopleths for both soil and sand samples are symmetrical about the plane of incidence, maxima and minima occuring in this plane. The curvatures of the isopleths reflect the relative forward and backward scattering properties of these two samples.

Vegetation. The bidirectional reflectance of grass has been measured for four incident wavelengths by Coulson[2]. The reflectance values are probably sensitive

TABLE 14.1. ρ' of turf: principal plane

Source elev.	Viewing elevation ϕ_r(rad)					
ϕ_i(rad)	0.07	0.105	0.14	0.175	0.21	0.245
0.26	0.0273	0.0264				
0.35			0.0255	0.0273		
0.435	0.0164	0.0182			0.0305	0.0296
0.52			0.0209	0.0228		
0.61	0.0100	0.0124			0.0236	0.0276
0.70			0.0157	0.0200		
0.785	0.0069	0.0091			0.0182	0.0240
0.87			0.0126	0.0173		
0.96	0.0047	0.0073			0.0149	0.0201
1.04			0.0113	0.0155		
1.13	0.0040	0.0066			0.0121	0.0175
1.21			0.0106	0.0142		
1.30	0.0035	0.0060			0.0102	0.0156
1.39			0.0096	0.0137		
1.48	0.0031	0.0047			0.0095	0.0140
1.56			0.0093	0.0124		
1.65	0.0034	0.0032			0.0104	0.0138
1.74			0.0109	0.0127		
1.83	0.0035	0.0032			0.0110	0.0154
1.92			0.0118	0.0135		
2.00	0.0040	0.0037			0.0138	0.0162
2.09			0.0126	0.0142		
2.18	0.0042	0.0047			0.0162	0.0169
2.26			0.0135	0.0147		
2.35	0.0055	0.0058			0.0197	0.0195
2.44			0.0158	0.0155		
2.53	0.0069	0.0064			0.0217	0.0219
2.61			0.0180	0.0167		
2.70	0.0104	0.0104			0.0236	0.0232
2.79			0.0175	0.0178		
2.88	0.0150	0.0158			0.0217	0.0219
2.96			0.0191	0.0191		
3.05	0.0177	0.0149			0.0134	0.0130

TABLE 14.2. ρ' of turf: outside principal plane

Incident elevation ϕ_i(rad)	Azimuth difference θ(rad) 0	$\pi/6$	$\pi/3$	$\pi/2$	$2\pi/3$	$5\pi/6$	π
				ϕ_r = 0.07 rad			
$\pi/12$	0.0220	0.0190	0.0113	0.0085	0.0110	0.0130	0.0145
$\pi/6$	0.0140	0.0125	0.0088	0.0065	0.0070	0.0075	0.0076
$\pi/4$	0.0086	0.0077	0.0065	0.0045	0.0049	0.0051	0.0053
$\pi/3$	0.0074	0.0071	0.0055	0.0031	0.0034	0.0036	0.0038
$5\pi/12$	0.0057	0.0047	0.0037	0.0025	0.0033	0.0041	0.0053
$\pi/2$	0.0034						
				ϕ_r = 0.105 rad			
$\pi/12$	0.0264	0.0208	0.0119	0.0103	0.0121	0.0140	0.0158
$\pi/6$	0.0153	0.0130	0.0098	0.0075	0.0079	0.0082	0.0084
$\pi/4$	0.0091	0.0080	0.0068	0.0050	0.0052	0.0055	0.0058
$\pi/3$	0.0082	0.0076	0.0058	0.0035	0.0038	0.0040	0.0042
$5\pi/12$	0.0060	0.0051	0.0040	0.0030	0.0032	0.0039	0.0058
$\pi/2$	0.0039						
				ϕ_r = 0.14 rad			
$\pi/12$	0.0335	0.0264	0.0139	0.0125	0.0136	0.0146	0.0183
$\pi/6$	0.0209	0.0153	0.0132	0.0110	0.0128	0.0142	0.0180
$\pi/4$	0.0142	0.0100	0.0085	0.0075	0.0087	0.0103	0.0147
$\pi/3$	0.0113	0.0100	0.0092	0.0080	0.0098	0.0100	0.0126
$5\pi/12$	0.0101	0.0090	0.0074	0.0065	0.0079	0.0099	0.0114
$\pi/2$	0.0093						
				ϕ_r = 0.175 rad			
$\pi/12$	0.0355	0.0285	0.0183	0.0142	0.0168	0.0166	0.0185
$\pi/6$	0.0228	0.0180	0.0150	0.0135	0.0142	0.0157	0.0167
$\pi/4$	0.0187	0.0156	0.0138	0.0120	0.0128	0.0138	0.0151
$\pi/3$	0.0155	0.0138	0.0121	0.0111	0.0119	0.0132	0.0142
$5\pi/12$	0.0139	0.0121	0.0109	0.0098	0.0106	0.0121	0.0131
$\pi/2$	0.0124						
				ϕ_r = 0.21 rad			
$\pi/12$	0.0356	0.0290	0.0187	0.0155	0.0175	0.0185	0.0198
$\pi/6$	0.0260	0.0210	0.0175	0.0155	0.0170	0.0180	0.0200
$\pi/4$	0.0200	0.0185	0.0165	0.0134	0.0145	0.0170	0.0180
$\pi/3$	0.0160	0.0145	0.0130	0.0120	0.0128	0.0140	0.0152
$5\pi/12$	0.0133	0.0126	0.0113	0.0100	0.0113	0.0126	0.0132
$\pi/2$	0.0129						
				ϕ_r = 0.245 rad			
$\pi/12$	0.0360	0.0280	0.0200	0.0185	0.0192	0.0209	0.0213
$\pi/6$	0.0285	0.0240	0.0195	0.0175	0.0188	0.0207	0.0216
$\pi/4$	0.0214	0.0203	0.0183	0.0148	0.0162	0.0188	0.0197
$\pi/3$	0.0175	0.0162	0.0145	0.0134	0.0142	0.0156	0.0168
$5\pi/12$	0.0138	0.0132	0.0117	0.0105	0.0119	0.0131	0.0136
$\pi/2$	0.0135						

to the condition of the sample. Coulson's measurements on a thick grass sample with upright blades 4 or 5 cm long show that the bidirectional reflectance generally increases with incident angle, with the backscattering peak developing more quickly than that of forward scattering. The results are not greatly different from those of the desert sand. With a sample of a broad-leafed, rather waxy plant, however, some particularly strong changes with both incident and viewing angle were observed. A third sample of grass, measured by SCRDE[16], showed prominent backscattering but very little forward scattering. SCRDE also measured two types of leaves. These samples were found to be very similar to the grass, except that bidirectional reflectance at small reflected angles was a little higher.

All the foregoing data concerning vegetation refer to measurement in the plane of incidence. The general features are the pronounced backscattering and smaller amount of forward scattering, these peaks becoming more prominent with increased angle.

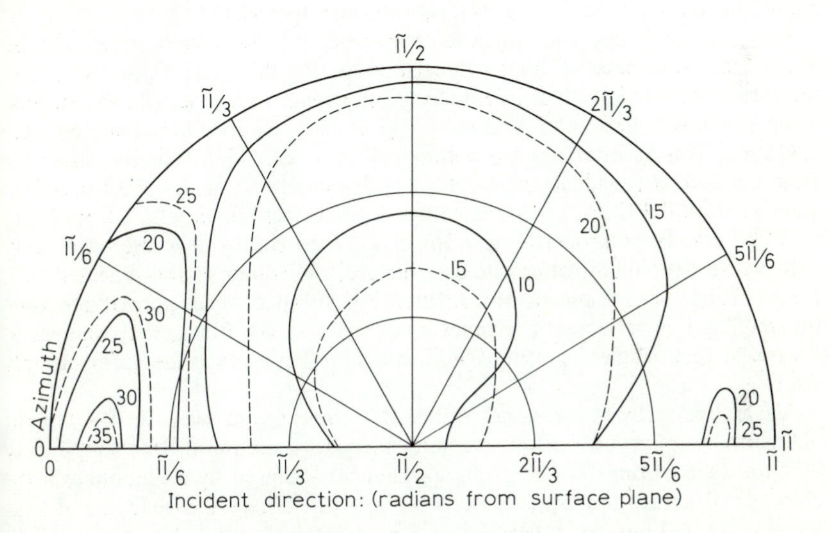

Fig. 14.7. Contours of bidirectional reflectance, (steradian)$^{-1}$ × 10^3, of turf as a function of incident and azimuth directions for two reflected angles – 0.14 rad (solid isopleths) and 0.245 rad (dashed isopleths), measured from the surface plane (from Crowther[6]).

Crowther[6] has measured the bidirectional reflectance of turf, both in the plane of incidence and outside it, for white light at a wide range of reflected angles. The results are given in Tables 14.1 and 14.2 for reflectance in the plane of incidence and outside it respectively. Typical isopleths are shown in Fig. 14.7. Field measurements[8,11], in which the sun defines the incident direction, exhibit similar features, but effects are softened due to skylight.

Snow and Ice. Middleton and Mungall[17] have measured the reflectanc properties of several snow and ice surfaces. The types of snow may be listed i order of increasingly large specular component as follows: surface hoar, settlin snow, new snow, rain crust, wind packed snow, glazed rain crust (ice). Apar from ice surfaces, for which virtually all the reflected light appears at th specular angle, diffuse reflection contributes significantly at small incidenc angles. This is particularly so for the least compacted surfaces.

14.3.2 Spectral Characteristics

The dependence of bidirectional reflectance on the wavelength of the inciden light has been determined for a few surfaces[2,3]. The measurements wer confined to angles in the plane of incidence, and generally show an increase i bidirectional reflectance with wavelength through the visible spectrum.

For bare soil and sand surfaces, increasing the wavelength raises the bi directional reflectance without appreciably altering the angular distribution. I the case of quartz sand it is essentially proportional to wavelength, the increas being about 16% between 0.492 and 0.643 μm and the same between 0.643 an 0.796 μm. This accounts for the yellow colour of sand. For red clay, under th same conditions (incident angle 0.92 rad), the change is of the order of a 25% increase from 0.492 to 0.643 μm, with a small decrease of about 5% from 0.64 to 0.796 μm. These properties give the clay its red colour. It is interesting that with white light illuminating the clay surface, the colour of the reflected ligh would change with the angle of incidence. For instance, in the normal direction the red light is reflected 2.5 times as strongly as the blue-green, whereas a 1.4 rad incident angle the ratio of bidirectional reflectances at the normal is only 1.8.

Vegetative surfaces are rather different from soil and sand. At 0.4 μm th bidirectional reflectance is low, reaching a small maximum in the region 0.5 to 0.55 μm (accounting for the green appearance) followed by a minimum in th chlorophyll absorption band centred on 0.68 μm. There is a sudden and larg increase at 0.7 μm to a level which is maintained into the infrared. Th angular distribution of bidirectional reflectance is, unlike that for soil and sand dependent on wavelength. At wavelengths greater than about 0.7 μm the forwar and backward scattering maxima are greatly enhanced.

The wavelength dependence of bidirectional reflectance means that any ligh source for laboratory measurements should be chosen with care. In a laborator simulation of natural reflectance properties the source of illumination shoul have approximately the same spectral composition as daylight. As the spectra content of skylight varies with position in the sky, weather conditions and tim of day, this simulation is difficult to achieve.

14.3.3 Directional-Hemispherical Reflectance

If, for a particular incident angle, the bidirectional reflectance is integrated over the upward hemisphere, the directional-hemispherical reflectance is obtained. This parameter serves as a useful comparison between different surfaces.

Kondratyev[18] claims that the directional-hemispherical reflectance of soils varies between about 5% and 45%, compared with a value of 100% for a perfect diffuser. The higher values are found for sands and dry soils, whilst the dark, heavier soils have the lower reflectances. In general, the smoother the surface (on the large scale) the higher the total reflectance. For example, a flat field reflects more than does a ploughed one. The directional-hemispherical reflectance of vegetation lies roughly between 10 and 25%, a smaller range than that of soil. The directional-hemispherical reflectances of all these surfaces increase with the angle of incidence. It should be noted that the term albedo strictly refers to the proportion of the total incident illumination which is reflected, and is derived from ρ' by integration over both downward and upward hemispheres. It is not very useful for comparative purposes because some distribution of illumination is implied.

14.3.4 Polarisation

For completeness, it is interesting to look briefly at the polarisation of reflected light. Natural surfaces exhibit their own characteristic polarisation effects. Coulson[3] summarises the data existing in 1965. More recent work is that of Chen and Rao[19]. The highest values of polarisation (70% to 95%) have been found for dark mineral surfaces. High values (65% to 70%) have also been observed for wet clay surfaces, whilst most dry soils and fine-grained materials exhibit low polarisation ($< 10\%$) at most angles. Black loam soil (up to 20%) is rather an exception. The behaviour of apparently smooth water surfaces can only be described as scattering by an 'optically rough, locally smooth' surface, according to Chen and Rao. In the light of the above, measured contrast when viewed by a polarisation sensitive 'receiver' will certainly be dependent on the plane of polarisation to which the 'receiver' is sensitive, a fact which may have important implications on acquisition tasks in the field.

14.3.5 Symmetry

Coulson[3] found the patterns of bidirectional reflectance of desert sand and black loam to be symmetrical about the plane of incidence. It seems likely that this is true of flat, amorphous surfaces in general. The data for grass have a larger scatter, although Coulson does not note any asymmetry. If the vegetation

TABLE 14.3. Summary of terrain reflectance

Surface	Bidirectional reflectance	Spectral characteristics	Total reflectance
Soils, sand, etc.	Back and forward scattering Sand has larger forward scattering Loam has small forward scattering Moisture increases forward and reduces back-scattering	Increasing to 1 μm Decreasing above 2 μm	5% to 45% Moisture decreases reflectance by 5% to 20% Smoother surfaces have higher reflectances
Vegetation	Backscattering Small forward scattering	Small below 0.5 μm Peak at 0.5 to 0.55 μm Chlorophyll absorption at 0.68 μm Sharp increase at 0.7 μm Decrease above 2 μm Variation with growing season	5% to 25% Diurnal effects; max. reflectance at large angles (from normal Marked annual variatior
Water	Back and forward scattering, large at large angles (from normal)	Maximum at 0.5 to 0.7 μm Dependent on turbidity	Small reflectance Maximum at grazing angles Dependent on turbitidy
Snow and Ice	Diffuse component plus specular component Specular component increases with incidence angle	Decreases slightly with increasing wavelength Large variability depending on purity, wetness and physical condition	Variable, 25%–100%

(leaves, grass blades, etc.) is randomly oriented, the average bidirectional reflectance, measured over a large enough area, should be symmetrical. If the distribution is not random, for example crops leaning in the wind, there will probably be some asymmetry about the incident plane. The striped effect of recently mown grass previously mentioned is a good illustration.

A further symmetry consideration arises from the reciprocity theorem[2]. If this theorem applies, then a reversal of incident and reflected directions leaves the bidirectional reflectance unchanged. The surfaces for which this theorem is valid are not known at present, though it probably does not hold for many natural surfaces.

A summary of the reflectance properties of the main surfaces for which data are available is to be found in Table 14.3.

14.4 MEASUREMENT OF REFLECTANCE

The method of measurement of reflectance employed must depend on the use to which the measurements are to be put. If a practical measurement of reflectance for normal viewing in a stable environment is required, then it may be simplest to measure the hemispherical-directional reflectance as a function of viewing angle, with the integrated light appropriate to the situation. If, however, the incident light conditions are not stable, it is necessary to measure the bi-directional reflectance as a function of angle of incident light. Then, knowing or estimating the distribution of incident light, the hemispherical-directional reflectance may, in theory, be computed for any given condition.

14.4.1 Reference Surfaces

In any method of reflectance measurements, one of the prime requisites is a standard reference surface. Ideally such a surface should be completely diffuse and should reflect 100% of the incident light. In practice there are no such surfaces. The closest approach is possibly a sublimation of magnesium oxide, but such a surface is very easily damaged and deteriorates quickly. Considerable efforts have been made to produce durable reference surfaces with not very great success. The nearest to a standard reference for everyday use is a ceramic tile of defined surface properties which can be calibrated at some Standards Bureau (such as the National Physical Laboratory in the UK or the National Bureau of Standards in the USA) (e.g. Crowther[6]). Such standard tiles can now be obtained in white or several standard colours.

14.4.2 Measurement of Bidirectional Reflectance

In order to measure bidirectional reflectance of a surface, it is necessary to have control of the direction of incident light and, ideally, the incident light should be collimated. It is thus hardly feasible to measure bidirectional reflectance out of doors. Even indoors the requirements are sufficiently complex for it to be uneconomic to try to make *ad hoc* measurements. A much more satisfactory approach is to construct a facility with the required degrees of freedom and bring samples to it. This is, in fact, what has been done at BAC (GW), amongst a small number of other establishments.

14.4.3 The BAC Polar Reflectometer

The BAC Reflectometer[6] is shown schematically in Fig. 14.8. It comprises a specimen table, adjustable in azimuth and elevation, a source mirror which can

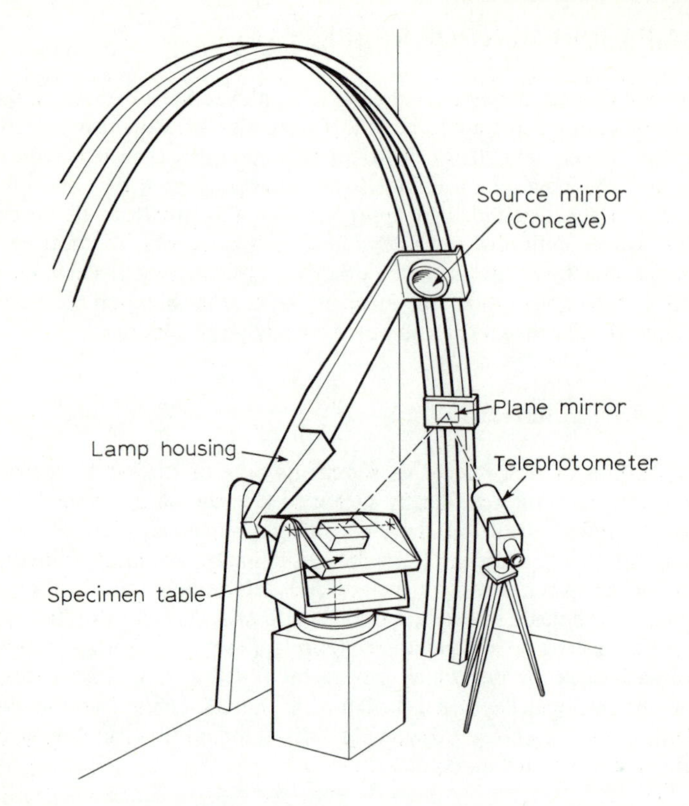

Fig. 14.8. Schematic representation of the B.A.C. (GW) Polar Reflectometer.

be moved in a fixed vertical arc (4 m diameter) centred on the sample, and a receiver mirror also constrained to move in the same arc. The specimen table ha a continuous adjustment and can be set with an accuracy of 1 or 2 mrad. The receiver elevation can be similarly adjusted, whilst the elevation of the source mirror is limited to intervals of 87 mrad.

The source is a quartz iodine lamp having a colour temperature around 3400°K. Provision is made for filtering the incident beam (with, for example polarising or colour temperature correction filters). The specimen table i illuminated with a parallel beam of light by locating the source in the focal plane of a 20 cm diameter concave mirror positioned on the vertical arc. A telephoto-meter is used to collect the reflected light after further reflection at a plane front-silvered mirror, which is also mounted on the arc.

Incident and reflected angles can be varied independently over a wide range o angles above the surface, provided the sample is of solid material. For loose samples, such as sand, the range of angles is limited by the amount of til

•ermitted, and for liquids the plane of incidence and reflectance can only lie •erpendicular to the surface.

In a typical measurement program the light source is filtered so that the :olour temperature is about 5500°K, which corresponds approximately to mean ıatural daylight illumination. The receiver system is also filtered so that its esponse is photopic.

The field of view of the telephotometer can be varied to suit the nature of the ample. For homogeneous surfaces a probe with an acceptance angle of l.75 mrad is generally used, whereas samples which have larger scale surface rregularities, such as grass, require a 17.5 mrad probe. This does not affect the neasured value of ρ', providing it is constant over the 17.5 mrad range. A standard ile must, of course, be measured with the corresponding probe to maintain the :orrect calibration factor. Calibration checks are, in fact, frequently carried out hroughout the measurement program.

4.4.4 Computation of Reflectance in Field Conditions

t is all well and good to have reliable data for the distributed bidirectional eflectance of various surfaces but, unless we know what forms of hemispherical llumination distribution typically exist in a field situation, and can compute ntereactions with the surface reflectances, the data are of very limited practical ıse. In an effort to rectify the situation, BAC (GW) have undertaken large scale tatistical measurements of sky luminance distributions as a function of time of lay, time of year and weather[20,21]. This work has provided a large data bank 'or statistical prediction of the hemispherical incident illumination distribution. At the same time a computer program, 'Polar Bear', has been written which akes statistical sky luminance data and bidirectional reflectance data and :omputes hemispherical-directional reflectances for all viewing angles[6].

14.5 IMPLICATIONS ON MODELLING

The important conclusion from the results of studies reported in this chapter is :hat, in any visual viewing situation where incident lighting and viewing angle are ıot absolutely stable, there is no such thing as basic local scene luminance. Each ocal portion of the scene – object of interest and background – will have its own characteristic law of luminance variation with time as the viewing con- litions change. We at BAC (GW) have found to our cost that such local temporal uminance signatures upset markedly the modelling of threshold trends as a function of range, since they can and do interact strongly with known range lependent factors such as size and atmospheric effects (see particularly Chapter 15). They also interfere with attempted photometry to establish atmospheric conditions.

TABLE 14.4. Model clear sky luminance distribution (cd/m^2) for a solar elevation $\pi/4$ rads, azimuth 0 rads

	Elevation (rad)						
Azimuth (rad)	0	$\pi/12$	$\pi/6$	$\pi/4$	$\pi/3$	$5\pi/12$	$\pi/2$
0	16 580	13 840	18 980	12 000	5 890	3 347	2 145
$\pi/4$, $7\pi/4$	9 250	6 990	5 824	5 070	3 940	2 994	–
$\pi/2$, $3\pi/2$	5 000	3 238	2 415	2 107	2 035	2 151	–
$3\pi/4$, $5\pi/4$	4 385	2 597	1 658	1 357	1 364	1 597	–
π	4 728	2 665	1 569	1 213	1 179	1 466	–

An example of the effects of polar luminance variations when viewing the ground obliquely from a low flying aircraft has been studied by Crowther[6] using the 'Polar Bear' computer program. The necessary input for this exercise comprised the set of bidirectional reflectance data for turf given in Tables 14.1 and 14.2 and a sky luminance distribution. Two different sky types have been used; a clear sky, with the sun at $\pi/4$ rads elevation, and a complete overcast. The clear sky model was based on the measurements of Jones and Condit[22] and the sky luminance values are set out in Table 14.4. Direct sunlight adds 58 850 lm/m^2 to the ground plane illumination. The overcast sky was modelled on Moon and Spencer's formula[23]

$$B_\phi = B_h(1 + 2 \sin \phi)$$

in which B_ϕ is the sky luminance at elevation angle ϕ. The scaling factor B_h, which equals the horizon sky luminance, was given the value 1000 cd/m^2.

TABLE 14.5. Reflected luminance of turf as a function of viewing direction under clear and overcast sky models.

		Luminance (cd/m^2)					
			Clear (azimuth indicated)				
Reflected angle (rad)	Range (km)	Overcast	0	$\pi/4$, $7\pi/4$	$\pi/2$, $3\pi/2$	$3\pi/4$, $5\pi/4$	π
0.070	8.7	47.7	619	519	350	379	398
0.087	7.0	49.7	639	533	370	393	417
0.105	5.8	51.7	659	546	389	408	436
0.122	5.0	66.6	834	622	485	557	730
0.140	4.3	81.4	1009	698	581	705	1023
0.156	3.85	91.7	1159	877	730	831	1046
0.174	3.5	101.9	1309	1055	879	958	1068
0.192	3.15	107.2	1357	1147	930	1040	1165
0.210	2.90	112.5	1405	1240	980	1121	1261
0.227	2.63	117.9	1455	1302	1031	1183	1320
0.244	2.45	123.4	1506	1365	1082	1245	1380

Table 14.5 contains all the computed luminance values under both clear and overcast skies as a function of the viewing direction. This direction has also been converted into effective slant range for an aircraft flying at 600 m. The luminances due to a clear sky are dependent on azimuth differences between the sun and the viewing direction. Therefore, values have been computed for a variety of azimuth directions relative to the sun. The underlying reflectance characteristics are clearly evident: backscatter is somewhat greater than forward scatter with a minimum at right angles to the sun's direction.

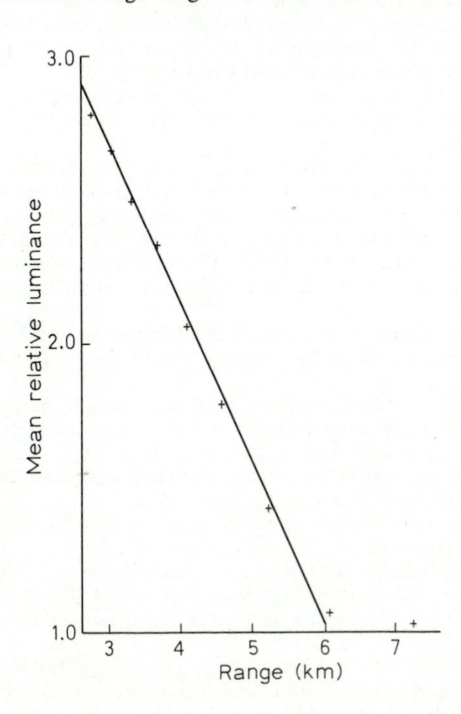

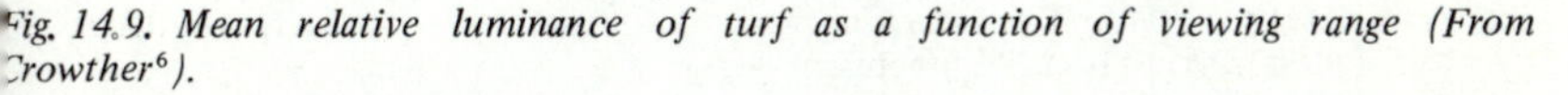
Fig. 14.9. Mean relative luminance of turf as a function of viewing range (From Crowther[6]).

If a mean relative luminance against viewing angle is determined then the esult, on plotting against range, is as shown in Fig. 14.9. Between 2.5 and 5.5 km the relative luminance can be seen to be approximated by a straight line.

This graph is of course *only* applicable to the sample of turf measured, but does serve to illustrate how markedly the intrinsic luminance of turf is dependent on viewing range when viewed from an aircraft at constant altitude. In practice this intrinsic luminance will itself be modified as a range function due to the effects of atmospheric attenuation to be discussed in the next chapter.

REFERENCES

1. Born, M. and Wolf, E. (1964). *Principles of Optics*, Chap. 4.8. Pergamon Press
2. Coulson, K. L. (1966). 'Effects of Reflection Properties of Natural Surfaces in Aeria Reconnaissance', *Applied Optics,* **5**, 905
3. Coulson, K. L., Bouricius, G. M and Gray, E. L. (1965). 'Optical Reflection Propertie of Natural Surfaces', *J. of Geophys. Res.,* **70**, 4601
4. Brennan, B. and Bandeen, W. R. (1970). 'Anisotropic Reflectance Characterisitics o Natural Earth Surfaces', *Applied Optics,* **9**, 405
5. Hodgson, E. W. and Overton, T. K. W. (1965). 'Texture and its Significance ir Camouflage', Directorate of Stores and Clothing Development, MOD
6. Crowther, A. G. (1972). 'Polar Luminance and Contrast', Annex A of *Final Report o Third Visual Studies Contract*, BAC (GW) Ref. L50/196/1535
7. Crowther, A. G. (1971). 'A note on Comparative Measurement of the Bidirectiona Reflectance of Black Nylon Velvet and Matt Black Paint' BAC (GW) Tech. Memo No. 7, Ref. L50/249
8. Duntley, S. Q., Gordon, J. I., Taylor, J. H., White, C. T., Boileau, A. R., Tyler, J. E. Austin, R. W. and Harris, J. L. (1964). 'Visibility', *Applied Optics,* **3**, 556
9. Boileau, A. R. and Gordon, J. I. (1966). 'Atmospheric Properties and Reflectances o Ocean Water and Other Surfaces for a Low Sun', *Applied Optics,* **5**, 803
10. Gordon, J. I. and Church, P. V. (1966). 'Overcast Sky Luminances and Directiona Luminous Reflectances of Objects and Backgrounds under Overcast Skies', *Applie Optics.,* **5**, 919
11. Gordon, J. I. and Church, P. V. (1966). 'Sky Luminances and Directional Luminou Reflectances of Objects and Backgrounds for a Moderately High Sun', *Applied Optics* **5**, 793
12. Born, M. and Wolf, E. (1964). *Principles of Optics*, Chap. 1.5, Pergamon Press, Londo
13. Webber, D. S. (1971). 'Surface Winds from Sun-glitter Measurements from a Space craft', in *Proceedings of SPIE, Vol. 27, Remote Sensing*, 93
14. Cox, C. and Munk, W. (1954). 'Measurement of the Roughness of the Sea Surface fro Photographs of Sun's Glitter', *J. Opt. Soc. Am.,* **44**, 838
15. Nicodemus, F. E. (1965). 'Directional Reflectance and Emissivity of an Opaqu Surface', *Applied Optics,* **4**, 767
16. Littlefield, T. A. 'Distribution of Light due to Texture', Ministry of Hom Security – Research and Experimental Dept., Rep. REN 477
17. Middleton, W. E. K. and Mungall, A. G. (1952). 'The Luminous Directional Reflectanc of Snow', *J. Opt. Soc. Am.,* **42**, 572
18. Kondratyev, K. (1965). 'Actinometry', NASA TTF–9712, Washington DC, Novembe
19. Chen, H. and Rao, C. R. N. (1968). 'Polarisation of Light on Reflection by som Natural Surfaces', *Brit. J. Appl. Phys. D.,* **1**, 1191
20. Brown, M. B. (1971). 'The Measurement of Sky Brightness – Final Report', BAC (GW Ref. L50/22/PHY/114/1296, February
21. Brown, M. B. (1974). 'The Measurement of Sky Radiance – Final Report', BAC (GW Ref. ST11930
22. Jones, L. A. and Condit, H. R. (1948). 'Sunlight and Skylight as Determinants o Photographic Exposure', *J. Opt. Soc. Am.,* **38**, 123
23. Moon, P. and Spencer, D. E. (1942). 'Illumination from a Non-uniform Sky', *Illum Eng. (NY),* **37**, 707

5 Atmospheric Attenuation

ı the foregoing chapters all the discussions have made the implicit assumption ıat the atmosphere between the object being viewed and the observer is ptically inert and perfect (i.e. it does not affect the luminance, contrast or uality of the object or scene of interest in any way). Whilst this is usually ısumed to be true in controlled laboratory experiments, it is readily seen not to e true in some outdoor viewing situations. For instance, when viewing over long aths it is readily observed that the prevailing meteorological visibility modifies ıe scene contrast as a function of range. This is, after all, why there is such a ıing as meteorological visibility. Similarly, on a hot day, it is not uncommon to ›serve objects at a distance, and close to the ground, 'shimmering'. This is a ›rm of degradation of quality due to atmospheric turbulence. These two optical nperfections of the atmosphere are the subject of the next two chapters. In this ıapter we shall deal in depth with the problem of atmospheric attenuation. urbulence will then be discussed in depth in Chapter 16.

In our considerations of atmospheric attenuation we shall be concerned with ıe outdoor viewing situation. However, it should always be remembered that ıy particles or vapour in an atmosphere will introduce attenuation effects, and such it is quite possible to meet atmospheric attenuation indoors. A good ‹ample of a violent indoor effect is the effect of steam from a kettle or bath in ı enclosed space. A less obvious, but significant, one is the effect of tobacco noke in a cinema.

5.1 BASIC ATTENUATION MECHANISMS

tmospheric attenuation is caused by scattering of light from small particles or oisture droplets suspended in the atmosphere and by absorption. This has a vo-fold effect on contrast transmittance. Firstly, light coming from an object interest and from its background is progressively removed from the viewing ıth and does not reach the observer. Such removal of light follows an ‹ponential law in homogeneous air.[1] Thus a collimated beam of light containing a flux F_0 at the source will have a residual flux at range R from the urce in homogeneous air given as

$$F = F_0 \exp[-(b + k_s)R] = F_0 \exp(-\sigma_e R) \tag{15.1}$$

general, where b and k_s are the scattering coefficient and absorption ›efficient respectively, and where σ_e is the extinction coefficient. Secondly, ght which has not come directly from the object or from its immediate ıckground is scattered into the viewing path. This additional light is called *air-light* or path radiance. Unfortunately, even on a clear day and in homo geneous air, the flux of scattered radiation is not isotropic. This is because th volume scattering function of air $\beta'(\phi)$, which is the intensity in a given directio per unit incident flux per unit volume of air, varies with scattering angle ϕ. Thus in order to determine the scattering of light into a viewing path, one must know the form of dependence of $\beta'(\phi)$ on ϕ, the absolute magnitude of $\beta'(\phi)$ and th distribution of energy around the entire background sphere. The form o dependence of $\beta'(\phi)$ on ϕ is very dependent on the type of atmospher considered, and will be discussed in the following sections. Equally th distribution of energy around the background sphere is, in practice, complex an will receive consideration later. The absolute level of $\beta'(\phi)$, on the other hand may be related directly to the scattering coefficient by the expression

$$2\pi \int_0^\pi \beta'(\phi) \cdot \sin\phi \cdot d\phi = b \qquad (15.2$$

Alternatively a relative scattering function $F(\phi)$ may be used, where $F(\phi)$ defined as

$$F(\phi) = \frac{\beta'(\phi)}{\beta'(0)} \qquad (15.3$$

$\phi = 0$ referring to the direction of light propagation.[1] Other workers[2] defin $F(\phi)$ as $4\pi\beta'(\phi)/b$, $4\pi\beta'(\phi)$ being the integral of equation 15.2 if $\beta'(\phi)$ is constant.

15.1.1 Rayleigh Scattering

The way in which the molecular constituents of the air (mainly oxygen an nitrogen) scatter radiation can be expressed by a theory first proposed by Lor Rayleigh. This is discussed by Middleton,[3] and it is shown that the relationshi between the molecular scattering coefficient $b_1(\lambda)$ and wavelength λ of inciden radiation can be stated as

$$b_1(\lambda) = C_1\lambda^{-4} \qquad (15.4$$

where C_1 is a constant determined by the gaseous composition of th atmosphere. For use at sea level a value of $C_1 = 1.1 \times 10^{-3}$ has been propose by Möller,[4] where the wavelength is expressed in μm. Rayleigh used this invers fourth power relationship to explain the blue colour of the sky and the fact th the sky is bluest when it is purest. It is also shown by Middleton[3] that the for of scattering is such that $\beta'(\phi)$ may be given as

$$\beta'(\phi) = k\,(1 + \cos^2\phi) \qquad (15.5$$

where k is a constant independent of ϕ. Thus, for Rayleigh scattering, it can be seen that $\beta'(\phi)$ is a maximum at $\phi = 0$ and π, with a minimum of just half the maximum value when $\phi = \pi/2$. Furthermore the entire distribution of scattered light may be fairly accurately predicted from Equations 15.4 and 15.5.

15.1.2 Aerosol Scattering

Unfortunately it is only in the driest and cleanest air that the scatteringoccurs solely from gaseous molecules. A real atmosphere usually comprises a suspended aerosol in addition to the air molecules, and the aerosol itself varies greatly with type of location. For example, in any industrial area there is often a pronounced haze due to a suspension of soluble and insoluble smokes, combustion particles and so on. Alternatively, over ocean surfaces, salt particles are present, whilst the most common aerosol suspension consists of small water droplets, contaminated with condensation nuclei. Now for Rayleigh scattering laws to be obeyed, the scattering particle radius must be much smaller than the wavelength of light and, whilst gaseous molecules conform to the requirements, most aerosol particles do not. This is particularly true in misty or foggy conditions, where the particle radii are frequently in the range 0.1 to 10λ. In such situations the scattering obeys a theory originally developed by Mie[5] for the field of colloidal chemistry. This appears to have been first applied to atmospheric attenuation in the 1930's.[6,7] The complete theory is given by Stratton,[8] whilst summarised treatments are given by Gambling[9] and by Middleton.[3] Tabulations of the intensity due to scattering have been performed for various particle radii between 0.1 and 70λ, and for particle refractive indices of 1.33, 1.44, 1.55 and 2.0.[10-12] Those for $\mu = 1.33$ are of very obvious interest in the field of atmospheric attenuation, since a great deal of scattering takes place due to water droplets in the atmosphere. A condensed version of these tables for $\mu = 1.33$ is to be found at Table 15.1, taken from Middleton,[3] as a function $S(\phi)$, $S(\phi)$ being dependent on particle refractive index, particle radius and angle ϕ. It can be seen from the table how, as particle size increases, the angular properties of scattering start to deviate massively from the Rayleigh theory, the majority of scattering being increasingly concentrated at small angles of ϕ. If the air contains N particles/cm^3, then

$$N \cdot S(\phi) = \beta'(\phi).$$

Now, for *extinction* of light due to Mie scattering, we require the value of the scattering coefficient b. In order to arrive at this, it is necessary to integrate $S(\phi)$ with respect to ϕ, the total energy scattered being then given as

$$F = \frac{\lambda^2}{2\pi} \int_0^{\pi} S(\phi) \cdot \sin\phi \cdot d\phi \qquad (15.6)$$

TABLE 15.1. $S(\phi)$ from the Mie theory, for water droplets, in units of $\lambda^2/4\pi^2$ cm^2

$\gamma = \phi$ (rad)	0.5	1.0	1.5	2.0	2.5	3.0	4.0	5.0	6.0
ϕ(rad)									
0	0.00071	0.05260	0.6465	3.937	14.20	41.69	197.7	585.8	1253.
0.17	0.00069	0.05152	0.6288	3.790	13.45	38.68	173.7	478.0	927.1
0.35	0.00066	0.04846	0.5790	3.382	11.42	30.85	116.6	251.8	349.0
0.52	0.00061	0.04383	0.5054	2.800	8.705	21.00	57.71	76.70	56.58
0.70	0.00055	0.03827	0.4196	2.155	5.938	12.02	19.48	13.168	27.61
0.87	0.00048	0.03248	0.3331	1.546	3.626	5.646	4.333	9.441	28.16
1.04	0.00042	0.02706	0.2551	1.038	1.984	1.1549	2.064	10.234	9.396
1.22	0.00037	0.02253	0.19087	0.654	0.9808	0.6463	2.744	4.872	3.863
1.39	0.00033	0.01915	0.14213	0.3867	0.4519	0.3214	2.375	2.635	5.914
1.57	0.00032	0.01700	0.10765	0.21354	0.2149	0.3936	1.230	1.909	3.573
1.74	0.00032	0.01598	0.08472	0.10831	0.13202	0.4806	0.4746	2.390	1.5505
1.92	0.00034	0.01588	0.07025	0.04918	0.12225	0.4630	0.4771	1.354	3.052
2.09	0.00037	0.01645	0.06138	0.02028	0.14755	0.35880	0.8279	0.5854	2.730
2.27	0.00042	0.01741	0.05605	0.01096	0.19170	0.23317	1.0006	1.300	0.8624
2.44	0.00046	0.01851	0.05284	0.01370	0.2462	0.14452	0.8518	2.222	2.555
2.61	0.00050	0.01956	0.05088	0.02262	0.3034	0.1188	0.6302	2.065	4.647
2.79	0.00054	0.02042	0.04970	0.03288	0.3540	0.1446	0.6083	1.593	3.456
2.96	0.00056	0.02096	0.04906	0.04068	0.3888	0.1855	0.7196	1.829	2.880
3.14	0.00057	0.02115	0.04887	0.04358	0.4013	0.2042	0.8709	2.155	3.507

(Reproduced from Middleton[3] by courtesy of the University of Toronto Press – adapted by Middleton from La Mer and Sinclair.[10])

This expression gives the lumens scattered by one particle per $\mathrm{lm/cm^3}$ of local path illuminance, and may be thought of as the 'effective area' or 'extinction cross-section' of the particle, that is, the area of the wavefront affected by it. Dividing by the cross-section area of the particle, where r_p is the particle radius, yields a pure number K, given as

$$K = \frac{\lambda^2}{2\pi^2 r_p{}^2} \int_0^\pi S(\phi) \cdot \sin\phi \cdot d\phi \tag{15.7}$$

b is then given as

$$b = NK\pi r_p{}^2 \tag{15.8}$$

for monodisperse aerosols. For aerosols having particles of various sizes we must write instead

$$b = \sum_{i=1}^{n} N_i K_i \pi r_{pi}{}^2 \tag{15.9}$$

The relation between K and $\gamma = 2\pi r_p/\lambda$ is shown in Fig. 15.1.

For aerosol scattering, experimental data indicate a wavelength dependence given by an empirical expression of the form

$$b(r_p\lambda) = C_2\lambda^{-z} \tag{15.10}$$

where r_p and λ are of approximately the same order. Values of z which best fit the observed results range between 0.5 and 2.0. Junge[13] suggests that a value of

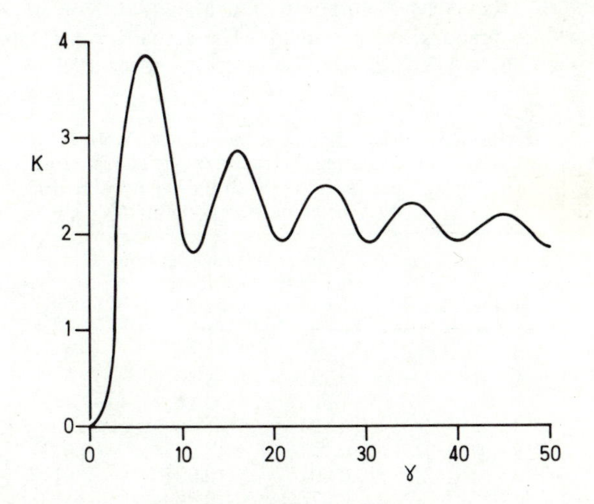

Fig. 15.1. Illustrating the relationship between K and γ for water droplets (Reproduced from Middleton[3] by courtesy of the University of Toronto Press).

1.3 is compatible with other studies of particle size distribution. C_2 is a constant whose magnitude has been put at 8×10^{-2} by Möller[4] for horizontal paths at sea level.

When the particle size is large compared to the wavelength of incident light the nature of the scattering becomes independent of wavelength. The process is called non-selective scattering. This situation can occur is some fogs and clouds, for example, where droplet radii can easily reach 50 μm.

15.1.3 Relative Importance of Scattering Mechanisms

Rayleigh and aerosol scattering may be superimposed to represent the overall scattering properties of a real atmosphere. In Elterman's formulation of a model for a 'clear standard atmosphere'[14] the scattering properties of a Rayleigh (molecular) atmosphere and of the added aerosol component are evaluated. The figures in Table 15.2 are calculated values of the ratio

$$\frac{\text{aerosol scattering coefficient}}{\text{Rayleigh scattering coefficient}}$$

at various altitudes and for three wavelengths in the lower, middle and upper parts of the visible spectrum. The graphs in Fig. 15.2 show the variation with height of both Rayleigh and aerosol scattering coefficients for the same three wavelengths.

Elterman's model is a combination of the US Standard Atmosphere,[15] which is a pure (Rayleigh) atmosphere, and a standard aerosol whose height profile is based on practical measurements. The actual aerosol density has been adjusted to give a meteorological visibility of 25 km at sea level.

TABLE 15.2. The variation of the ratio (aerosol scattering coefficient/Rayleigh scattering coefficient) with altitude and wavelength for a metereological visibility of 25 km.

	Wavelength (μm)		
Altitude (km)	0.4	0.55	0.7
0	4.65	13.59	30.9
1	2.25	6.59	15.0
2	1.08	3.14	7.14
3	0.50	1.46	3.33
5	0.124	0.362	0.823
10	0.0018	0.005	0.0119
20	0.0275	0.080	0.183
30	0.0294	0.086	0.195

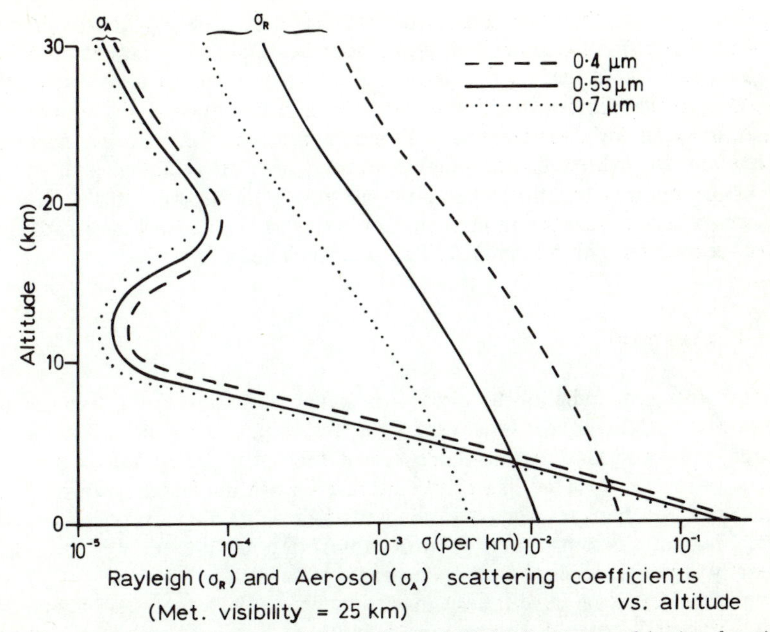

Fig. 15.2. Computed scattering coefficients for Rayleigh (σ_R) and aerosol (σ_A) as functions of altitude and wavelength based on a standard atmosphere and a meteorological visibility of 25 km.

It is evident from Fig. 15.2 and Table 15.2 that, in this atmosphere, the aerosol attenuates an optical beam more effectively than does the Rayleigh component up to altitudes of 2 to 3 km, due to its greater scattering coefficient. The height at which Rayleigh and aerosol attenuation become equal is wavelength dependent, but varies between 2 and 5 km over the visible spectrum. At ground level the aerosol scattering coefficient is about an order of magnitude greater than the Rayleigh scattering coefficient for the same wavelength.

The figures in Table 15.2 should not be taken to indicate the relative importance of the two modes of scattering. Anticipating a result of Section 15.2, the attenuation produced, for example, by Rayleigh scattering of radiation at 0.55 μm along a path at zero altitude is 0.87 of the attenuation caused by the aerosol.

It is important to note that these results apply to conditions where the visibility is about 25 km. When the visibility is considerably reduced, due to a denser aerosol component, then the relative importance of Rayleigh scattering is less. However, even when the visibility is reduced to 2 km, Rayleigh scattering is still accounting for more than 10% of the total attenuation.

The results of Waldram's experiments[16] indicate that the total attenuation coefficient over various non-industrial areas of England is roughly an order of magnitude greater than the theoretical Rayleigh scattering coefficient for pure air. This conclusion is compatible with Elterman's model of a clear atmosphere. Waldram also found that, in two industrial areas, the attenuation coefficient at ground level was almost a further order of magnitude up on the corresponding values in clean air. However, as Waldram points out, 'there is no necessary correlation; the industrial atmospheres were measured in weather which was on the whole thicker than that when clean air measurements were made'. It must also be expected that the modern use of 'smokeless' fuels will have had some influence on attenuation in industrial areas since Waldram's work.

15.1.4 Absorption

The absorption of radiation by clean atmosphere is greatest in the infra-red and ultra-violet regions of the spectrum. The absorbing agents are mainly water vapour, carbon dioxide, nitrous oxide, ozone, molecular and atomic oxygen, and nitrogen. In the infra-red there are discrete absorption bands, whilst in the ultra-violet the transmission at wavelengths below 0.3 μm is very low. In the visible region, however, there is little absorption on a clear day, and it is generally negligible when the water content is low.

The situation is rather different in regions where the atmosphere is polluted with industrial waste. A suspension of black soot particles is quite a strong absorber. In fact, with such atmospheres, absorption has been found to be very dependent on the particular conditions, both in magnitude and spectral properties.[3,16] A great deal of practical field measurements of attenuation in such conditions as industrial haze suggests that it is difficult to predict the absorption properties in a given situation. Thus it has become fashionable to assume that, in most viewing situations, it can be ignored as a separable function. This is probably satisfactory where one is dealing with extinction alone, since one may then assume the exponential laws of Equation 15.1 to apply to the sum of scattering and absorption. However, as will be seen later, it can be dangerous to ignore absorption without due thought where light is being added to a viewing path as well as removed from it, (the usual daylight viewing situation), since in this case light added to the path is only added by scattering, whilst that removed *may* include a major contribution due to absorption.[16]

15.2 BASIC ATTENUATION LAWS FOR HORIZONTAL VIEWING

In general, whatever the form of scattering prevalent in a given viewing situation, there will be two effects which contribute to the change of object luminance and contrast with viewing range. Firstly some light will inevitably be scattered out of the viewing path and some will be absorbed, thus reducing the amount of light received by an observer from the object of interest and its surroundings.

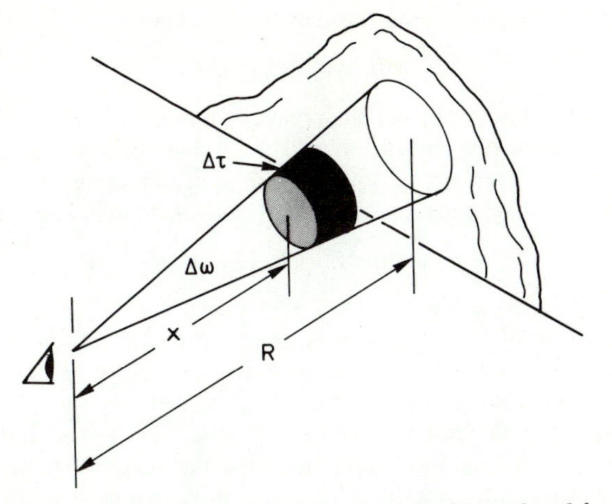

Fig. 15.3. Illustrating Koschmieder's theory of extinction. (Reproduced from Middleton[18] by courtesy of the University of Toronto Press).

Secondly there willl be light scattered into the viewing path from all other sources of light – the sun (if visible), the sky, clouds and the ground. This light scattered into the path is usually called *air-light*. In order to attempt to model the effects of atmospheric scattering, Koschmieder[17] developed a simple theory for viewing through a controlled atmosphere. This theory, discussed and developed at length by Middleton,[18] considered a viewing path through a homogeneous atmosphere, with restrictions on the form of illumination such that the whole viewing path should be illuminated equally. Under such conditions one may consider the viewing situation to be as in Fig. 15.3.

It is then shown by Koschmieder and Middleton that any element $\Delta\tau$ will have an intensity ΔI given as

$$\Delta I = \Delta\tau \cdot k \cdot b \tag{15.11}$$

where k is a constant of proportionality defined by the prevailing illumination conditions. From this it is shown that the incremental contribution to apparent luminance of an intrinsically black object from $\Delta\tau$ at range x will be

$$\Delta B = kb \exp(-bx)\,\mathrm{d}x. \tag{15.12}$$

Then, by integration, the apparent luminance due to the whole path will be

$$B_b = \int^R kb \exp(-bx)\,\mathrm{d}x = k[1 - \exp(-bR)] \tag{15.13}$$

For horizontal viewing, and assuming effectively a flat earth for purposes of computation, it is then assumed that the black object can be moved infinitely far away, with the result that $B_b \to k$. But, under such viewing conditions, B_b is effectively the luminance of the horizon sky B_h. Thus Equation 15.12 becomes

$$B_b = B_h\,[1 - \exp(-bR)]. \tag{15.14}$$

For other than black objects the apparent luminance due to the light from the object itself, ingnoring absorption, will be $B_0 \exp(-bR)$, where B_0 is the intrinsic luminance (see Equation 15.1). Thus the apparent luminance of any object viewed along a homogeneous horizontal path may be closely specified as

$$B_R = B_0 \exp(-bR) + B_h[1 - \exp(-bR)] \tag{15.15}$$

if absorption may be ignored.

If absorption must be taken into account, then this may be done, with the same limitations of homogeneity, by replacing b by σ_e as in Equation 15.1, so long as no strong colour contrasts are involved. If there are strong colour effects, these may equally be taken into account by considering the spectral composition of the object and background, together with the spectral dependence of the scattering and absorption functions. Thus

$$B_{R,\lambda} = B_{0,\lambda} \exp(-\sigma_\lambda R) + B_{h,\lambda}\,[1 - \exp(-\sigma_\lambda R)] \tag{15.16}$$

and

$$B_R = \int B_{0,\lambda} : \exp(-\sigma_\lambda R) \cdot \mathrm{d}\lambda + \int B_{h,\lambda} \cdot [1 - \exp(-\sigma_\lambda R) \cdot \mathrm{d}\lambda \tag{15.17}$$

where $B_{R,\lambda}$, $B_{0,\lambda}$, $B_{h,\lambda}$ and σ_λ are the various parameters previously defined at wavelength λ; $B_{R,\lambda}$, $B_{0,\lambda}$ and $B_{h,\lambda}$ being measured in *illumination* units. For a full treatment of the colour of distant objects see Middleton.[19]

If we now consider contrast of an object against its immediate background we may say, from Equation 15.15, that

$$(B_R - B_R') = (B_o - B_o') \exp(-\sigma_e R) \tag{15.18}$$

where B_R and B_R' are the apparent luminances of the object and its background at range R and B_o and B_o' are the intrinsic luminances of the object and its background.

But apparent contrast C_R is defined as

$$C_R = \frac{(B_R - B_R')}{B_R'} \tag{15.19}$$

and intrinsic contrast C_0 is defined as

$$C_0 = \frac{(B_0 - B_0')}{B_0'} \tag{15.20}$$

Equation 15.18 may then be written

$$C_R = C_o \left(\frac{B_o'}{B_R'}\right) \exp(-\sigma_e R) \tag{15.21}$$

For a fuller discussion the reader is referred to Middleton[18] and Duntley.[20]

In general, Equation 15.2 requires knowledge of the factor (B_o'/B_R) for its solution. However, in the particular case of viewing against the horizon sky (subject to the adequacy of Koschmieder's original assumptions), B_o' and B_R' both refer to the horizon sky and are approximately equal. Under such circumstances Equation 15.21 reduces to

$$C_R = C_o \exp(-\sigma_e R) \tag{15.22}$$

Equation 15.22 is a widely used simple expression for attenuation of contrast by the atmosphere. However, whilst in many situations it may provide a good guide to the practical effects, the user is nevertheless warned strongly to take adequate note, in its use, of the restrictions implied in its derivation and of the practical effects discussed later in this chapter.

Whilst concerning ourselves with this expression, its is appropriate to consider the derivation of meteorological visibility, which comes directly from the formula. Meteorological visibility is defined as that range at which a large, intrinsically black object viewed against the horizon sky has its contrast reduced either to 0.05 or to 0.02, both definitions being in common use. But a black object has a contrast of -1. Therefore, from Equation 15.21, the meteorological visibility must be such that $\exp(-\sigma_e R_v) = 0.05$ or 0.02, where R_v is the meteorological visibility. Hence $R_v = 3/\sigma_e$ for an 0.05 residual or $3.9/\sigma_e$ for an 0.02 residual. For search modelling purposes (c.f. Chapter 8) the author has found a residual of 0.03 to be most appropriate, this being in agreement with the median value obtained by Middleton and Mungall (see Section 15.5.1).

Having derived the basic laws of atmospheric attenuation for ideal horizontal paths, it should be stressed again that their use is very restricted as they stand. A homogeneous atmosphere restricts the viewing path to horizontal, and even then the earth's curvature reduces the validity of the modelling for long paths. Any broken cloud situation immediately violates the assumption of constant path illumination. Local terrain variations will influence ground reflectance and also will affect the homogeneity of the atmosphere, even for horizontal paths.

The implications of some of the foregoing are discussed in the following paragraphs.

15.3 SLANT PATH VIEWING

As soon as we introduce a differential height between the observer and the object of interest, we immediately violate the law of homogeneity necessary for use of the Koschmieder equations. The extent to which this is so depends largely on particular meteorological conditions.

In order to attempt to overcome the basic problems of altitude dependency of σ_e, Duntley[20] introduced the concepts of an *optical standard atmosphere* and an *optical slant range*. The *optical standard atmosphere* is analagous in concept to the internationally agreed standard atmosphere defined in terms of temperature, pressure and height. From temperature and pressure one may determine the relative number of molecules per unit volume at various heights, which, for heights up to 10 km, is found to be expressed closely as

$$N/N_0 = \exp\left(\frac{-h}{9\,200}\right) \tag{15.23}$$

where N_o is the number of molecules/unit volume at sea level and N is the number of molecules/unit volume at h metres above sea level.

Duntley defines the *optical standard atmosphere* as one in which the scattering particles are of the same kinds and dimensions at all altitudes, but vary in number according to Equation 15.23. Under such conditions he shows that, providing attenuation is predominantly by scattering and not by absorption, the exponential function $(-\sigma_e R)$ implied in Equation 15.15 and shown in Equation 15.22 may be replaced by $(-\sigma_o \bar{R})$, where σ_0 is the attenuation coefficient at the lower end of an inclined path. $\bar{R}$ is the *optical slant range* defined as

$$\bar{R} = \int_0^{R_s} F(R) \cdot \mathrm{d}R \tag{15.24}$$

where R_s is the actual length of the slant path. But $F(R)$ for the *optical standard atmosphere* is given as N/N_0. Also along any slant path at height h the distance from the intersection with the sea level plane is $R \sin\theta$, where θ is the elevation of the path. Thus Equation 15.24 becomes

$$\bar{R} = \int_{R_1}^{R_2} \exp\left(\frac{-R \sin\theta}{9200}\right) \cdot \mathrm{d}R$$

$$= 9200 \operatorname{cosec}\theta \left[\exp\left(\frac{-R_1 \sin\theta}{9200}\right) - \exp\left(\frac{-R_2 \sin\theta}{9200}\right)\right] \tag{15.25}$$

A thorough discussion of the foregoing is given by Middleton.[18]

Unfortunately the concept of an *optical standard atmosphere* on many occasions proves to be far removed from the practical situation. Not only is absorption often a significant part of σ_e, resulting in an immediate invalidation of the concept due to the mechanisms of addition of light to the viewing path (pure scattering) and extinction of light (scattering plus absorption) being different, but also the atmosphere is frequently 'non-standard'. This latter fact can be appreciated by considering the appearance of ground mists, which exhibit a marked discontinuity in σ_e a few metres above the ground, and what might be considered as low cloud, where the value of σ_e actually *increases* rather than decreases with increasing altitude.

A number of attempts have been made to measure σ_e as a function of height, of which those of Waldram,[16,21] Siedentopf[22,23] and more recently the Visibility Laboratories of the University of California[24–26] are amongst the most important. Waldram's work, in particular, illustrates how far removed most real atmospheres in England are from the ideal, even when the air is 'clean'. An extensive summary of Waldram's and Siendentopf's work is given by Middleton,[3] whilst the general form of the studies carried out by the Visibility Laboratories is well covered by Boileau.[24]

Notwithstanding the serious limitations of the *optical standard atmosphere*, it is still useful to consider how far the various attenuation equations can be applied to viewing along inclined paths.

15.3.1 Upward Slant Paths

In the case of upward viewing, it will be usual that $B_R{}'$ is the apparent luminance of the sky near an object of interest, $B_o{}'$ being its luminance as seen by an observer at the object and looking in the same direction. However, the concept of a homogeneous atmosphere and a horizon sky luminance, as necessary for use of Equation 15.15, no longer applies. In order to generalise the situation, Duntley[20] introduced an alternative formulation of the basic attenuation equation by use of *two* constants to describe a lamina of air of unit thickness – σ_e as before and the luminance of the lamina, B_a, as viewed by a nearby observer looking in the direction of interest. He was then able to redefine Equation 15.15, subject to B_a and σ_e being similar functions of range R, as

$$B_R = \frac{B_a(0)}{\sigma_0} [1 - \exp(-\sigma_0 \bar{R}) + B_0 \exp(-\sigma_0 \bar{R})] \qquad (15.26)$$

where $B_a(0)$ and σ_{eo} refer to the values of B_a and σ_e at the bottom end of an inclined path (see Middleton[18] for a summary of the development of Equation 15.26).

If we now write $\bar{R}_{0,\infty}$ for the *optical slant range* corresponding to the integration of equation 15.24 from zero to infinity and $\bar{R}_{R_s,\infty}$ for that obtained by integrating from R_s to ∞ then, from Equation 15.25, we get, for upward viewing against a clear sky

$$B_R{}' = \frac{B_a(0)}{\sigma_0} [1 - \exp(-\sigma_0 \bar{R}_{0,\infty})] \qquad (15.27)$$

and

$$B_0{}' = \frac{B_a(0)}{\sigma_0} [1 - \exp(-\sigma_0 \bar{R}_{R_s,\infty})] \qquad (15.28)$$

Substituting these values in Equation 15.21 we get

$$C_R = C_0 \exp(-\sigma_0 \bar{R}) \left\{ \frac{1 - \exp(-\sigma_0 \bar{R}_{R_S,\infty})}{1 - \exp(-\sigma_0 \bar{R}_{0,\infty})} \right\} \qquad (15.29)$$

which is independent of B_a and is seen to depend only on σ_0 and $\bar{R}$. Since, in a given atmosphere, if horizontally stratified, $\bar{R}$ depends only on R_s and viewing path elevation θ, it follows that an object of fixed inherent contrast, moving at constant altitude in a circular path having its centre vertically above the observer, will present the same apparent contrast at all points of its path, even if the sun is shining. However, it should be noted that the intrinsic contrast of an actual target is likely to be far from fixed under such circumstances, and the only object for which the hypothesis is really justified is an ideal black object.

Other theories of contrast attenuation and acquisition range when looking upward have been attempted by Löhle[27] and Foitzik[28] but Duntley's is considered to be the closest to practical applicability. However, for the special but important case of viewing an object such as an aircraft at constant altitude, it is considered by the author that a useful practical simplification may be made by use of the relationship

$$\sigma_0 \bar{R} \equiv \bar{\sigma}_e R_s \tag{15.30}$$

The true range can then be used instead of the optical slant range $\bar{R}$. This necessitates the introduction of $\bar{\sigma}_e$, which is the equivalent attenuation coefficient of the air between ground level and the object of interest, $\bar{\sigma}_e$ being itself defined by the equation

$$\bar{\sigma}_e = \frac{1}{H} \int_0^H \sigma_e(h) \cdot \mathrm{d}h \tag{15.31}$$

where H is the height of the object above the observer and $\sigma_e(h)$ is the value of σ_e at height h above the observer. For constant altitude difference between observer and object of interest, and assuming, as usual, a stratified atmosphere, this function $\bar{\sigma}_e$ is a constant. From Equations 15.27, 15.28 and 15.29 it then follows that, for this specific situation, the apparent contrast may be expressed as

$$C_R = C_0\, k \exp(-\bar{\sigma}_e R_s) \tag{15.32}$$

where $k = B_0'/B_R'$. Although in general k will vary with viewing direction, for situations where sight line elevation with respect to the sun does not change markedly it is often adequate to consider k a constant. Alternatively, providing that $F(\phi)$ is known, it is possible to describe k as an analytic function. For a full treatment of this form of upward viewing the reader is referred to Crowther.[29]

Strictly the preceding paragraphs only refer to the situation of viewing against clear sky – that is, with an infinity of air behind the object of interest. When the sky is overcast there is no longer only air-light behind the object – the clouds act as an effective solid body of finite intrinsic luminance. However, the effective background is not at the same range as the object of interest and so the equations of extinction for object and background are dissimilar. This leads to complex equations which will not be quoted here. The reader wishing to pursue this is referred to Crowther[29] and Overington.[30] The case of broken cloud will be discussed separately in Section 15.4.

15.3.2 Downward Slant Paths

For downward viewing it is possible to consider the general case for a layered atmosphere, according to Duntley,[20] to be similar to the Koschmieder approach for horizontal paths, except for two important differences. Firstly it is necessary to consider the horizon sky brightness to be that measured in such a direction that the angle between the sun's direction and the sight line to the object of interest is the same as the angle between the sun's direction and the sight line to the necessary part of the horizon. This is in order for the scattering angles for direct sunlight to be similar for the viewing path and the horizon sky brightness measurement. Duntley does not explain what we are to do when there is *no* intersection with the horizon of the viewing cone of angle equal to that between the sun's direction and the object sight line. Such a situation can frequently be met when the sun is behind the observer and the sight line is depressed by several tens of milliradians.

The second difference to the horizontal path situation, as in the case of upward slant paths, is the need to use a modified variable in place of the product $\sigma_e R$. Duntley choses again to use the optical slant range $\bar{R}$, whilst at BAC(GW) we have again chosen to employ the concept of $\bar{\sigma}_e$ for the special case of constant altitude viewing. The two approaches are discussed by Middleton[18] and the present author[30,31] respectively.

15.4 EFFECT OF STRUCTURED ILLUMINANCE OF THE VIEWING PATH

15.4.1 Broken Cloud Situations

In the previous section it was stated that the formulae derived were only strictly applicable to cases of clear sky or a uniform overcast. Now over much of the world such sky conditions are the exception rather than the rule. It is thus necessary to consider what the implications of broken cloud cover are.

To appreciate the situation, let us first assume the object to be fixed and the broken clouds to be effectively stationary. Then the illumination of some parts of the viewing path will be by direct sunlight plus general sky light, whilst other parts of the viewing path will be illuminated by sky light only. Thus the *addition* of air-light at local parts of the path will be dependent on the illumination conditions of that part of the path, whilst the *extinction* of such components of added air-light will still generally be controlled by the exponential function governed by σ_e or $\bar{\sigma}_e$ as described previously. This leads to a need to restate B_b

and B_h in Equations 15.13 and 15.14 as

$$B_b = \int_0^R KE_i b \exp(-bx)\, \mathrm{d}x \tag{15.33}$$

and

$$B_h = \int_0^\infty KE_i b \exp(-bx)\, \mathrm{d}x \tag{15.34}$$

where KE_i replaces k in the former equations, E_i being the local incident illumination on the viewing path at range x. Thus, in general, the apparent luminance at range R for horizontal viewing may be given, in place of Equation 15.15, as

$$B_R = B_0 \exp(-bR) + \int_0^R KE_i b \exp(-bx)\, \mathrm{d}x \tag{15.35}$$

Similarly expressions for other viewing situations may be modified.

If the object of interest or observer is now allowed to move, or if the clouds are moving across the sky, then not only will the air-light be modulated but also the object and/or the observer will be moving in and out of sunlight. This essentially means that B_0 may be a variable and E_i may be a variable function of range as time or viewing distance change. Detailed discussions of the case for downward viewing of fixed objects by a moving observer have been given by the present author,[31] whilst the case for upward viewing of a moving object by a fixed observer has been discussed at length by Crowther.[29]

15.4.2 Typical Cloud Structure

In order to assess the importance of the functions discussed in Section 15.4.1., it is necessary to determine the statistically most probable cloud patterns. To this end two exercises have been carried out at BAC(GW).

Firstly, some years ago, as an initial feel for the subject, the variation of illuminance of a diffuse, near horizontal surface was recorded throughout all daylight periods when there was no rain over a period of some six weeks. During these recording periods the wind speed and direction were recorded by the local meteorological office every half hour. By invoking the concept that the clouds would, on average, move at the same rate and in the same direction as the ground wind (although realising that there would be very considerable discrepancies from such a law) a first order transform of temporal variation of local ground plane illumination to spatial variation of ground plane illumination (and hence cloud patterns) could be achieved. Using this concept a rough guide to frequency of occurrence of various spatial periods and modulation amplitudes was

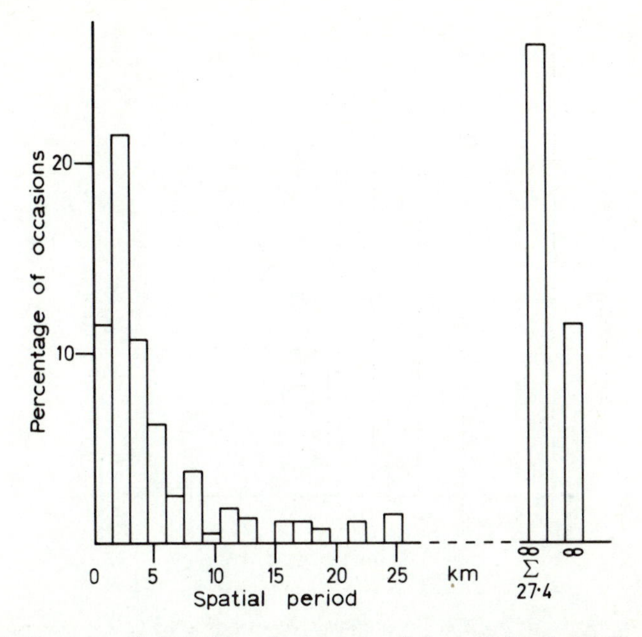

Fig. 15.4. Histogram of frequency of occurance of the spatial periods of major ground plane illumination cycles as measured by B.A.C. (GW). $\Sigma_{27.4}^{\infty}$ *indicates the sum total of frequencies of occurance of all spatial periods above 27.4 km.* ∞ *indicates frequency of occurance of clear skies or uniform overcast.*

determined.[30,32] The results seemed to show a most probable spatial period of around 3 km and a most probable modulation of around 2:1, both with a strongly skewed distribution (see Figs. 15.4 and 15.5). When applied to the modelling of Section 15.4.1 it is found that such typical periodicities and modulation depths can produce a very serious distortion of the exponential laws for contrast attenuation. For convenient presentation of the effects an alternative form of definition of apparent contrast C_R, as derived by Middleton,[18] is utilised, – viz.

$$C_R = C_0 \left[1 - \frac{B_h}{B_0'} (\exp \sigma_e R - 1)\right]^{-1} \tag{15.36}$$

This equation, which may be derived directly from Equations 15.15, 15.19 and 15.20, may alternatively be written

$$\frac{1}{C_R} = \frac{1}{C_0} \left[1 - \frac{B_h}{B_0'} (\exp \sigma_e R - 1)\right]$$

$$= \frac{1}{C_0} \cdot \left[1 - \frac{B_h}{B_0'}\right] - \frac{B_h}{C_0 B_0'} \cdot \exp \sigma_e R \tag{15.37}$$

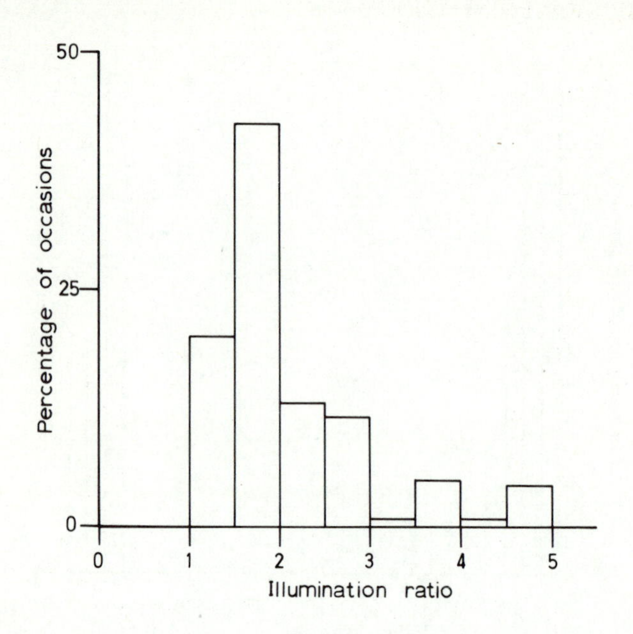

Fig. 15.5. Histogram of frequency of occurance of various ground plane illumination ratio as measured by B.A.C.(GW).

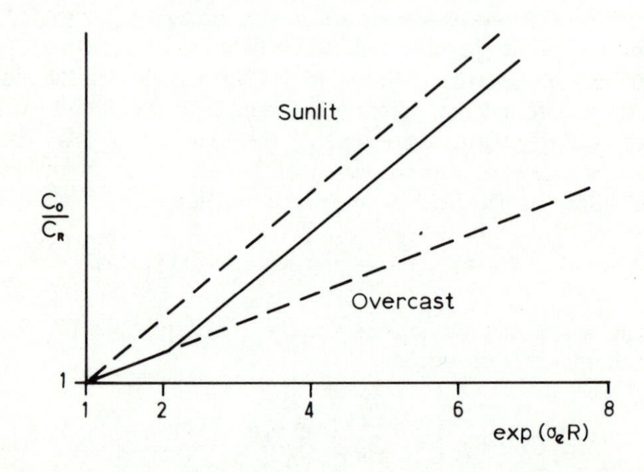

Fig. 15.6. The form of reciprocal contrast versus exp($\sigma_e R$) function where the target and its immediate vicinity are in cloud shadow with the rest of the path in sunlight.

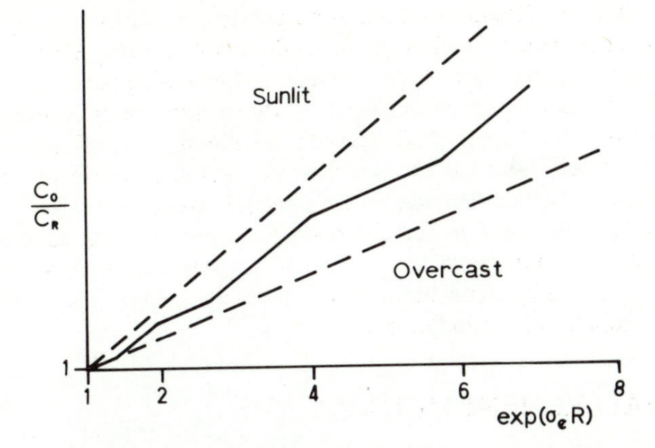

Fig. 15.7. The form of reciprocal contrast versus exp($\sigma_e R$) function where the viewing path has alternate portions in cloud shadow and sunlight.

where it will be seen to be comprised of a *positive* exponential and a constant term. If then C_0/C_R is plotted against exp $\sigma_e R$, for conditions of stable illumination the result should be a straight line. However, a change from sunlight to shadow will change the value of $B_h/C_0 B_0{}'$ and hence the slope of the line.

Figures 15.6 and 15.7 illustrate two simple cases of C_0/C_R trend for non-uniform path illumination and fixed objects of interest. Figure 15.6 shows a situation where the region around the object of interest is in cloud shadow and the rest of the path is sunlit, whilst Fig. 15.7 shows the effect due to a periodic cloud pattern causing a path illumination ratio of 2:1 between sunlit and shadow parts.

In recent years a much larger program of direct measurement of broken cloud structure on a statistical basis has been undertaken.[33,34] This program has provided a large bank of data on statistical cloud distribution as a function of weather type, time of day and season of year for the United Kingdom. It is believed that the statistical distributions with weather types may be extrapolated, within reason, to other parts of the world, although the frequency of occurance of various weather patterns may well vary considerably. At this time no attempt has been made to apply the statistical findings to the modelling of Section 15.4.1.

15.4.3 Local Path Illumination

An alternative and important condition of air-light is when there is little or no general illumination but a strong local source at a *finite* distance from the viewing path. Examples are viewing at night under local artificial illumination and many interior situations where atmospheric attenuation is a problem. The major difference here from that of the sunlit outdoor scene is the fact that the local source (or sources) are relatively close to the viewing path. Thus, instead of being able to assume the illumination of the viewing path constant along its length and at a fixed angle, both intensity of illumination of the path and the angle are variable. This leads to important changes in the air-light function and corresponding important changes in the contrast attenuation laws. For instance, under flare light illumination at night, instead of apparent contrast reducing with range whilst scene luminance stays constant, apparent contrast tends to remain relatively stable whilst scene luminance reduces with range. For some discussion of local illumination see Middleton.[35]

15.5 MEASUREMENT METHODS

Methods of measurement of atmospheric attenuation may be divided into four classes. These are the visual range measurements as commonly used by meteorological observers, the extinction measurements based on the attenuation of energy over a fixed path from a known source, local air sampling techniques (nephelometers) and the direct instrumental recordings of reduction of contrast over a given range. None of the methods can give the entire story about prevailing attenuation along a given sight line, although each has its uses. The limitations and advantages will be discussed in the following paragraphs.

15.5.1 Visual Range Estimates

The observations made by meteorological observers are usually known as visual range *estimates*. It is considered by the author that *estimate* is the important word here. In general such measurements are based on the ability of a trained meteorological observer to see various prominent objects against the horizon sky at known ranges from his observation post. But it is well known from detection threshold studies that the threshold contrast for an object depends considerably on its size, shape and the level of background luminance (e.g. Chapter 4). Hence the apparent visual range (or visibility) recorded will depend on what object is looked at. Furthermore the threshold is variable from observer to observer and from time to time. Middleton and Mungall[36] carried out a field trial using meteorological observers to look at standard test objects typical of those used in practice. They found that the spread of threshold contrasts was from 0.01 up to 0.1 with a median at 0.03. This spread is equivalent to a spread of 2:1 in visual range or quoted visibility – a rather large tolerance when attempting to model acquisition in the field. Further, the measurement tells one nothing about the fluctuation of the atmospheric attenuation coefficient along the viewing path. The large spread on visual range estimates is leading to an increasing use of instrumental methods at the present time.[37]

15.5.2 Extinction Methods

At night it is, of course, not possible to use the visual range estimate method. Thus methods have been developed which depend on the estimation or measurement of the attenuation of light from a self-luminous source over a fixed range (e.g. Middleton[38]). Some of these methods (using light modulation) can also be used in daylight. They suffer from the disadvantages that they require a known and controlled source of light and that the measurement system has to be 'double-ended'. However, they are basically capable of giving an accurate reading of the *average extinction coefficient* over the measurement path.

15.5.3 Local Air Sampling

The two methods described to this point both suffer from the disadvantage that they provide *no* information about fluctuations of σ_e along a viewing path. For other than horizontal viewing this is, of course, a serious limitation because of the inevitable change of σ_e with altitude. In order to overcome this, a range of sampling instruments known as nephelometers have been developed over the last few decades (e.g. Middleton[38] and Overington *et al*[39]). The basic principle of most of these instruments is to take a sample of air at some point and to measure its *volume scattering coefficient*. Typically this is achieved by passing an adequately baffled light beam through a controlled column of the sampled air and detecting the amount of light scattered out of the beam as shown schematically in Fig. 15.8. It is usual to calibrate such a device by injecting gases

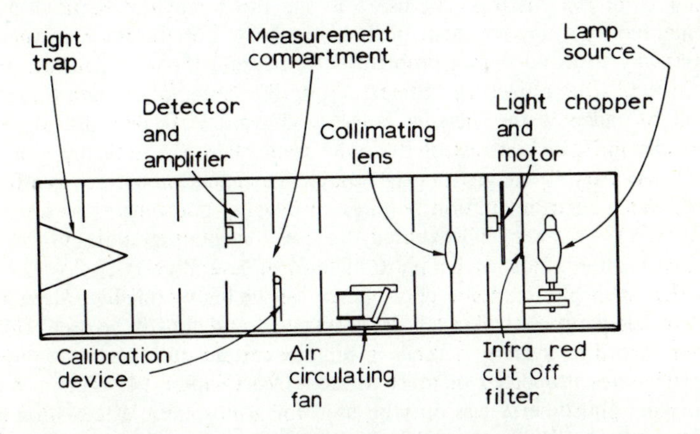

Fig. 15.8. Schematic diagram of a typical nephelometer[39]. A sample of air is sucked into the instrument and the light scattered from the collimated light beam in the measurement compartment is measured. Strategically placed baffles and the light trap prevent measurement of light scattered elsewhere than in the measurement compartment.

of known vapour pressure and scattering properties.[40] It should be noted that such instruments *only* measure the volume *scattering* coefficient and therefore take no account of absorption, it being assumed, in their use, that absorption is usually small compared to scattering. How far this assumption is true is open to conjecture (see Section 15.1). It would seem, from the author's interpretation of the literature, that such an assumption is reasonably valid for mists in areas away from industry, but that there is considerable doubt as to its validity in the case of industrial fogs and in other circumstances where solid particles are a prime attenuator (e.g. snow flurries, dust storms, smoke, etc.).

Subject to the foregoing limitations it is theoretically possible to study the local σ_e over any viewing path in sufficient detail to provide a complete attenuation profile. In practice, other than for very stable atmospheric conditions, it is not very practical to carry out the necessary sampling with range because of the time necessary to transport the nephelometer from point to point and establish a stable sample. It should also be noted that, since such instruments of necessity contain a light source (which also provides heat), unless great care is taken in design and use, the air samples tested may attain a different temperature to that in free air, with attendant changes in their scattering properties (see Section 15.5.5).

15.5.4 Direct Recording of Contrast Reduction

In viewing through the atmosphere it is rare that one is looking at objects silhouetted against the horizon sky. Unless this is the case, all measurements of σ_e obtained from the methods discussed in the three previous sections must be supplemented by a measurement of the appropriate horizon sky brightness. One potential way to avoid such a problem is to measure the local contrast of scene detail around the object of interest from the observation point (and as a function of range if the viewing range is changing). If one also arranges to measure the intrinsic contrast of the same scene then, theoretically, one has all the data necessary to reconstruct the complete attenuation pattern as a function of range. Such instrumentation requires an imaging photometer of some form. At BAC(GW) we have endeavoured to use the photographic process in a controlled fashion as just such a photometer over several years.[31,41,46] We have shown that such photographic photometry can be highly reliable. Alternatively, subject to adequate control, a television system could equally be used. However, whatever recording system is used, in practice certain problems arise due to the polar reflectance properties of most surfaces (see Chapter 14). This means that such imaging photometry can only be used for atmospheric attenuation studies with the greatest care. For instance, it is imperative that the viewing angles for measuring intrinsic contrast and apparent contrast at range are correctly related. Otherwise anomolies are introduced due to effects of polar reflectance and also possibly of $\beta'(\phi)$. However, on the other hand, where sufficient detail can be recorded it has been possible to illustrate vividly the importance of obtaining reliable data on atmospheric attenuation related to the *viewing path*. For instance, in the course of a series of field trials concerned with viewing of approaching aircraft from the ground, a photographic photometer was used to record the slant path attenuation by measurement of aircraft contrast against its

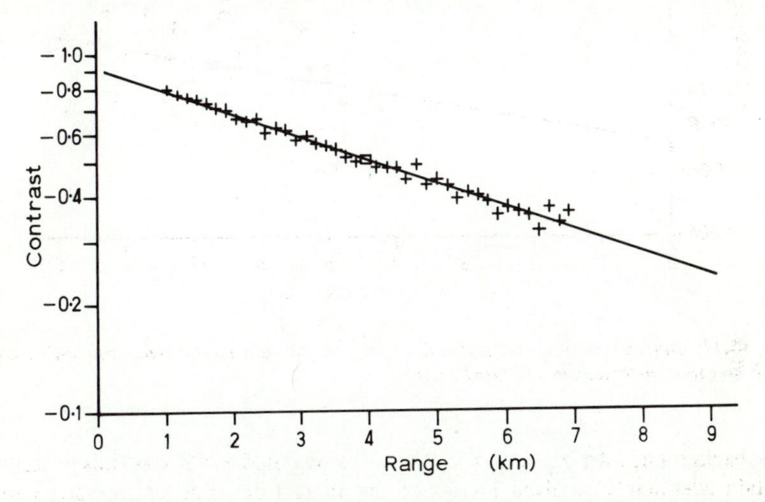

Fig. 15.9. A typical plot of apparent/contrast versus range on a log-linear scale for slant path attenuation where the ideal exponential attenuation law holds.

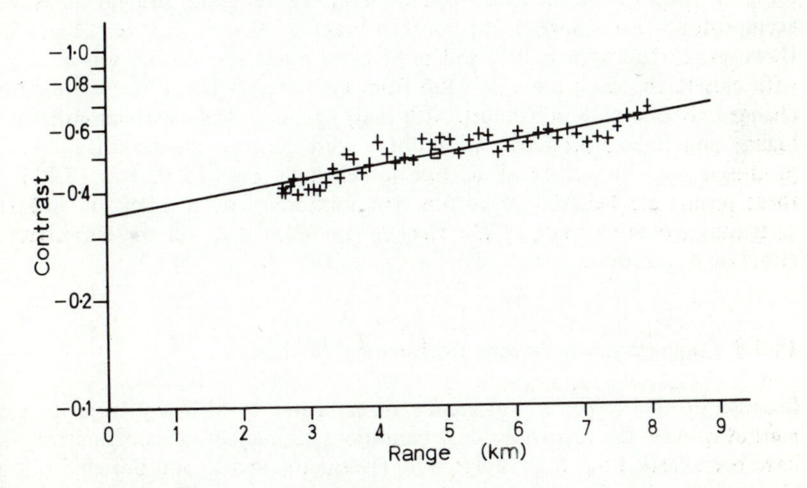

Fig. 15.10. Illustration of a contrast/range function for slant path attenuation which suggests a 'negative' attenuation coefficient.

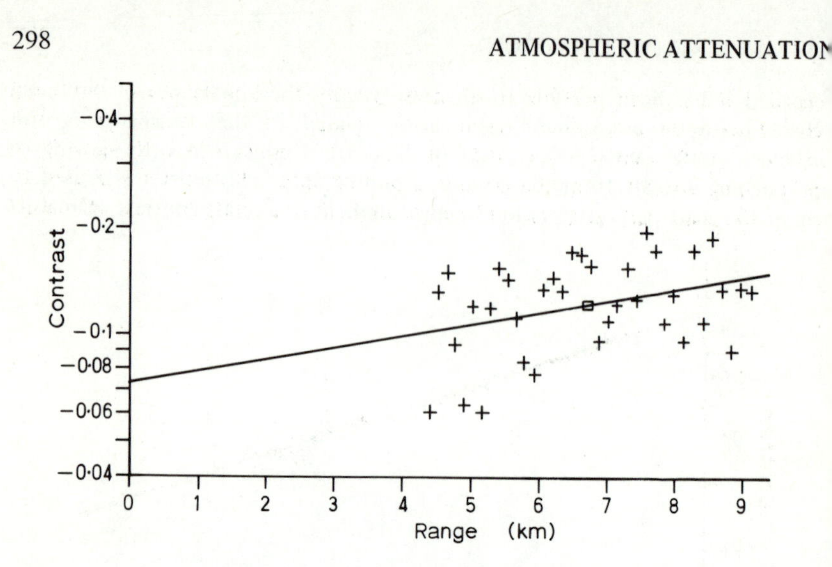

Fig. 15.11. Illustration of a contrast/range function for slant path attenuation which shows both 'negative' attenuation and gross scatter.

sky background. An extreme telephoto lens was used and it was thus possible to obtain adequately resolved images of the aircraft detail at ranges out to about 8 km, such that the apparent luminance of a particular part of the aircraft surface could be monitored with closing range. For direct approaches the typical plots of logarithmic contrast against range were good straight lines with asymptotes at zero range in the contrast bracket −0.4 to −1.0 (e.g. Fig. 15.9). However, certain approaches, and most cases where the aircraft was on a path with closest approach some 2–4 km from the observer (i.e. where the sight line changed considerably in azimuth with range) produced plots which either were highly non linear, produced asymptotes with contrast greater than −1.0 or produced negative values of σ_e (see for example Figs. 15.10 and 15.11). All these results are believed to be due to a mixture of polar reflection and $\beta'(\phi)$ variations *which are part of the viewing condition* and will therefore have an effect on acquisition.

15.5.5 Comparisons of Various Measurement Methods

Because of the suspected difference in estimates of visibility from different methods, even for relatively ideal conditions, a number of comparative trials have been carried out (e.g. Hodgson,[47] Overington *et al*[48] and Brown[49]). It has already been indicated that visual range estimates are subject to considerable variability due to observer differences and effects of size, shape and scene luminance. We shall concentrate here, by the way of illustration, on a particular comparison between a nephelometer and photographic photometers.

In order to compare a nephelometer with a photographic photometer, a field trial was organised by BAC(GW) on Pendine Sands, South Wales.[48] In this trial, large contrasting boards were set up at one end of a long flat stretch of the sands and photographed from various ranges along the sands up to 6.4 km. To ease instrumentation, camera equipment was driven across the sands, recordings being taken at selected stations along the sands. To obtain a better average set of data, since setting up and sighting of equipment took around ten minutes at each station, records were taken on the outward journey away from the boards and on the return, a complete series of records occupying about two hours. At each station, whilst the photographic record was being taken, a nephelometer was set up at the same height above the ground as the camera and a sample reading of local σ_e obtained. This trial procedure was repeated on several occasions, both by day and by night, to obtain a statistical set of data covering a variety of visibilities. An encouraging stability of the atmosphere as recorded by the nephelometer was observed during most trial periods.

The results of the exercise showed that there was good agreement between the two methods when visibility was greater than 10 km, but as visibility

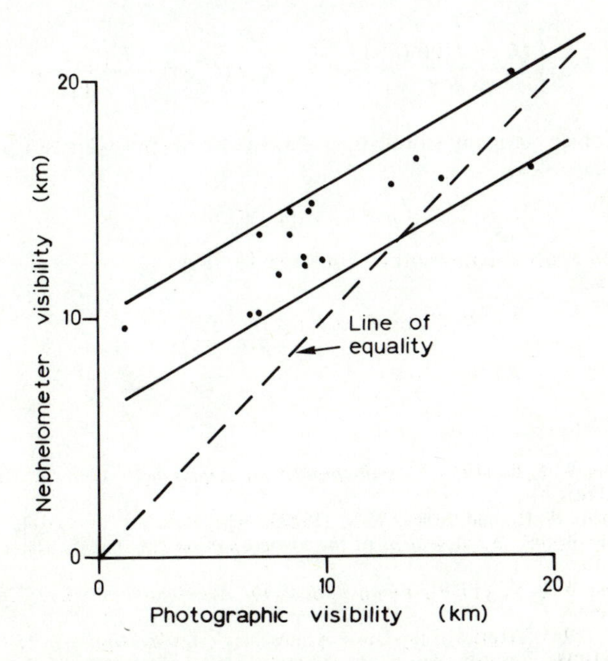

Fig. 15.12. Comparison of nephelometer and photographic visibilities as measured over identical horizontal near-ground paths.

reduced the nephelometer proceeded to read progressively higher than the photographic photometer. In addition there was a lot of scatter in the results from period to period. More detailed analysis showed that the scatter could be attributed to the photographic method at high visibilities, whilst at low visibilities, where the photographic method became very sensitive, the nephelometer was unreliable. To confirm this an opportunity was taken to make a nephelometer reading when the visibility was only a few hundred metres (visually). The nephelometer reading was in this case 10 km (believed due to the effect of small temperature changes within the instrument). The collected results are shown in Fig. 15.12.

15.6 IMPLICATIONS ON MODELLING OF VISUAL THRESHOLD

Provided that it is possible to measure or estimate atmospheric attenuation along a viewing path as a function of intrinsic contrast, the effects of atmospheric attenuation in a field situation can be modelled directly in terms of any of the equations in previous chapters by replacing ϵ by $F(C_o)$ where C_o is the intrinsic contrast. Thus, for single glimpse field detection Equation 7.7, for instance, becomes

$$\log_e \left[\frac{(K_2 + K_3)F(C_0) + 1}{K_3 F(C_o) + 1} \right] \approx K_2 F(C_o) = \frac{K_1}{n(n-1)} + \delta \qquad (15.38)$$

Equally, since for many situations it is adequate to approximate C_R by Equation 15.22, i.e.

$$C_R = C_o \exp(-\sigma_e R)$$

a simpler and approximate form of Equation 15.38 is

$$K_2 C_o \exp(-\sigma_e R) \approx \frac{K_1}{n(n-1)} + \delta \qquad (15.39)$$

REFERENCES

1. Middleton, W. E. K. (1958). *Vision through the Atmosphere*, Chap. 2, University of Toronto Press
2. Chesterman, W. D. and Stiles, W. S. (1948). 'The Visibility of Targets in a Naval Searchlight Beam', in *Proceedings of the Symposium on Searchlights, Illum. Eng. Soc.*, London, 75
3. Middleton, W. E. K. (1958). *Vision through the Atmosphere*, Chap. 3, University of Toronto Press
4. Möller, F. (1964). 'Optics of the Lower Atmosphere', *Applied Optics*, 3, 157
5. Mie, G. (1908). 'Contribution to the Optics of Turbid Media, Special Colloidal Metal Suspension', (in German), *Ann der. Phys.*, **25**, 377

6. Stratton, J. A. and Houghton, H. G. (1931). 'A Theoretical Investigation of the Transmission of Light through Fog', *Phys. Rev.*, **38**, 159
7. Sinclair, D. (1947). 'Light Scattering by Spherical Particles', *J. Opt. Soc. Am.*, **37**, 475
8. Stratton, J. A. (1941). *Electromagnetic Theory*, McGraw Hill, New York
9. Gambling, D. J. (1966). *The Visibility of Distant Objects*, MSc Thesis, University of Adelaide
10. LaMer, V. K. and Sinclair, D. (1943). 'Verification of the Mie Theory', *O.S.R.D. Rep. No. 1857* (Dept, of Commerce, Office of Publication Board, No. 944), Washington
11. Gumprecht, R. O. and Sliepcevich, C. M. (1951). 'Tables of Light Scattering Functions for Spherical Particles', University of Michigan Eng. Res. Inst., Special Publ.
12. Gumprecht, R. O., Sung, N. L., Chin, J. H. and Sliepcevich, C. M. (1952). 'Angular Distribution of Intensity of Light Scattered by Large Droplets of Water', *J. Opt. Soc. Am.*, **42**, 226
13. Junge, C. (1955). 'The Size Distribution and Ageing of Natural Aerosols as determined from Electrical & Optical Data on the Atmosphere', *J. Meteorol.*, **12**, 13
14. Elterman, L. (1963). 'A Model of a Clear Standard Atmosphere for Attenuation in the Visible Region and Infra-Red Windows', *Rep. No. AFCRL–63–675*, Air Force Cambridge Research Lab., USA
15. Anon. (1962). ICAO Standard Atmosphere, ICAO Document 7488, US Government Printing Office, Washington DC
16. Waldram, J. M. (1945). 'Measurement of the Photometric Properties of the Upper Atmosphere', *Trans. Illum. Eng. Soc.*, **10**, 147
17. Koschmieder, H. (1924). 'Theory of the Horizontal Range of Vision', (in German), *Beitr. Phys. freien. Atm.*, **12**, 33 & 171
18. Middleton, W. E. K. (1958). *Vision through the Atmosphere*, Chap. 4, University of Toronto Press
19. Middleton, W. E. K. (1958). *Vision through the Atmosphere*, Chap. 8, University of Toronto Press
20. Duntley, S. Q. (1948). 'The Reduction of Apparent Contrast by the Atmosphere', *J. Opt. Soc. Am.*, **38**, 179
21. Waldram, J. M. (1945). 'Measurement of the Photometric Properties of the Upper Atmosphere', *Quart. J. of the Royal Met. Soc.*, **71**, 319
22. Siedentopf, H. (1947). 'On the Dispersion of Light by Water Droplets' (in German), *Zeits. für Meteorol.*, **1**, 342
23. Siedentopf, H. (1947). 'On the Optics of Atmospheric Haze', (in German), *Zeits für Meteorol.*, **1**, 417
24. Boileau, A. R. (1964). 'Atmospheric Properties', *Applied Optics*, **3**, 570
25. Gordon, J. I. (1969). 'Model for a Clear Atmosphere', *J. Opt. Soc. Am.*, **59**, 14
26. Duntley, S. Q., Johnson, R. W. and Gordon, J. I. (1972). 'Airborne Measurements of Optical Atmospheric Properties in Southern Germany', Visibility Laboratory, Scripps Institution of Oceanography, San Diego, Rep. SIO.72–64, July
27. Löhle, F. (1935). 'About Vision over Slant Paths' (in German), *Meteorol. Zeits.*, **52**, 435
28. Foitzik, L. (1947). 'Theory of Slant Path Vision' (in German), *Zeits für Meteorol.*, **1**, 161
29. Crowther, A. G. (1971). 'Contrast Attenuation by the Atmosphere along a Slant Path', BAC(GW) Ref. L50/22/PHY/186/1197
30. Overington, I. (1969). 'Local Variations of Atmospheric Attenuation', *Optics Technol.*, **1**, 78
31. Overington, I. (1967). 'Photographic Techniques for the Study of Atmospheric Attenuation in Air to Ground Viewing', *J. Phot. Sci.*, **15**, 164
32. Murphy, M. J., Overington, I. and Williams, D. G. (1965). 'Final Report on Visual Studies Contract', App. 6.3, BAC(GW) Ref. R41S/11/VIS

33. Brown, M. B. (1971). 'The Measurement of Sky Brightness – Final Report', BAC(GW) Ref. L50/22/PHY/114/1296
34. Brown, M. B. (1974). 'The Measurement of Sky Radiance – Final Report', BAC(GW) Ref. ST.11930
35. Middleton, W. E. K. (1958). *Vision through the Atmosphere*, Chap. 7, University of Toronto Press
36. Middleton, W. E. K. (1958). *Vision through the Atmosphere*, Chap. 10, University of Toronto Press
37. Anon (1969). *Observers Handbook*, Chap. 3, Meteorological Office, HMSO
38. Middleton, W. E. K. (1958). *Vision through the Atmosphere*, Chap. 9, University of Toronto Press
39. Overington, I., Duncan, I. and Brown, M. B. (1970). 'Final Report on Visual Studies II Contract', App. 3.1.II, BAC(GW) Ref. L50/22/PHY/158/1164
40. Fielding, W. F. and Austin, K. E. (1969). 'The Use of Standard Gases as a Sub-standard for the Calibration checking of the W.R.E. Visibility Meter', Int. Dept. Note No. 299, Weapons Dept., RAE
41. Overington, I. (1963). 'The Study of Time and Space Variable Surface Temperatures of Self-luminous Bodies and Other Brightness Parameters by Colour Densitometry', *Light and Heat Sensing*, Chap. 12, (AGARDograph 71), Pergamon Press
42. Overington, I. (1967). 'The Stability of the Photographic Process for Absolute Measurement of Radiation', *J. Phot. Sci.*, **15**, 11
43. Overington, I. (1967). 'The Accuracy and Stability of Photographic Equipment for use as part of a Photographic Detector System', *J. Phot. Sci.*, **15**, 84
44. Overington, I. (1967). 'The Measurement of Contrast and Range in Air to Ground Viewing', *J. Phot. Sci.*, **15**, 277
45. Newton, A. M. and Lavin, E. P. (1968). 'Photographic Detectors V: Automatic Data Handling of Film Records', *J. Phot. Sci.*, **16**, 148.
46. Overington, I. (1968). 'Photographic Detectors VI: Photographic Pyrometry', *J. Phot. Sci.*, **16**, 199
47. Hodgson, I. S. (1969). 'Report of a Trial to assess Methods of measuring Visibility', *TERA Rep. 12024*
48. Overington, I. Duncan, I. and Brown. M. B. (1970). 'Final Report on Visual Studies II Contract', BAC(GW), Ref. L50/22/PHY/158/1164, Apps. 3.2.I & 3.2.II.
49. Brown, M. B. (1969). 'Comparison of Nephelometer and Inclined-Path Visibilities', Study Note No. 6 – Second Visual Studies Contract, BAC(GW) Ref. L50/20/PHY/158/1074

16 Atmospheric Turbulence

As briefly mentioned in the introduction to Chapter 15, a second atmospheric optical effect which can be troublesome in viewing through any significant atmospheric path is the presence of turbulence. Extreme examples of this are familiar to all of us – the shimmering of objects near the ground on a hot day or the twinkling of stars and distant lights. Unfortunately turbulence may be a problem in viewing even when it is not so readily detectable. After all, the fact that objects shimmer is due to local refraction by the atmosphere which varies with time. Even when they do not shimmer it is highly likely that there are temporal fluctuations in apparent object position which must interfere with viewing. More importantly there are also likely to be changes of refraction across the wavefronts being received by the two eye pupils, with resultant degradation of retinal images in much the same way as by aberrations in the optics.

When using visual aids such as binoculars this local refraction becomes much more serious. In such cases the entrance pupil is much enlarged compared with that of the eye (M times the eye pupil where M is the magnification of the visual aid) with the result that any tendency to shimmer is largely integrated. However, this results in a degraded image quality replacing the image motion. In addition any resultant degradation is magnified in its effect by the instrument. Thus turbulence becomes potentially very important when viewing through visual aids. It is particularly a problem in astronomy.

It is the purpose of this chapter to bring to the reader's attention the forms of turbulence, the available data on modelling of turbulence in terms of image quality and the methods available to attempt to measure it.

16.1 THE NATURE OF TURBULENCE

The normal state of the atmosphere is a turbulent one due to the presence of winds and thermal currents which continuously create eddy air currents. For simplicity, the atmosphere may therefore be thought of as a large number of regions of various dimensions called eddies. Each eddy is then looked upon as a parcel of air, or turbulent blob, over which the temperature, refractive index, etc. deviates from the average. In general, turbulence effects are considered to be isotropic for eddy sizes less than some value L_0 called the outer scale of turbulence.

Energy is introduced into the turbulence via the largest of these eddies, the two major sources of energy being the wind shear and convective heating from the ground. If the energy in the eddy exceeds a critical level determined by the Reynold's number, then the eddy is no longer stable but breaks up into, and transfers energy to, smaller eddies. Further break up then occurs until eventually, for very small eddies, viscous effects become important and the energy is dissipated. This viscous dissipation begins for eddies smaller than some value l_0, called the inner scale of turbulence. The region between the energy receiving eddies and the energy dissipating eddies is called the inertial sub-range (Kolmogoroff[1]). In this range, which extends from eddies of the order of hundreds of metres down to millimetres, energy is assumed to be passed on without loss.

Thus, the turbulent atmosphere is a randomly inhomegeneous medium in which the refractive index is a function of position and time. When an optical beam propagates through such a medium its interaction with the eddies produces random variations in the amplitude and phase of the signal. These variations lead to several effects which tend to degrade the performance of an optical system.

16.2 FORMS OF DEGRADATION DUE TO TURBULENCE

Variations in the angle of arrival of a received optical wave-front will cause an image to be focussed at different points in the focal plane of the receiving optics. This results in a continuous, rapid movement of the image about a mean point, and is known as 'image motion' or 'dancing'. Particularly slow oscillatory motions of this form are called 'wanderings' or 'beam steering' and are due to deviations of the entire beam from the line of sight. In astronomical telescopes of large aperture these two effects combine to form a diffuse, enlarged 'tremor-disc'. The terms 'pulsation' or 'breathing' are used to describe a (fairly rapid) fluctuation of the cross-section of a propagated light beam, particularly laser beams, due to small angle scattering by the atmospheric inhomogeneities. This scattering also produces destructive interference within the beam, which in turn produces local fluctuations in amplitude and therefore intensity (scintillation). These intensity fluctuations appear as areas of bright and dark compared to the average intensity over the beam cross-section, as illustrated by the framing sequence from a cine record of the time variant beam structure introduced into a pencil beam by the atmosphere shown in Fig. 16.1.

'Scintillation' in its strictest sense refers only to (rapid) fluctuations in intensity, and in particular applies to point light sources (e.g. the 'twinkling' of stars). 'Image distortion' or 'blurring' is the integrated effect of image motion and pulsation involving many points which form the details of an extended object. It is clear that if such points move independently of, and out of phase with, one another, the details become blurred, lose contrast and cannot be distinctly recovered from the image. This is often loosely referred to as 'boiling'; more specifically, however, 'boiling' means the time-varying non-uniform illumination in a large spot image. 'Shimmer' is a general term often given to the tremulous appearance, apparent distortion and motion of an object seen, for instance, through a layer of air immediately above a heated surface. The quality

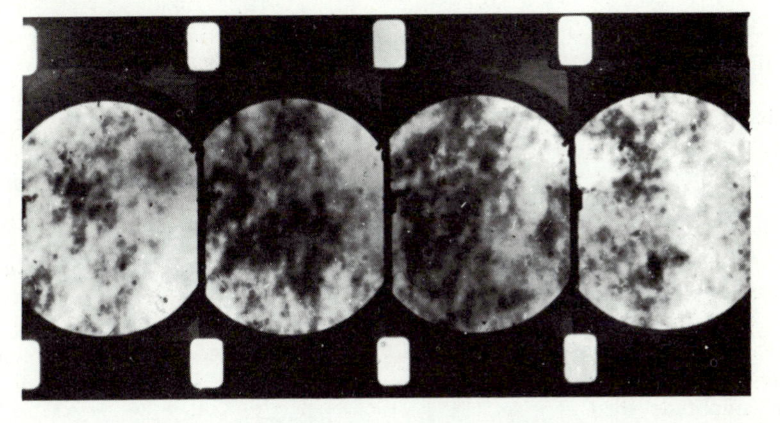

Fig. 16.1. The temporally varying induced structure in luminance distribution across a uniform beam of light projected through the atmosphere as recorded by a 16 frames/s cine record.

of image transmission through the atmosphere is frequently referred to as the 'seeing' quality.

16.3 THEORETICAL STUDIES

All the above effects are produced by amplitude and phase fluctuations within the propagating optical beam, and to give a quantitative discussion requires the solution of the electromagnetic wave equation with the appropriate boundary conditions. Unfortunately, there do not yet exist general analytical techniques for the exact solution of this equation to include the random refractive index fluctuations of the atmosphere. However, three approximation techniques have been developed and these will now be discussed.

16.3.1 Rytov Approximation Methods

The analysis of the effect of atmospheric turbulence on optical properties rests primarily on Tatarski[2], who based his work on the atmospheric theories of Kolmogoroff[1]. The latter author presented evidence for an inertial sub-range in the atmosphere, within which fluctuations of the properties of interest are isotropic. Tatarski applied this inertial sub-range concept to the effects of atmospheric turbulence on electromagnetic radiation, solving the resulting wave equation by employing the Rytov approximation[3], i.e. expressing the wave equation in terms of the logarithm of the amplitude (log-amplitude) and then applying a perturbation technique. A comprehensive discussion of this analysis technique is given by Strohbehn[4]. From this analysis, Tatarski arrived at a expression for the wave-structure function, $D(p)$.

The importance of this wave-structure function to atmospheric turbulenc theory (Fried[5]) lies in its ability to represent all the statistics of phase an log-amplitude fluctuations of a propagated wave, and it therefore contain information concerning the 'shape' of the deformed wavefront. By definitior

$$D(p) = D_\phi(p) + D_l(p) \tag{16.1}$$

where $D_\phi(p)$ is the phase structure function and $D_l(p)$ is the log-amplitud structure function, these being defined as follows:

If, at the points x and x' in the cross-section of the propagated wave, th phase and log-amplitude variations associated with the wavefront deformatio are denoted by $\phi(x)$ and $\phi(x')$ for the phase and $l(x)$ and $l(x')$ for th log-amplitude, then

$$D_\phi(p) = \langle [\phi(x) - \phi(x')]^2 \rangle \tag{16.2}$$

$$D_l(p) = \langle [l(x) - l(x')]^2 \rangle \tag{16.3}$$

where $p_s = x - x'$ and the brackets $\langle\rangle$ denote an ensemble (i.e. space) average Knowledge of the 'shape' of the deformed wavefront is necessary if the effect on an optical system are to be predicted.

Since the atmosphere must be considered as part of any optical system use under field conditions, it is convenient to talk of the Modulation Transfe Function (MTF) of the atmosphere as a measure of the effect of atmospheri turbulence on optical properties. Fried[6] distinguishes between two differen cases:

(1) the MTF obtained when a short exposure is used
(2) the MTF when a long exposure is used

although he does not specifically define what he means by a long or shor exposure. However, the two distinct cases arise because part of the wav distortion can be attributed to a random tilt of the wavefront (i.e. imag motion) which is time averaged for the long exposure situation. For very shor exposures, wavefront tilt only displaces an image, and so does not contribute t the short exposure MTF. Evaluating the MTF's for the two cases, Fried[6] foun

$$\langle \tau(f_s) \rangle_{LE} = \tau_0(f_s)\exp\{-\tfrac{1}{2}D(p)\} \tag{16.4}$$

for long exposures, and

$$\langle \tau(f_s) \rangle_{SE} = \tau_0(f_s)\exp\left\{-\tfrac{1}{2}D(p)\left[1 - \frac{\gamma p_s{}^{5/3}}{D}\right]\right\} \tag{16.5}$$

for short exposures, where $\langle \tau(f_s) \rangle$ is the transfer function of a complete imagin system, $\tau_0(f_s)$ is the transfer function of a diffraction-limited lens, p_s is th ample point separation in the aperture plane ($=\lambda X_L f_s$), f_s is the spatial 'requency in the image plane (cycles per unit length), D_L is the diameter of the ens, X_L is the focal length of the lens, λ is the wavelength and γ is a constant.

It can be seen that these two equations are identical apart from the factor [$1 - \gamma p_s^{5/3}/D_L$] in the short exposure case. The constant γ lies between ½ and , depending on the range of propagation. Coulman[7] discusses the exposure imes for which the above relationships hold and concludes that 'integration over).04 s does not constitute a 'short exposure' in the sense of Equation 16.5. It is of practical value to know that, in daytime conditions, an exposure appreciably horter than 0.04 s would be needed to remove entirely the effects of image novement.'

For infinite plane waves and horizontal propagation paths, Tatarski obtained esults for the wave-structure function $D(p)$, concluding that

$$D(p) = 2.91\, k_\lambda^2 p_s^{5/3} \int_0^L C_\mu^2(s)\,ds \qquad (16.6)$$

or $\sqrt{(\lambda L)} \gg l_0$ (the inner scale of turbulence) where k_λ is the wavenumber $= 2\pi\lambda - 1$), $C\mu$ is the refractive-index structure coefficient, representing the optical) 'strength' of the turbulence and s is the position along the propagation path.

Following Tatarski, Schmeltzer[8] applied the Rytov approximation to the study of the propagation of a laser beam through a turbulent atmosphere. However, since Schmeltzer presents his results in terms of multiple integrals with complex integrands, it is not immediately apparent what are the magnitudes of he effects indicated by his results. Fried[9] simplified the situation by evaluating he limiting case of Schmeltzer's results. He was thus able to derive the vave-structure function for the propagation of a spherical wave over a horizontal path, i.e.

$$D(p) = 2.91\, k_\lambda^2 p_s^{5/3} \int_0^L C_\mu^2(s) \left(\frac{s}{L}\right)^{5/3} ds \qquad (16.7)$$

or $\sqrt{\lambda L} \gg l_0$ which, aside from the factor $(s/L)^{5/3}$, is equivalent to the corresponding result for the propagation of an infinite plane wave as expressed by Tatarski (see Equation 16.6).

It will be seen, however, that it is often more convenient to express the wave tructure function in the following manner:

$$D(p) = Kp^{5/3} \qquad (16.8)$$

where K is a constant whose value depends on the propagation path length, the wavelength, the 'strength' of the turbulence along the path and the nature of the unperturbed wave front. Fried[5] notes that, for horizontal propagation, the spherical wave coefficient K is exactly 3/8 of the corresponding coefficient for an infinite plane wave propagating over the same path.

Before proceeding further, it must be stated that the validity of the Rytov approximation (also known as the 'Method of Smooth Perturbations') is the subject of much dispute. Theoretical calculations by Brown[10], Hufnagel and Stanley[11] and Strohbehn[12] indicate that the Rytov approximation is only of limited validity and not generally applicable as claimed. However, other authors (De Wolf[13] and Heidbreder[14]) believe that the limitation can be removed when a constant (chosen from energy conservation considerations) is added to the Rytov solution. In support of the view that no limitation exists, Fried[15] uses experimental data of Coulman[16] to verify the Rytov approximation outside the domain of validity defined by these theoretical calculations. However, Strohbehn[12] in a summary of recent work by Russian authors, particularly Tatarski, shows that as a result of this work the data of Coulman are at the limit of the valid range, not outside it.

This limit is determined from scintillation measurements (see Gracheva and Gurvich[17] and is usually expressed in terms of the propagation path length. From their experiments, Gracheva and Gurvich found that the limit on the path length was often as small as 1 km for strong daytime turbulence, but could be extended to distances of 100 km or more at night.

16.3.2 Averaging Methods

While they were expressing their doubts about the Rytov approximation, Hufnagel and Stanley[11] were formulating their own plane-wave solution to the electromagnetic wave equation. They reasoned that since image-degradation fluctuations are averaged to some degree over any exposure time it is only necessary to compute an averaged value of the degradation. By applying this idea to the solution of the wave equation, Hufnagel and Stanley arrived at an expression for the system MTF

$$\langle \tau(f) \rangle = \tau_0(f) M(\lambda X_L f_s) \tag{16.9}$$

for horizontal propagation, where $M(\lambda X_L f_s)$ is the averaged atmospheric MTF and is given by

$$M(\lambda X_L f_s) = \exp\{-\tfrac{1}{2} k_\lambda^2 \langle [S(p)]^2 \rangle\} \tag{16.10}$$

Assuming C_μ to be constant along the propagation path

$$\langle [S(p)]^2 \rangle = 2.91 p_s^{5/3} L C_\mu^2 \tag{16.11}$$

Comparing this result with the long exposure MTF calculated by Fried (Equation 16.4), it can be seen that the two results exactly agree for plane waves. Fried[6] comments on this exact agreement and concludes: 'The Hufnagel and Stanley and Rytov approximations are so fundamentally different that, before seeing the solutions they lead to, no one could have reason to expect that they would result in the same solution – unless the approximations are sufficiently accurate that they both yield the exact solution'. He expressed this conclusion even though Chase[18] has indicated that Hufnagel and Stanley have an unevaluated approximation in their results.

16.3.3 Geometrical Optics Approximations

Duntley, Culver, Richey and Preisendorfer[19] considered that a perturbation approach to the solution of the wave equation yielded results in an unattractive form for direct application to the transmission of optical images. They therefore decided to study the problem of image transmission through an inhomogeneous medium by using a geometrical optics approximation. Duntley *et al* followed Chandrasekhar[20] in assuming that the atmosphere consists of a series of discrete eddies whose refractive indices differ from some mean value. Only small-angle deflections are considered and diffraction effects are ignored, i.e. geometrical optics are limited to the region in which the effect of amplitude fluctuations is small compared to the effect of phase fluctuations (Strohbehn[4]). The latter assumption leads to a restriction on range, first given by Tatarski as

$$\sqrt{\lambda L} \ll l_0 \tag{16.12}$$

i.e. diffraction effects must be small compared to even the smallest eddy sizes (see Strohbehn[4] and deWolf[13]). Alternatively, this restriction may be thought of as placing an upper limit on the number of eddies allowed in the propagation path for a geometrical optics solution to be valid. The true picture, however, is one of *many* intermixed eddies and so diffraction effects should not be ignored.

Many authors have accepted the above restriction on range. Nevertheless, Beckman[21] notes that formulae derived from simple geometrical optics often appear to agree with other methods for propagation paths outside these limits. This led Taylor[22] to investigate the accuracy of the restriction. He found that within the single scatter region, the restriction on the propagation path length can be significantly relaxed, the new limitation being expressed as

$$l_0{}^{11/3} L_0{}^{1/3} \gg (\lambda L/27)^2 \tag{16.13}$$

Within these limits, Duntley *et al*[19] calculated that a continuum of light rays starting with a given intial direction and traversing a turbulent medium will give rise to a normal distribution of intensity, i.e. the angular spread of the image forming rays is normally distributed. Rogers[23] shows that this means that a line object will have a spread function which is gaussian. Mathematically, Duntley *et al* express the intensity profile as

$$P(\psi) = [\sigma \sqrt{(2\pi)}]^{-1} \exp\left\{-\frac{\psi^2}{2\sigma^2}\right\} \tag{16.14}$$

Rogers shows that this is equivalent to a normalised transfer function of

$$M(\lambda X_L f_s) = \exp\{-2\pi^2 \sigma^2 f_s{}^2\} \tag{16.15}$$

or, in angular units

$$M(f_s)_i = \exp\{-2\pi^2 \langle \Delta\theta^2 \rangle (f_s)_i^2\} \tag{16.16}$$

where: ψ is an angular variable, σ is the standard deviation, $\Delta\theta$ is the angular deviation, $(f_s)_i$ is the spatial frequency in the image plane (c/mrad.)

From the derivation of the intensity profile, it is evident that the quantity $\langle \Delta\theta^2 \rangle^{1/2}$ should exhibit a square-root dependence on range.

A similar approach (i.e. using geometrical optics to explain wave propagation through an atmosphere of discrete turbulent eddies) was also made by Weiner[24]. By assuming the eddy size of the turbulence to be L_0 and each eddy to act as though it were wedge shaped, Weiner arrived at an expression for $\langle \Delta\theta^2 \rangle$, given by

$$\langle \Delta\theta^2 \rangle = 2\overline{\Delta\mu^2} \frac{L}{3L_o} \tag{16.17}$$

where $\overline{\Delta\mu^2}$ is the mean square (temporal) of refractive-index fluctuations.

In general, for a random variable (such as μ), time and space averaging can be considered identical. Thus

$$\langle \Delta\theta^2 \rangle = 2\langle \Delta\mu^2 \rangle \frac{L}{3L_o} \tag{16.18}$$

This agrees well with the result of Chandrasekhar[20], which, expressed in our notation, becomes

$$\langle \Delta\theta^2 \rangle = \text{constant} \langle \Delta\mu^2 \rangle \frac{L}{L_o} \tag{16.19}$$

(Note that the results of both Weiner and Chandrasekhar predict a square-root dependence on range for $\langle\Delta\theta^2\rangle^{1/2}$).

16.4 PARAMETER EFFECTS

It can be seen that two of the above solutions are identical for plane-waves propagating over a horizontal path. Both predict a five-thirds power law of spatial frequency. In the published literature, development of the theory of wave propagation through turbulent media to include the effects of receiver altitude, receiver aperture, slant paths, etc. is based on this five-thirds relationship.

16.4.1 Non-Horizontal Paths

For an object located above ground level, and viewed at a zenith angle ϕ by a lens system at an altitude H, Fried[25] replaces the integration variable s of the wave-structure, Equations 16.6 and 16.7, by $(h - H) \sec \phi$ where h is the new (height) integration variable, i.e.

$$h = H + s\cos\phi \tag{16.20}$$

Hufnagel and Stanley[11] and Chandrasekhar[20] also suggest a cosine relationship for the effect of turbulence along slant paths. Although this cosine relationship ignores the effect of the curvature of the earth, Evvard[26] shows that while the radial density gradient associated with this curvature may shift the apparent position of an object, it does not decrease the resolution of the atmosphere.

For plane-waves, the wave-structure function now becomes

$$D(p) = 2.91 k_\lambda^{\,2} p_s^{\,5/3} \sec \phi \int_H^{H_1} C_\mu^{\,2}(h)\mathrm{d}h \tag{16.21}$$

where H_1 is the height of the object. Alternatively, from Equation 16.8,

$$D(p) = K_p^{\,5/3}$$

where K is now a constant whose value depends on the propagation path length, zenith angle, the wavelength, the 'strength' of the turbulence along the (slant) path and the nature of the unperturbed wavefront.

The relationship expressed by Equation 16.21, however, must be treated with care for large values of ϕ (i.e. near horizontal paths), since sec ϕ rapidly approaches infinity and H_1 tends to H. For the vertical propagation of plane-waves, the wave structure function simplifies to

$$D(p) = 2.91 k_\lambda^{\,2} p_s^{\,5/3} \int_{\substack{\text{vertical} \\ \text{propagation path}}} C_\mu^{\,2}(h)\mathrm{d}h \tag{16.22}$$

The situation is not so clear for spherical-waves because the form of the wave-structure function depends on the direction of propagation (or, more exactly, the point from which the path length is measured). For a spherical wave propagated from the ground, Fried[6] finds the wave-structure function measured at an altitude L directly above the propagating source (i.e. the vertically downward viewing situation) to be of the form

$$D(p) = 2.91 k_\lambda^{\,2} p_s^{\,5/3} \int_0^L C_\mu^{\,2}(h) \left(\frac{h}{L}\right)^{5/3} \mathrm{d}h \tag{16.23}$$

When the positions are reversed (i.e. the vertically upward viewing case), the wavefront position is no longer measured from the originating source, but from the observer. Sutton[27] shows that, for this case, $(p_s h/L)^{5/3}$ should be replaced by $p_s(1 - h/L)^{5/3}$. This is then equivalent to diminishing the effect of those fluctuations at high altitudes furthest from the receiver optics, i.e. allowance is made for the location of the turbulence along the optical path.

16.4.2 Aperture Dependence

We have already seen in Equation 16.8 that the structure function can be written as

$$D(p) = Kp^{5/3}$$

where the constant K has now been redefined to include also the direction of propagation as one of its variables. Following Fried[6], it is now convenient to define two new quantities p_0 and $\mathcal{R}$, where

$$p_o = \left(\frac{6.88}{K}\right)^{3/5} \tag{16.24}$$

and

$$\mathcal{R} = \int_{\text{all } f_s} \langle \tau(f_s) \rangle \, \mathrm{d}f_s \tag{16.25}$$

The quantity p_0 is thus another constant which depends on the same variables as K, while $\mathcal{R}$ is an alternative measure of the performance of the imaging system and is referred to by Fried as the resolution. (The relation of this 'resolution' to usual resolution criteria is discussed by Birch[28]). It can be seen that $\mathcal{R}$ is defined to be the integral over spatial frequencies of the system's (i.e. atmosphere plus lens) ensemble-average MTF.

Fried evaluated the resolutions from the long-exposure MTF (Equation 16.4) and the short-exposure MTF for the two extreme values of γ (Equation 16.5), referred to as the near-field case for $\gamma = 1$ and the far-field case for $\gamma = ½$. In these equations, $\tau_0(f_s)$ is the MTF of a diffraction-limited lens and is given by Fried[6] as

$$\tau_0(f_s) = \begin{cases} 2\{\cos^{-1}(\lambda x_L f_s/D_L) - \lambda x_L f_s[1 - (\lambda x_L f_s/D_L)^2]^{1/2}/D_L\}/\pi & \text{if } \lambda x_L f_s \leqslant D_L \\ 0 & \text{if } \lambda x_L f_s > D_L \end{cases} \tag{16.26}$$

If the lens diameter is allowed to become large, then, for $\lambda x_L f_s \leqslant D_L$, $\tau_0(f_s)$ tends to unity and so $\langle \tau(f_s) \rangle$ becomes the MTF of the atmosphere alone. Thus, it is now possible to determine the atmospheric resolution limit, i.e.

$$\mathcal{R}_{max} = \lim_{D_L \to \infty} \mathcal{R}(\text{Long exposure}) \tag{16.27}$$

This value $\mathcal{R}_{max}$ implies that atmospheric turbulence places an absolute upper limit on the resolution that can be obtained with a long exposure through the atmosphere. (The long exposure MTF is used in this context, since any viewing system necessarily has a long exposure time – see Coulman[7]. Fried calculates the value of $\mathcal{R}_{max}$ as

$$\mathcal{R}_{max} = \pi p_0{}^2/4(\lambda L)^2 \quad c^2/m^2 \text{ on the ground} \tag{16.28}$$

Therefore, according to Fried, the 'minimum resolvable length' Δl, due to atmospheric turbulence alone is given by $1/2\mathcal{R}_{max}{}^{½}$

$$\Delta l = \frac{\lambda L}{p_o\sqrt{\pi}} \tag{16.29}$$

The dependence of the normalized resolution, $\mathcal{R}/\mathcal{R}_{max}$, on the normalized lens diameter, D_L/p_o, for the three exposure cases is shown in Figure 16.2

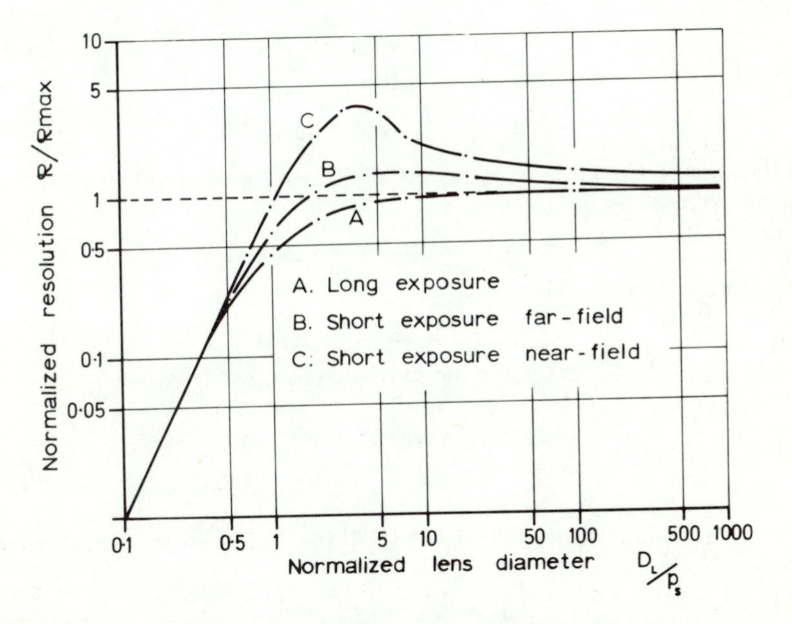

Fig. 16.2. The dependence of normalised resolution on normalised lens diameter for the limiting exposure and path length situations (Reproduced from Fried[6] by courtesy of the Optical Society of America).

(reproduced from Fried[6]). Two additional effects are also clearly indicated:–

(1) Significantly better resolutions are obtained when short exposures are used.
(2) For long path lengths (i.e. the far-field situation), resolution becomes essentially independent of exposure time.

However, Fried notes that it is often difficult to determine whether an experiment is carried out under near-field or far-field conditions. It is therefore convenient that most imaging experiments for studying propagation theory turn out to be applicable to the long exposure theory.

16.4.3 Optical Turbulence 'Strength'

In theoretical derivations, the optical 'strength' of turbulence is represented by $C\mu$, the refractive-index structure coefficient (Tatarski[2]). The significance of this structure coefficient to turbulence theory is illustrated by the following reasoning, as discussed by Hodara[29].

There is considerable evidence (both experimental and theoretical) to suggest that, in the inertial sub-range

$$C\mu^2 = D\mu(p)p_s^{-2/3} \tag{16.30}$$

where $D\mu(p)$ is the refractive-index structure function defined as

$$D\mu(p) = \langle [\Delta\mu(x+p_s) - \Delta\mu(x)]^2 \rangle \tag{16.31}$$

(cf. Equations 16.2 and 16.3). Assuming locally homogeneous turbulence within the inertial sub-range, i.e.

$$\langle \Delta\mu^2(x+p_s) \rangle = \langle \Delta\mu^2(x) \rangle \equiv \langle \Delta\mu^2 \rangle \text{ (say).} \tag{16.32}$$

then, from Equation 16.31

$$D\mu(p) = 2[\langle \Delta\mu^2 \rangle - \Gamma\mu(p)] \tag{16.33}$$

for $L_0 \geqslant p_s \geqslant l_0$, where $\Gamma\mu(p)$ is the correlation function defined by

$$\Gamma\mu(p) = \begin{cases} \langle \Delta\mu(x+p_s) \cdot \Delta\mu(x) \rangle & \text{for} \quad l_o \leqslant p_s < L_o \\ 0 & \text{for} \quad p \geqslant L_o \end{cases}$$

It therefore follows from Equations 16.30 and 16.33 that, as an approximation

$$C\mu^2 \approx \langle \Delta\mu^2 \rangle \cdot L_0^{-2/3} \tag{16.34}$$

This expresses the relationship between the refractive-index structure coefficient $C\mu$ and the refractive-index fluctuations $\langle \Delta\mu^2 \rangle$ (see also Strohbehn[4]).

The refractive-index structure coefficient, through its dependence on the vertical temperature gradient (see Hulett[30]) shows a complete dependence on the time of day, season, terrain, height and meteorological conditions. The effect of these variables on the magnitude of turbulence in the atmosphere will therefore be automatically covered by discussions of C_μ.

In practice, the enormous number of possible combinations of the above parameters (if only from meteorological considerations) precludes the prediction of a single value of C_μ for a particular experiment. For this reason, any experimental determinations of the wave-structure function should always be accompanied by sufficient data to allow C_μ to be evaluated. (Tatarski[2] shows that a knowledge of the vertical temperature gradient profile along the path satisfies this requirement for optical wavelengths). The normal procedure followed in the literature has, therefore, been simply to represent average values of C_μ. Hufnagel and Stanley[11], note, however, that in individual situations significant departures from the average can be expected and that even the averages themselves may be significantly in error.

There is no simple rule for estimating C_μ. Davis[31] characterizes average daytime conditions a few metres above the earth's surface in the following way:

$$\text{Weak turbulence } C_\mu = 8 \times 10^{-9} \text{ m}^{-1/3}$$

$$\text{Intermediate turbulence } C_\mu = 4 \times 10^{-8} \text{ m}^{-1/3}$$

$$\text{Strong turbulence } C_\mu = 5 \times 10^{-7} \text{ m}^{-1/3}$$

These categories agree well when compared to the experimental data for this region quoted by Deitz and Wright[32], Goldstein, Miles and Chabot[33], Ochs, Bergman and Snyder[34], Fried, Mevers and Keister[35], Newton *et al*[36], Wright and Schutz[37]. The data of Wright and Schutz show clearly

(1) the diurnal variation of C_μ,
(2) the large variation (nearly 100:1) that can be typically expected,
(3) the maximum value corresponding to the period of maximum solar flux.

Determinations of C_μ require either a knowledge of the vertical temperature gradient along the entire optical path (but normally only measured at one point close to the receiver optics due to cost and complexity considerations) or an experimental measurement of the magnitude of scintillation effects. These latter measurement are usually made from laser beam propagation experiments, and the value of C_μ obtained is therefore an average value for the whole propagation path. However, care must be taken when using this method because there is evidence to suggest that the scintillation 'saturates', i.e. an increase in turbulence or propagation path length produces no further increase in scintillation (see Gracheva and Gurvich[17]). This 'saturation' phenomenon has since been reported by several workers, particularly by Newton *et al*[36] and Deitz *et al*[32].

Measurement of how C_μ varies with altitude is not easy and has to be approached indirectly. Hufnagel and Stanley[11] overcame the problem by first

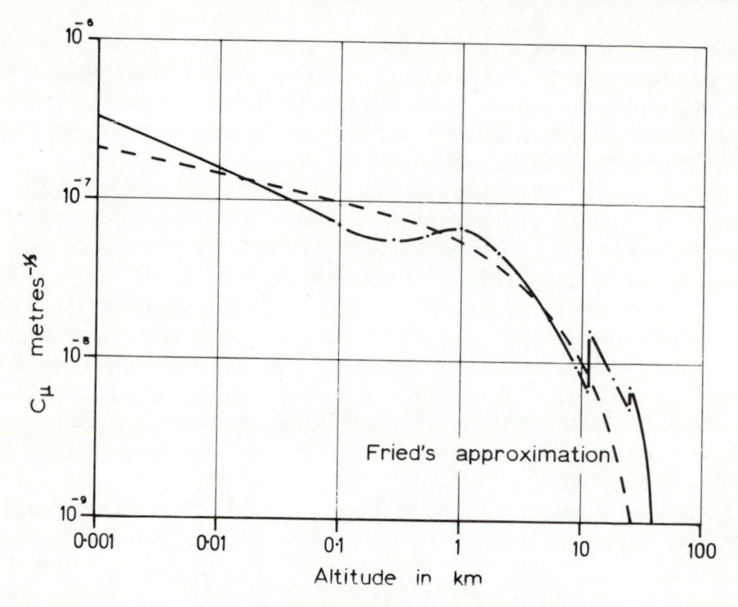

Fig. 16.3. The variation of the refractive index structure coefficient C_μ with altitude as determined from experimental data for 0 to 1 km, region and data from a model atmosphere. (Reproduced from Hufnagel and Stanley[11] *by courtesy of the Optical Society of America). Fried's exponential approximation is also shown.*

expressing the structure constant in terms of meteorological parameters and then using available experimental data for the 0 to 1 km region plus data from a model atmosphere above this region. The net result of this analysis was a graph of C_μ versus height, reproduced in Figure 16.3. (The discontinuities shown by this C_μ/height curve result from the model atmosphere and are not physically significant). For his analytical models of turbulence, Fried[9] approximated this C_μ curve of Hufnagel and Stanley by an experimental function of the form

$$C_\mu^{\,2} = 4.2 \times 10^{-14} h^{-1/3} \exp\{h/h_0\} \mathrm{m}^{-2/3} \tag{16.35}$$

where $h_0 = 3\,200$ m.

Obviously this equation cannot be valid down to ground level and so can only be strictly applied if the receiver optics are above any serious ground effects. The form of this exponential is also included in Fig. 16.3.

An alternative attitude function of C_μ for small values of h, as recently proposed by Brooker[38], is of the form

$$C_\mu^{\,2}(h) = C_{\mu_0}^{\,2} h^{-z} \exp\left(-\frac{h}{h_o}\right) \tag{16.36}$$

where $h_0 = 320$ m and z and $C_{\mu_0}{}^2$ are both variables according to time of day. Brooker suggests values of z and $C_{\mu_0}{}^2$ as shown in Table 16.1, where it will be seen that z, in particular, is considerably different for all times of day to the value of 1/3 implied in Equation 16.35.

TABLE 16.1. Values of z and $C_{\mu_0}{}^2$ for various times of day and conditions.

Conditions	z	$C_{\mu_0}{}^2$
Sunny day	5/6	3.6×10^{-13}
Clear night	1	1.6×10^{-13}
Dawn/dusk (minimum turbulence period)	2/3	8.7×10^{-15}

Hulett[30] attempted to verify the C_μ/height relationship of Hufnagel and Stanley by estimating values of C_μ from the many measurements of stellar scintillation available in the literature. This investigation led Hulett to the conclusion that the values given by Hufnagel and Stanley are too large and suggests that, as a first approximation, they should be multiplied by a factor of 0.19. He also quotes a private communication which shows that Hufnagel has since revised the C_μ curve (again using scintillation data) and that the new values approximately agree with his own calculations. When the above correction is applied to the Hufnagel and Stanley data, the quoted value of C_μ close to the ground then corresponds to the case of intermediate turbulence as expressed by Davis[31].

The accuracy of Fried's approximation is demonstrated by a calculation performed by Fried himself[9]. Using his experimental relationship, the worst ground resolution Δl_0 that could be expected for spherical waves propagating along a vertical path through the atmosphere is found to be approximately 4½ cm. Repeating the calculation using numerical integration of the Hufnagel and Stanley data, however, gives a value of approximately 10 cm. Fried attributes this difference to a combination of the weight given to the high altitude values of C_μ by the (spherical wave) factor $(s/L)^{5/3}$ with the inaccuracy of the analytic approximation at these altitudes.

This exponential function suggested by Fried greatly simplifies any calculation of the way the minimum resolution length Δl varies with altitude. The height dependence for the case of spherical wavefronts and vertically downward viewing paths given by Fried[9] is clearly shown in Fig. 16.4. The angular resolutions implied by this curve are always better than 0.006 mrad and consequently can be considered to be negligible for most circumstances.

The minimum resolution length $\Delta l'$ for spherical waves propagating in the reverse direction, however, is shown by Farhat and DeCou[39] to be related to Δl

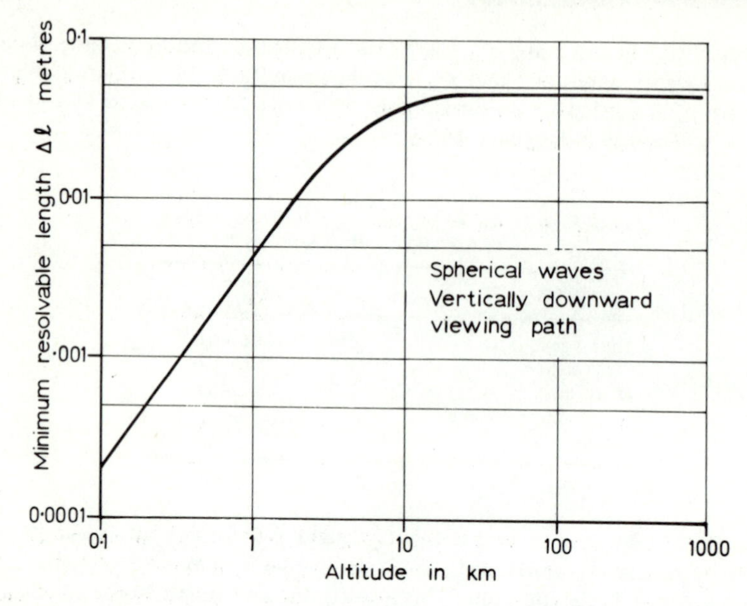

Fig. 16.4. The dependence of minimum resolvable length on altitude. (Reproduced from Fried[9] by courtesy of the Optical Society of America).

by

$$\left(\frac{\Delta l'}{\Delta l}\right)^{5/3} = \frac{\int_0^L C_\mu{}^2(h) \cdot (L-h)^{5/3} \cdot \mathrm{d}h}{\int_0^L C_\mu{}^2(h) \cdot h^{5/3} \cdot \mathrm{d}h}$$

Thus, $\Delta l'$ is greater than Δl.

In an effort to simplify the estimation of resolution achievable in a given situation, Titmuss[40] has generated a series of graphs based on the major work reported above. With the aid of these graphs (too bulky to reproduce here) it becomes a comparatively simple matter to determine the theoretical effect of any given turbulence situation for which a minimum set of parameters are known. These parameters must include the slant viewing range, the elevation of the path, the path length over which turbulence is effectively generated (if less than the complete path), the turbulence strength C_μ and the diameter of the receiving optics.

16.5 EXPERIMENTAL MEASUREMENTS OF TURBULENCE EFFECTS

The majority of papers in the published literature have been primarily concerned with the development of a theory of atmospheric turbulence and its effect on the performance of optical systems. Relatively few experimental data have been quoted for resolution measurements, probably because few such data exist due to the emphasis placed on scintillation measurements, both stellar and laser propagation, for theoretical verifications.

Early experimental work in this field was mainly concerned with the accuracy with which telescopes could be pointed at distant objects under conditions of high turbulence. For example, Washer and Williams[41] determined the probable error of a single telescope-pointing, under average daylight conditions, over a terrain consisting partly of grassland and partly of residential area. They found that for a path high above the ground the average error was of the order of 0.003 mrad.

First attempts to measure the resolution through a turbulent atmosphere were made by Riggs *et al*[42], who photographed a black and white line target from a distance of several hundred metres. They then determined the root mean square angular deviation, $[\langle \Delta\theta^2 \rangle]^{1/2}$, suffered by light travelling from a point on the target to the camera. For a viewing path at an average height of about 12 m above a predominantly level, grassy terrain, this deviation was often found to be in excess of 0.014 mrad, with a maximum approaching 0.05 mrad in bright sunshine. Under these experimental conditions Riggs *et al* state that, although no measurements were made, the probable error of a single pointing of a telescope would be greatly in excess of the 0.003 mrad reported by Washer and Williams[41].

Using a similar technique, the experiments of Smith *et al*[43] have shown that the extent to which turbulence degrades the photographic resolution is markedly dependent upon the location of the disturbance along the optical path. Turbulence near the camera significantly reduces the mean resolution, while the same turbulence located at the target end of the optical path has little or no effect. The results of a laboratory experiment to show this effect are summarized by Fig. 16.5. Similar measurements over a level, grassy plain under average daytime conditions showed that the resolution can be improved at short distances (i.e. less than 200 m) by using short exposures, but that, when the optical path is very long, the resolution becomes essentially independent of exposure time (Fig. 16.6). This again supports the interpretation that it is turbulence near the receiver optics which causes the most loss of resolution, since an increase in range can be considered as simply increasing the turbulence at the target end of the optical path. The curves of Fig. 16.6 also provide some verification of the near-field and far-field results of Fried[6] illustrated in Fig. 16.2.

One of the first experiments designed to measure the transfer function of the

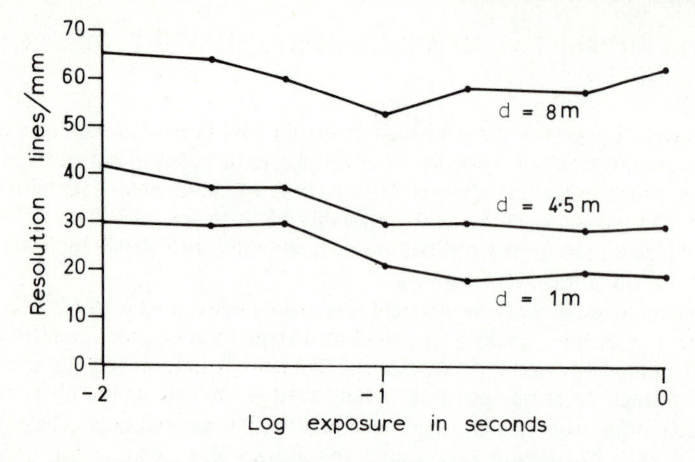

Fig. 16.5. The dependence of resolution on the distance d *of a local patch of turbulent air from a recording camera and on the exposure time. (Reproduced from Smith* et al[43] *by courtesy of the Optical Society of America).*

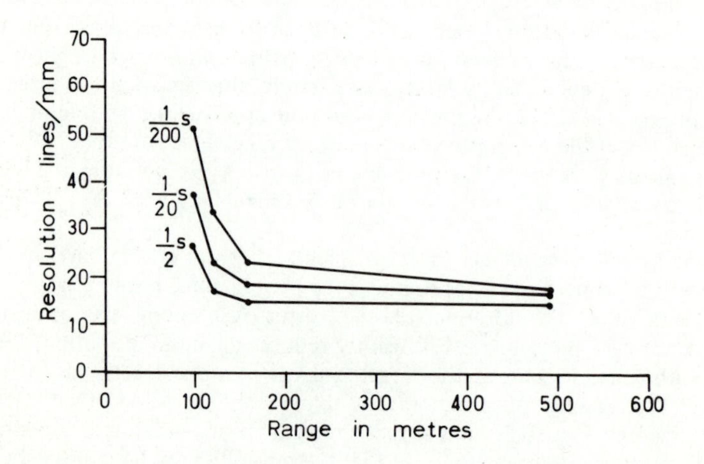

Fig. 16.6. The dependence of resolution of path length for various exposure times when the viewing path is over a grassy plain. (Reproduced from Smith et al[43] *by courtesy of the Optical Society of America).*

atmosphere was carried out by Djurle and Bäck[44]. The experimental arrangement consisted of a slit light source scanned by a photo-electric telephotometer placed 7 miles away. Hufnagel and Stanley[11] have compared the results of this experiment with their own plane-wave calculations and found at least an order of magnitude agreement. Hulett[30] shows that, when the results of Hufnagel and

Stanley are corrected for a point source (i.e. spherical waves) and then applied to the experimental conditions of Djurle and Bäck, the agreement is even better.

A similar light propagation experiment (i.e. a slit light source scanned with a telephotometer) was later performed by Coulman[45] over a grass range of 150 m. Measurements were made only on days when the sky remained either clear or uniformly overcast for periods of several hours. Coulman found that the transfer function fluctuated rapidly about a mean level which itself varied slowly according to the mean meteorological conditions. The form of the transfer function was also found to be similar to that deduced by Hufnagel and Stanley. In an improved version of this experiment, Coulman[16] set out to investigate quantitatively the transfer function of the atmosphere and its dependence on the propagation path length. By fitting his experimental data to empirical formulae, Coulman was able to determine three important relationships which govern image transmission through the atmosphere. These were:

(1) the power law of spatial frequency,
(2) the dependence of the averaged transfer function on the propagation range,
(3) the dependence of the root mean square angular deviation, $\langle \Delta\theta^2 \rangle^{1/2}$, on on the propagation range.

The calculations of the power law of spatial frequency produced a mean value of 1.65, which is remarkably near to the 5/3 value predicted by the theories of Tatarski, Fried and Hufnagel and Stanley. The dependence of the averaged transfer function on the propagation range was also found to satisfy these theoretical predictions, the value of 1.06 for the power of the range again being close to the theoretical value of unity. Unfortunately the third result does not continue this agreement, since $\langle \Delta\theta^2 \rangle^{1/2}$ was found to vary as the power 0.43 of range, which compares well with the predictions of Duntley *et al* based on Gaussian laws instead.

In an extensive set of experiments, Rogers[23,46,47] devised a photographic technique from which the atmospheric MTF could be derived and from which the effect of such parameters as viewing range, zenith angle, camera height and target height could be easily examined. His method involved photographing a line target from a range of several hundreds of metres with a telephoto lens to produce a line image on the film. The intensity distribution in the film image was then measured with a microdensitometer to give the spread function of the imaging system (including the atmosphere). The Fourier transform of this function was the transfer function of the entire optical system, from which Rogers retrieved the required atmospheric MTF. Rogers showed that his experimental MTF data could be fitted to the theoretical expression of Equation 16.16, i.e.

$$M(f_s)_i = \exp\{-2\pi^2 \langle \Delta\theta^2 \rangle \nu^2\}$$

i.e. the atmosphere behaves in a truly random manner as suggested by Duntley *et al.* This then enabled him to express each of the variables under investigation in terms of $\langle \Delta\theta^2 \rangle^{1/2}$, the root mean square angular deviation, for simplicity. By repeating his experiment over different viewing distances, Rogers[46] obtained a series of MTF curves of the above form but with different values of $\langle \Delta\theta^2 \rangle$. He found that the range dependence could then be expressed as

$$\langle \Delta\theta^2 \rangle^{1/2} = KL^{0.44} \tag{16.37}$$

(where K is a constant), which compares well to the square root variation predicted by Duntley *et al*[19]; c.f. also the experimental data of Coulman[16]. Extending his investigations to sightlines inclined to the horizontal, Rogers[23] found that $\langle \Delta\theta^2 \rangle^{1/2}$ decreases rapidly with elevation for the first few tens of milliradians, after which it tends to reach a constant level (Fig. 16.7). In another similar experiment, however, Rogers[46] shows that the improvement in resolution that results from raising the target is negligible in comparison to the improvement achieved when the camera (or receiving optics) is raised to the same height. This again supports the idea that it is the turbulence nearest the receiving optics which causes the most loss of resolution.

In an attempt to clarify the theoretical situation, Newton *et al*[36] were able to make use of data from a series of atmospheric propagation trials. Although these trials were primarily concerned with the ability of the atmosphere to propagate

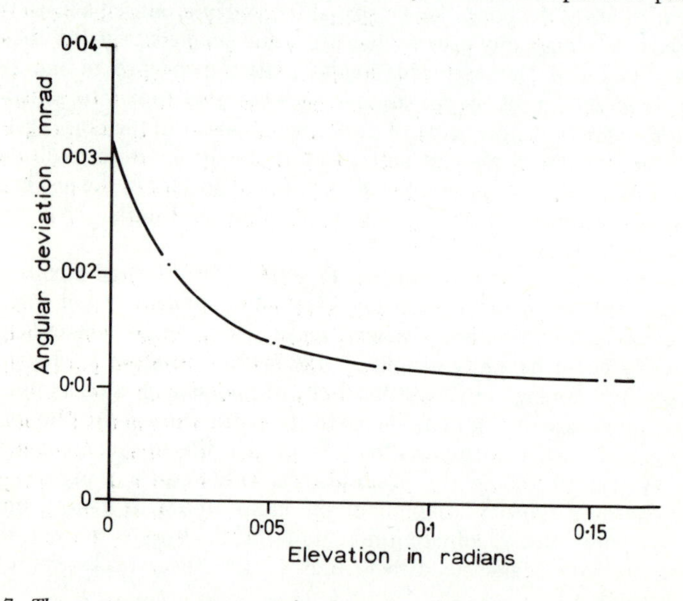

Fig. 16.7. The root mean square angular deviation $\langle \Delta\theta^2 \rangle^{1/2}$ as a function of sight line elevation. (Reproduced from Rogers[23] by courtesy of the Weapons Research Establishment Australian Department of Defence).

spatially modulated optical beams in both the visible and infra-red regions of the spectrum, viewing resolution experiments were also included to enable direct comparisons to be made with the experimental work of Rogers. Film recording techniques were again used. For both cases, the atmospheric MTF was characterized by

$$M(f_s) = \exp\{-F(w)f_s^{\,z}\} \tag{16.38}$$

where $F(w)$ is a numerical factor influenced by the path length, the wavelength, the 'strength' of the turbulence, the focal length of the optics and the ambient weather conditions, z is an unknown constant (not necessarily an integer).

It was found that fitting the experimental data to an equation of this form yielded a value for z of 1.91, with no significant variation over a very wide range of measuring conditions. This value was also found to be independent of the method of measurement (i.e. projection or viewing). Although this result is at variance with all the theoretical studies, it does show some agreement with the work of Duntley *et al*[19] (see also the experimental data of Rogers[47]).

16.6 CALCULATION OF C_μ FROM TEMPERATURE GRADIENT DATA

If the mean temperature varies with height in a logarithmic manner[2], i.e.

$$\bar{T}(h) = \text{constant} + T_* \log_e(h/h_0)$$

where $\bar{T}(h)$ is the temperature at a height h above some reference height h_0, then T_*, which is constant for any given atmospheric conditions irrespective of the sensor height, represents the vertical temperature gradient. Therefore, for two sensors at heights h_1 and h_2 above ground level, the temperature difference between them is given by

$$\bar{T}(h_1) - \bar{T}(h_2) = T_* \log_e(h_1/h_2)$$

Thus

$$T_* = \frac{\bar{T}(h_1) - \bar{T}(h_2)}{\log_e(h_1/h_2)} \tag{16.39}$$

For normal daytime turbulence close to the ground, the required relationship between C_μ, the refractive index structure coefficient, and T_* is given by Tatarski[2] as

$$C_\mu = 2.40K^{2/3}h^{-1/3}10^{-6}T_* \tag{16.40}$$

where K is a constant of approximate value 0.4. Thus

$$C_\mu \approx 1.3h^{-1/3}10^{-6}T_* \tag{16.41}$$

The above expressions for C_μ are only valid up to heights of the order of several tens of metres (i.e. $h = 30$ to 50 m). For heights in excess of this, C_μ is suggested by Fried to be given by Equation 16.35.

C_μ can also be determined from measurements of C_T, the temperature structure coefficient[2]. This is defined as

$$\overline{(T_1 - T_2)^2} = C_T{}^2 x^{2/3} \tag{16.42}$$

where T_1 and T_2 are the temperatures at two points a horizontal distance x apart (the bar denotes a time average). Thus, C_T is the rms temperature difference divided by the cube root of the distance between the sensors.

C_μ is then related to C_T by

$$C_\mu = \frac{69 \times 10^{-6}}{T^2} \rho C_T \tag{16.43}$$

where ρ is the pressure in millibars and T is the temperature in degrees Kelvin.

C_T can be measured directly from Equation 16.42 using two hot wire probes a known distance apart. The difference in the rates of cooling of the hot wires gives a measure of the temperature difference.

16.7 DISCUSSION

There have been three main approaches to the solution of the wave equation for optical waves propagating through a turbulent medium. Two have predicted (with no restrictions) a 5/3 power law for the spatial frequency dependence of the transfer function and have given identical plane-wave solutions for the atmospheric MTF along horizontal paths, i.e.

$$M(\lambda X_L f_s) = \exp\left\{-5.82\pi^2 \Lambda_L{}^{5/3} \lambda^{-1/3} f_s{}^{5/3} \int_0^L C_\mu{}^2(s)\, \mathrm{d}s\right\}$$

(Spherical waves simply include a further factor of $(s/L)^{5/3}$ in the integration). The third study, a geometrical optics derivation known to be of limited validity, predicts a square law of the following form:

$$M(\lambda X_L f_s) = \exp\{-2\pi^2 \sigma^2 f_s{}^2\}$$

However, it has been seen that derivations of the 5/3 relationship have been the centre of much critical discussion, with the result that their range of validity is now questioned.

Beckman[21] made the observation that geometrical optics predictions often appear to agree with other apparently more rigorous studies outside the limits imposed by Tatarski[2]. These observations, together with the discussions concerning the validity of the Rytov approximation, led Taylor[22] to investigate the whole theoretical situation from a range-of-application viewpoint. We have already seen that he was able to relax significantly the geometrical optics restrictions. In another paper, however, Taylor[48] discusses the other theoretical studies and concludes that all appear to be valid within exactly the same limits, each one being restricted to those regions where only single-scattering effects are important (i.e. the region where geometrical optics are applicable). If this is so, it therefore limits all theoretical predictions to those path lengths determined for the geometrical optics derivations (i.e. to propagation paths less than 1 km). This restriction, however, appears to be incompatible with the experimental evidence, which suggests an increase in path length has little effect on the performance of an optical system.

Even though doubts have been expressed about the usefulness of the theoretical studies, further development of the theory is based on the 5/3 relationship. The effect of slant paths and various receiver apertures on the optical resolution through the atmosphere have been considered separately, whereas it was found possible to group all other parameters within the discussions of C_μ.

The slant path dependence is expressed by a cosine rule of the form

$$M(\lambda X_L f_s) = \exp\left\{-5.82\pi^2 \Lambda_L{}^{5/3} \lambda^{-1/3} f_s{}^{5/3} \sec\phi \int_0^L C_\mu{}^2(h)\,\mathrm{d}h\right\}$$

(plane waves), since C_μ is known to vary with altitude. Whilst vertical paths are simply a special case of slant paths (i.e. $\phi = 0$), care must be exercised when near-horizontal paths (i.e. large ϕ) are considered, since sec ϕ rapidly approaches infinity as ϕ increases. For spherical waves, the direction of propagation (or viewing) must also be included in any calculations.

The dependence of the resolution on the lens diameter was found to be of the form illustrated by Fig. 16.2 which clearly shows that provided the path is short, i.e. the near-field case, the exposure has a significant effect. This range/exposure interaction is verified by the experimental data of Smith *et al*[43] reproduced in Fig. 16.6, and shows that the near-field interpretation only applies to paths of less than 150 m.

The most significant parameter in turbulence theory is the one which represents the (optical) 'strength' of the turbulence (i.e. the refractive-index structure coefficient, C_μ). To predict values of C_μ is difficult, with the best estimations probably coming from vertical temperature gradient data. Such data concerning the vertical temperature gradient encountered in different parts of the world do exist[49], while the probability distribution of the vertical temperature gradient in the UK with different sites, times of day and seasons is also available[50]. Whilst these data will undoubtedly be useful, they cannot provide the complete story. It has already been stated that cost and complexity considerations restrict vertical temperature gradient measurements to one point along the propagation path. This position is therefore usually chosen so that the vertical temperature gradient is measured in the region where the turbulence has its most optical effect (i.e. near the receiver optics). Thus, in this way, the effect of any variation in the turbulence strength along the optical path can be minimized to a small extent.

Experimental data in the literature are scarce. The most comprehensive studies have been made by Rogers[23,46,47] and Newton *et al*[36,51], whilst Titmuss[52] has reported on limited attempts to model the effects of turbulence in a shallow angle ground to air situation.

Rogers[23] shows qualitatively that, as expected, the MTF improves as the elevation angle of the sight path is increased (i.e. as ϕ decreases). Characterizing the MTF by $\langle \Delta\theta^2 \rangle^{1/2}$ allows a quantitative estimation of this improvement to be given, this being shown in Fig. 16.7. It can be seen that the horizontal-path value of $\langle \Delta\theta^2 \rangle^{1/2}$ found by Rogers is somewhat less than the maximum value of 0.05 mrad found by Riggs *et al*[42] but it does lie well within the range of values quoted by them.

Both Rogers[46] and Smith *et al*[43] have shown that it is turbulence near the receiving optics which causes the most degradation. Theoretical studies, however, suggest that this is only true for spherical wave propagation. Rogers[47] also claims to verify the theoretical predictions of Duntley *et al*[19] for the MTF shape and for the square-root range dependence of the root mean square angular deviation, $\langle \Delta\theta^2 \rangle^{1/2}$. This latter agreement is also obtained by Coulman[16]. Unfortunately, in his MTF calculations, Coulman found a 5/3 power law of spatial frequency and a unit power of range, both of which satisfy the theoretical studies of Hufnagel and Stanley and Fried and Tatarski. To add to this confusion, Newton *et al*[36] determined the power of spatial frequency, and obtained a value of 1.91, which does not agree with any of the theories although it is nearer to the square law predicted by Duntley *et al.*

In all experimental determinations it has been assumed that the known film-plus-lens MTF could be divided out of the measured system MTF (i.e. atmosphere plus lens plus film) to leave the required atmospheric MTF. However, such a method of separating the various MTF components of a system can only be strictly applied to incoherent or diffraction limited systems, as discussed in Chapter 9. Therefore, in general, there are difficulties implied in extracting effective atmospheric MTF from photographic records unless the lens used may be assumed to be diffraction limited. This is particularly emphasised and illustrated by Titmuss[52]. Equally the *effects* of a given measured atmospheric MTF viewed by any aberrated system must be open to some question.

REFERENCES

1. Kolmogoroff, A. H. (1941). 'Dissipation of Energy in Locally Isotropic Turbulence', (in Russian) *Dokl. Akad. Nauk. SSSR.*, **32**

2. Tatarski, V. I. (1961). *Wave Propagation in a Turbulent Medium*, McGraw-Hill
3. "The Rytov approximation was introduced and so attributed by Tatarski[2] though we have been uable to find such an approximation discussed in Rytov's rather lengthy 'source' paper" – D. L. Fried
4. Strohbehn, J. W. (1968). 'Line-of-Sight Wave Propagation through the Turbulent Atmosphere', *Proceedings of the IEEE,* **56**, 1301
5. Fried, D. L. (1965). 'Statistics of a Geometrical Representation of Wavefront Distortion', *J. Opt. Soc. Am.,* **55**, 1427
6. Fried, D. L. (1966). 'Optical Resolution through a Randomly Inhomogeneous Medium for Very Long and Very Short Exposures' *J. Opt. Soc. Am.,* **56**, 1372
7. Coulman, C. E. (1968). 'Effect of Averaging Time on MTF in a Turbulent Atmosphere', *J. Opt. Soc. Am.,* **58**, 1668
8. Schmeltzer, K. A. (1967). 'Means, Variances and Covariances for Laser Beam Propagation through a Random Medium', *Quart. Appl. Math.,* **24**, 339
9. Fried, D. L. (1966). 'Limiting Resolution looking down through the Atmosphere', *J. Opt. Soc. Am.,* **56**, 1380
10. Brown, W. P. (Jr.). (1967). 'Validity of the Rytov Approximation', *J. Opt. Soc. Am.,* **57**, 1539
11. Hufnagel, R. E. and Stanley, N. R. (1964). 'Modulation Transfer Function associated with Image Transmission through Turbulent Media', *J. Opt Soc. Am.,* **54**, 52
12. Strohbehn, J. W. (1968). 'Comments on Rytov's Method', *J. Opt. Soc. Am.,* **58**, 139
13. De Wolf, D. A. (1965). 'Wave Propagation through Quasi-Optical Irregularities', *J. Opt. Soc. Am.,* **55**, 812
14. Heidbreder, G. R. (1967). 'Multiple Scattering and the Method of Rytov', *J. Opt. Soc. Am.,* **57**, 1477
15. Fried, D. L. (1967). 'Test of the Rytov Approximation', *J. Opt. Soc. Am.,* **57**, 268
16. Coulman, C. E. (1966). 'Dependence of Image Quality on Horizontal Range in a Turbulent Atmosphere', *J. Opt. Soc. Am.,* **56**, 1232
17. Gracheva, M. E. and Gurvich, A. S. (1965). 'On Strong Fluctuations of the Intensity of Light in a Near-Earth Layer', *Radiofizika,* **8**, 717
18. Chase, D. M. (1965). 'Coherence Function for Waves in Random Media', *J. Opt. Soc. Am.,* **55**, 1559
19. Duntley, S. Q., Culver, W. H., Richey, F. and Preisendorfer, R. W. (1963). 'Reduction of Contrast by Atmospheric Boil', *J. Opt. Soc. Am.,* **53**, 351
20. Chandrasekhar, S. (1952). 'A Statistical Basis for the Theory of Stellar Scintillation', *Roy. Astron. Soc., Monthly Notices,* **112**, 475
21. Beckman, P. (1965). 'Signal Degeneration in Laser Beams Propagated through a Turbulent Atmosphere', *Radio Science Journal of Research,* **690**, 629
22. Taylor, L. S. (1968). 'Validity of Ray-Optics Calculations in a Turbulent Atmosphere', *J. Opt. Soc. Am.,* **58**, 57
23. Rogers, C. B. (1963). 'The Variation of the Optical Frequency Response of the Atmosphere with Sightline Elevation', Dept. of Supply, Australian Defence Scientific Service, Weapons Research Est., Tech. Note OID 35
24. Weiner, M. M. (1967). 'Atmospheric Turbulence in Optical Surveillance Systems', *Applied Optics,* **6**, 1984
25. Fried, D. L. (1967). 'Optical Heterodyne Detection of an Atmospherically Distorted Signal Wavefront', *Pro. IEEE,* **55**, 57
26. Evvard, J. C. (1968). 'Atmospheric Limits on the Observational Capabilities of Aerospacecraft', National Aeronautics and Space Administration, Washington DC NASA Technical Note; NASA TN D–4940
27. Sutton, G. W. (1969). 'Limiting Resolution looking upward through the Atmosphere', *J. Opt. Soc. Am.,* **59**, 115
28. Birch, K. G. (1969). 'A Survey of OTF Based Criteria used in the Specification of Image Quality', *Optical Metrology*, No. 5, NPL, April

29. Hodara, H. (1968). 'Effects of a Turbulent Atmosphere on the Phase and Frequency of Optical Waves', *Proc. IEEE,* **56**, 2130
30. Hulett, H. R. (1967). 'Turbulence Limitations in Photographic Resolution of Planet Surfaces', *J. Opt. Soc. Am.,* **57**, 1335
31. Davis J. I. (1966). 'Consideration of Atmospheric Turbulence in Laser System Design', *Applied Optics,* **5**, 139
32. Deitz, P. H. and Wright, N. J. (1969). 'Saturation of Scintillation Magnitude in Near-Earth Optical Propagation', *J. Opt. Soc. Am.,* **59**, 527
33. Goldstein, I., Miles, P. A. and Chabot, A. (1965). 'Heterodyne Measurements of Light Propagation through Atmospheric Turbulence', *Proc. IEEE,* **53**, 1172
34. Ochs, G. R., Bergman, R. R. and Snyder, J. R. (1969). 'Laser-Beam Scintillation over Horizontal Paths from 5.5 to 145 Kilometres', *J. Opt. Soc. Am.,* **59**, 231
35. Fried, D. L., Mevers, G. E. and Keister, M. P. (Jr.) (1967). 'Measurements of Laser-Beam Scintillation in the Atmosphere', *J. Opt. Soc. Am.,* **57**, 787
36. Newton, A. M., Lavin, E. P., Williams, R. B. and Titmuss, G. E. (1968). 'Optical Effects of Atmospheric Turbulence', BAC (GW) Rep. Ref. No. R44X/PHY/30/133/808
37. Wright, N. J. and Schutz, R. J. (1967). 'Measurement of the Refractive Index Structure Coefficient C_n', Ballistic Research Laboratories, Aberdeen Proving Ground, Maryland, Memo. Rep. No. 1885
38. Brookner, E. (1971). 'Improved Model for the Structure Constant Variations with Altitude', *Applied Optics,* **10**, 1960
39. Farhat, N. H. and DeCou, A. B. (1969). 'Relations between Wave Structure Function looking up and looking down through the Atmosphere', *J. Opt. Soc. Am.,* **59**, 1489
40. Titmuss, G. E. (1971). 'Curves for the Prediction of Resolution in the Presence of Turbulence', BAC (GW) Rep. Ref. No. L50/22/PHY/186/1300
41. Washer, F. E. and Williams, H. B. (1946). 'Precision of Telescope Pointing for Outdoor Targets', *J. Opt. Soc. Am.,* **36**, 400
42. Riggs, L. A., Mueller, C. G., Graham, C. H. and Mote, F. A. (1947). 'Photographic Measurements of Atmospheric Boil', *J. Opt. Soc. Am.,* **37**, 415
43. Smith, A. G., Saunders, M. J. and Vatsia, M. L. (1957). 'Some Effects of Turbulence on Photographic Resolution', *J. Opt. Soc. Am.,* **47**, 755
44. Djurle, E. and Bäck, A. (1961). 'Some Measurements of the Effect of Air Turbulence on Photographic Images', *J. Opt. Soc. Am.,* **51**, 1029
45. Coulman, C. E. (1965). 'Optical Image Quality in a Turbulent Atmosphere', *J. Opt. Soc. Am.,* **55**, 806
46. Rogers, C. B. (1963). 'The Influence of Camera Station Height on Optical Resolution for Horizontal Sight Lines', Dept. of Supply, Australian Defence Scientific Service, Weapons Research Est., Tech. Memo. OID 69
47. Rogers, C. B. (1965). 'Variation of Atmospheric 'Seeing' Blur with Object-to-Observer Distance', *J. Opt. Soc. Am.,* **55**, 1151
48. Taylor, L. S. (1967). 'Decay of Mutual Coherence in Turbulent Media', *J. Opt. Soc. Am.,* **57**, 304
49. Johnson, N. K. and Heywood, G. S. P. (1938). 'An Investigation of the Lapse of Temperature in the Lowest 100 Metres of the Atmosphere', *Geophysical Memoirs,* **77**, 12
50. Best, A. C., Knighting, E., Pedlow, R. H. and Stormonth, K. (1952). 'Temperature and Humidity Gradients in the first 100 metres over South-East England', *Geophysical Memoirs,* **89**, 8
51. Newton, A. M., Lavin, E. P., Titmuss, G. E. and Williams, R. B. (1968). 'Optical Effects of Atmospheric Turbulence', BAC (GW) Rep. Ref. K03/20/PHY/133/896
52. Titmuss, G. E. (1972). 'Turbulence Measurements at TERA, Ty Croes', BAC (GW) Rep. Ref. 820/186/1498

17 Field Thresholds

Having considered the properties of the human eye, copious data on the threshold functions for progressively more complex stimulus presentation in the laboratory, the properties of visual aids and simulation media and the optical properties of the atmosphere, it is time to look at threshold visual performance in the field and attempts to correlate this with laboratory threshold data. In addition, in this final chapter, we shall be considering attempts to bring field situations into the laboratory by simulation. In carrying out such a survey, the author has chosen to draw primarily from data concerned with the acquisition of objects on the ground from the air and of aircraft from the ground, since such studies are close to his own experience. However, much of the literature concerned with such field thresholds is subject to some restriction on general availability. It is only possible, therefore, to provide limited references and examples. Nevertheless, it is hoped that the examples and experience summarised will permit the reader to extrapolate to other situations possibly closer to his own field of interest.

For convenience, the studies to be discussed may be divided into two distinct categories – air/ground approach studies (characterised, in general, by size growth of an object of interest in cluttered surroundings with fixed scene luminance), and ground/air approach studies (characterised by size growth of an object of interest in relatively uncluttered surroundings and with possible temporal luminance distribution variations with respect to the object of interest). Of the foregoing the ground/air studies most closely approach the simple laboratory threshold situations, although even here there are very considerable differences.

17.1 FIELD INSTRUMENTATION

A prime requisite of field trials studying visual threshold performance is the provision of adequate instrumentation of the prevailing conditions and of the observer responses. Many trials results have been almost useless for predictive and comparative purposes due to inadequate recording of such data as prevailing atmospheric attenuation *along the viewing path*, object intrinsic contrast, terrain screening, background structural details and accuracy of observer response. Other problems arise when carrying out field trials involving optical aids or viewing through canopies. When viewing through visual aids such factors as eyepiece focus (see Section 9.1.3), eccentricity of eye and sight pupils (see Section 9.1.2) and sight/eye quality interactions (see Section 9.1.5) frequently cause potential problems and are rarely measured at all. In addition, for both viewing through visual aids and viewing through a canopy, local veiling glare (ghost images, etc. – see Section 9.1.6) can cause trouble, often without the observer being consciously aware of the problem. It is all to easy to look through any optics and ignore what one is not looking for, despite the fact that unwanted images distort the scene being viewed.

Of recent years attempts have been made to avoid these problems of instrumentation (e.g. Erickson and Gordon,[1] Murphy *et al*,[2] Overington and Duncan[3] and Kelly and Overington[4]) but all too frequently problems still remain. It is the author's opinion[5] that anybody planning a field trial involving visual observation should arrange to instrument the trial as fully as possible for scene luminance distribution, atmospherics and any optical interface effects, whether or not it is expected in advance that the data will be required. It is too late after the trial if some data are needed which were not adequately recorded.

17.2 GROUND/AIR FIELD TRIALS

17.2.1 Aircraft Acquisition

Most ground/air field trials are associated with ability to detect and/or recognise aircraft against a sky or cloud background in a search situation. As such it might be assumed that the task is relatively simple and that the answers should be predictable. However, an aircraft is a highly structured object bearing little resemblance to the simple shapes studied in laboratory experiments. Furthermore, as aircraft acquisition is usually accomplished through visual aids of some kind, the problem of combined performance of observer and aid (Section 9.1) becomes of prime importance. Finally, the presence of specularly reflecting surfaces, such as the canopy of the aircraft, leads to a very considerable possibility of 'glint' occuring if the aircraft is at any time sunlit. Should such glint occur in a wide field search situation it is more than likely that the temporary stimulus is so strong that attention will be drawn to it, even if the glint is viewed at a position many degrees from the fovea (see Section 13.3). Attention having thus been drawn, its is probable that a foveal glimpse in the direction of the glint will permit detection of the aircraft, even if glint no longer persists. In such circumstances the normal wide scale search problem is temporarily converted to little more than a single glimpse foveal detection problem – a vastly different threshold situation (see Chapter 8).

Another problem of considerable importance in ground to air viewing of aircraft is the prevailing visibility. It is frequently not good enough to estimate standard horizontal meteorological visibility. Due to the usual tendency for σ_e to reduce with height, the *effective* slant path visibility for viewing aircraft from the ground is usually much greater than the horizontal ground visibility (see Chapter 15). That this is so has been amply illustrated by comparative measurements of ground and slant path visibility reported by Overington *et al*[6]

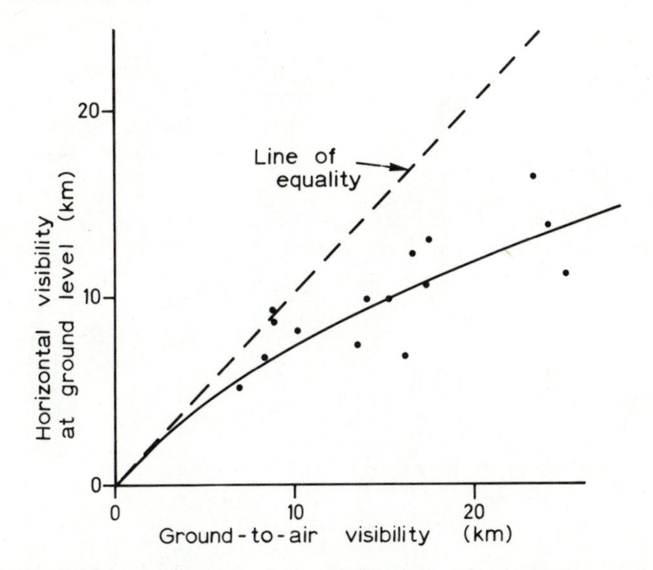

Fig. 17.1. Comparison of horizontal visibility at ground level with slant path visibility measured between ground level and 100–200 m altitude.

The results of this trial, where horizontal ground visibilities measured with a nephelometer were compared statistically with ground/air slant path visibilities measured by the photographic technique described in Chapter 15.5, are shown in Fig. 17.1. In this figure the ground visibilities measured by the nephelometer have been 'corrected' to photographic horizontal visibilities by use of the results of the Pendine Sands trial (Fig. 15.12). It will be seen that the general trend of slant path visibilities is considerably higher than that of the horizontal ground visibilities.

A further problem of measurement concerning ground to air viewing of aircraft is the determination of the intrinsic contrast of the aircraft. This cannot readily be estimated from ground measurements of surface reflectance because the apparent luminance of the aircraft depends not only on viewing direction and type of lighting but also, owing to the air-light scattered upwards from the thicker lower layers of the atmosphere (see for instance Gordon and Church[7,8]), on the altitude of the aircraft. A typical trend of intrinsic contrast against sky and cloud backgrounds obtained from a series of studies by BAC(GW)[9] is shown in Fig. 17.2. Here it can be seen that the probable aircraft contrast falls from –1.0 near ground level to around –0.4 at 3000 m. Such a trend curve is a useful guide to aircraft contrast within the limits of the group of trial measurements. However, the trials covered a restricted range of approach angles relative to the sun and an incomplete permutation of sky and cloud backgrounds. Thus the contrast trend shown should not be generalised beyond the conditions specified

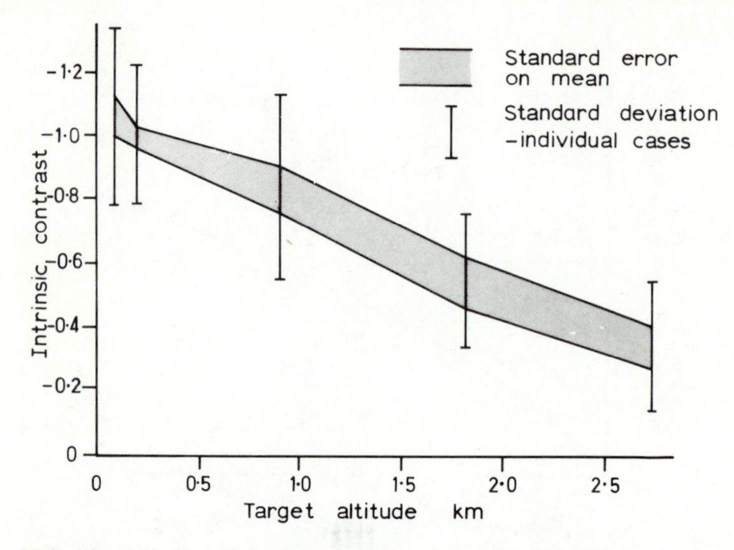

Fig. 17.2. The variation of intrinsic contrast of aircraft viewed from the ground as a function of height. Individual values are derived by extrapolation of apparent contrast/range graphs for directly approaching aircraft.

by Titmuss.[9] It is theoretically possible to apply some of the results of air luminance studies quoted by Gordon and Church, McClatchey *et al*,[10] Gordon[11] and Duntley *et al*[12] to prediction of aircraft contrast, provided care is taken to allow correctly for such factors as aircraft surface reflectivity (see Chapter 14).

A final problem of instrumentation concerns the effects of turbulence. As shown in Chapter 16, turbulence has the greatest effect on quality near the receiving optics. Also turbulence tends to be strongest close to the ground. Hence, particularly when viewing through aids where the magnification is high (and the aperture correspondingly large), the near ground turbulence can be a serious problem. Figure 17.3 illustrates the problem by comparison of the scene imaged by a high magnification system with low and moderate turbulence present. However, the more serious problems, from a field trial point of view, are when some significant turbulence is suspected (which can degrade system optical quality) but it is *not* visually obvious. Under such conditions it is imperative that *some* attempt be made to measure the turbulence present and to allow for it. This is not easy, since local turbulence depends on local ground surface conditions as well as altitude, but at least some estimate of its likely effect may be made by a series of measurements of vertical temperature gradient (VTG) in the vicinity of the observer (see Chapter 16).

With the above set of problems can one expect any success in controlling field trials to such an extent that a genuine comparison can be made with predictions

Fig. 17.3. Illustrating the effects of turbulence when viewing through a high power telescopic sight: (a) View of groups of resolution charts set at ranges of 1 km and 2 km through a high power telephoto lens under conditions of low turbulence. (b) The same view under conditions of moderate turbulence (note:– under some conditions of high turbulence most of the boards disappeared altogether).

from acquisition models? The answer, to the best of the author's knowledge, is that the majority of field trials on aircraft detection and recognition have been essentially exercises in obtaining empirical data on the detectability of a particular type of aircraft flying at a particular altitude with a given *horizontal ground* visibility and a given visual aid (or naked eye viewing conditions). However, in the last few years BAC(GW) have assisted in a highly controlled set of field trials where the prevailing conditions were fully instrumented. From the trials results those which produced reliable, stable data on slant path visibility, aircraft intrinsic contrast, etc. have been extracted. Using a development of the modelling of visual performance for the dynamic situation as discussed in Section 7.8, in conjunction with the search modelling of Chapter 8 and allowances for viewing through visual aids (Section 9.1.7), the field detection performance for these conditions has been successfully predicted. The trials conditions modelled ranged from head-on approaching aircraft (with motion at only 1 to 2 mrad/s in eye space), through crossing aircraft (with moderate rates of motion – 12 to 60 mrad/s in eye space), to mechanical elevation search of the visual aid (introducing aircraft rates in eye space of up to 600 mrad/s). They also involved slant path visibilities from some 12 km to 40 or 50 km and aircraft altitudes from 75 m to 3000 m. To have been able to predict over the whole range of conditions with small errors is taken by the author as evidence that one can expect ground to air field trials to give results as predicted from laboratory experiments, provided that the trials are thoroughly and correctly instrumented. The problem, in predicting ground to air acquisition in the field is thus believed to be not one of finding correction factors to apply to laboratory data, as sometimes suggested, but fairly and squarely one of adequate and correct instrumentation. This is supported by comparison of the aircraft acquisition trials results of Hoffman[13] with the theoretical predictions for a range of conditions based on laboratory thresholds and search modelling reported by the present author.[14] Making due allowance for the relative horizontal and slant path visibilities from Fig. 17.1 it has been found that agreement is very good indeed.

17.2.2 Astronomy

Many of the factors discussed in Section 17.2.1 also apply to astronomy from the earth's surface. Particularly important is the effect of turbulence, since once again the observer is generally in the region of greater turbulence. The matter is further aggravated in this case owing to the very large objective apertures of typical astronomical telescopes. What one can resolve, which in an ideal stable atmosphere would be limited by the aperture ratio of the system and its magnification, is in practice often limited by the atmosphere instead. For this reason man has sought to build his giant telescopes on mountains – thus avoiding the worst of the atmosphere – and more recently in satellites such as

Skylab – where there are no atmospheric problems. In this latter case, size is limited and viewing can only be carried out during the manning of Skylab.

In an attempt to overcome some of the problems of viewing from the ground level a system of image recording through telescopes has recently been studied which substantially removes the atmospheric effects. This involves the recording of a photographic time exposure of the blur disc of a star of interest, and then triggering the camera shutter for future exposures only when the image is centred on the integrated blur disc. In this way, exposures are made only when the atmospheric blur is low. Further improvements of such a technique may be possible by making use of the known spread function of the telescope and by detecting first differentials of image illuminance. However, the more refined such a technique is made, the less frequent will be the instants when the shutter is triggered, and hence such astronomy is restricted to indirect viewing from a photographic record.

17.3 AIR TO GROUND ACQUISITION

In air to ground acquisition, in contrast to the ground to air situation, the object of interest and its surroundings are usually constantly illuminated and the visual task is most frequently carried out with unaided vision. In addition, the downward viewing path means that the predominent atmospheric effects are close to the object of interest rather than the observer. This has the double result of minimising effects of turbulence and maximising the component of contrast attenuation due to air-light. Typically the apparent contrast as seen from the air will be considerably lower than the apparent contrast over an equivalent horizontal path at ground level. It thus becomes of utmost importance in air/ground field trials to be able to measure the air/ground slant path attenuation constants. In order to achieve this, BAC(GW) have on a number of occasions attempted to employ the photographic photometry technique (Section 15.5), but in this case the application is by no means as straight forward as in the ground/air case. In general the camera must be carried in the aircraft with the field trials observer. If a telephoto system is to be used, in order to provide a large image of the object of interest and its surroundings (so that contrast may be monitored directly), this requires that someone in the aircraft knows exactly where to point the camera from ranges well beyond visual acquisition range. Although such a requirement might sound relatively simple with the aid of modern electronic tracking equipment, it usually proves unmanageable or uneconomic in practice (e.g. Murphy *et al*[15] and Overington and Duncan[16]). The alternative is to provide a wider angle photographic coverage and find alternative large objects on the ground which can be used to monitor attenuation. In practice the sort of objects found desirable have been fairly large contrasting fields,[15,17] preferably on ground sloping gently away from the observer. However, the fields will have polar reflectance characteristics (see Chapter 14), which may distort the measurements of attenuation unless they can be allowed for.

In addition to the problems of measuring atmospheric attenuation it is necessary to monitor intrinsic contrast and viewing range. Both these are frequently subject to considerable error in air to ground trials. Firstly, unless the polar reflection characteristics of the object of interest and also the distribution of illumination are known, it is difficult to define the intrinsic luminance from a given direction (Section 14.4.4). Secondly, unless the sight line is greatly depressed – an unusual situation except for very high altitude viewing – the background against which an object is viewed will probably vary due to perspective effects. Allowance for this requires full knowledge of the terrain structure and the geometry of the situation. To make matters more difficult, the fact that the background will often be at a different range to the object of interest can make allowance for atmospheric attenuation extremely tedious. Finally, it is not particularly easy to arrange for accurate ranging of an aircraft from particular ground features when a field trial on acquisition must involve approaches to many different ground features. The alternative is to estimate ground position from the aircraft, correlating the ground fixes with the record of observations and the attenuation camera, or to estimate slant range from a cine record.[18]

With the foregoing in mind, what can be achieved with air/ground field trials? Well, several trials have been carried out on air to ground acquisition of typical ground features. Of these, some of the most fully instrumented are those carried out jointly by BAC(GW) and the Royal Aircraft Establishment,[19,20] and as such these will be reported here as being typical. These trials were concerned with the ranges at which aircrew were able to detect (and in some cases recognise) a variety of 'typical' ground features, such as large buildings in various surroundings, bridges of various types, railway stations, airfields, small isolated villages, radio stations, etc.. Samples of the various forms of 'target' and surroundings are to be found in Figs. 17.4–17.11. It will be seen immediately how far removed the situation is from idealised laboratory detection stimulus presentations. However, as already discussed in Chapers 12 and 13, this is the realistic situation. The briefing for the trials was usually the provision of an Ordnance Survey map (1:63 000 and 1:250 000), with the target, route and one or more intermediate reference points (IP's) marked on it, the request being for the observers to press a button when they located the target (thus localising search) and to press a second time when they could see something where the target should be (our definition of detection in Chapter 1). It was noticeable that, in general, the target area was detected considerably before target detection. Thus one might expect detection performance to approximate to a no search situation. On some occasions a third response was elicited when the aircrew could confirm their target (as the definition of recognition in Chapter 1).

During all these trials photographic records were taken and analysed to provide data on target intrinsic contrasts and atmospheric conditions. The

Fig. 17.4. Example of a simple block target in open country – a close approach to a simple laboratory stimulus in a plain background.

Fig. 17.5. Example of well-defined targets (the bridges) but which are difficult to model in terms of 'size'.

Fig. 17.6. Example of an isolated, well-defined target (the Motorway bridge) which i. nevertheless in a relatively structured local background.

Fig. 17.7. Example of a basically isolated target, but one which presents considerabl difficulties in 'modelling' due to detail luminance structure.

Fig. 17.8. Example of a relatively well-defined target but set in a very complex local background.

Fig. 17.9. Example of a target which, as well as being in a highly complex background, is difficult to define in terms of size and contrast owing to local background interactions.

Fig. 17.10. Example of a target which, owing to its height (the chimney), is seen against widely different local backgrounds dependent on viewing range and altitude.

Fig. 17.11. Example of a target (the oil storage tanks) which, in addition to being viewed against a complex background, possesses multiple signatures (one tank or the group of tanks), making predictive modelling difficult.

records were also analysed, where possible, to determine the angular size of the target as a function of range. In many cases, owing to perspective effects and screening, the variation of size with range was by no means a simple inverse square function in terms of area. Also, owing to extreme complexity, the 'size' as measured was open to considerable errors. Attempts were made to determine the probable errors in estimation of size, contrast and atmospheric factors for those cases which lent themselves to analysis at all (about 50% of the trial runs).[19,21]

From the set of trial results which were analysable, those which contained relatively small estimated errors were extracted, the size and contrast at 'detection' being plotted in the same manner as the Tiffany data (Section 4.1). The results are shown in Fig. 17.12. In this figure they are compared with the laboratory threshold curve for detection of circles when given infinite viewing time, since it appeared that there was little search in the field. It will be seen that

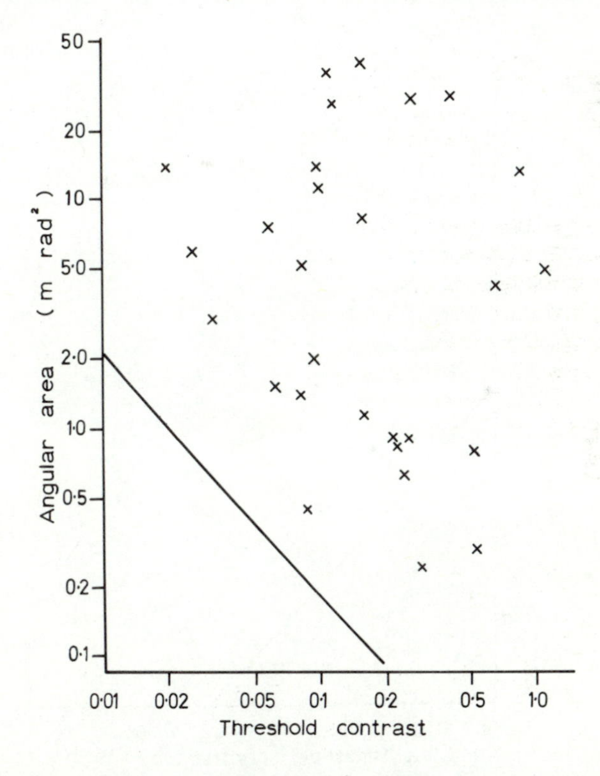

Fig. 17.12. Collected detection thresholds for air/ground acquisition of a variety of targets in the field. Typical laboratory forced-choice threshold curve for simple disc targets and infinite viewing time (Blackwell–Tiffany) for comparison.

some of the results cluster quite close to those for the laboratory detection of circles, whilst others form a cluster removed a long way from the Blackwell threshold curve. Further analysis of the *types* of targets constituting each group shows that most of those giving detections close to the laboratory curve are simple, isolated targets (such as Fig. 17.4–17.7). Conversely, most of those clustering in the group a long way from the laboratory curve are complex targets such as Figs. 17.8–17.11.

A further striking feature was revealed by analysis of the results for *recognition*. It was found that the *recognition* results for simple, isolated targets clustered in a very similar position on the size/contrast graph to the *detection* results for complex targets. The detection results for *simple* targets are to be found in Fig. 17.13. The comparative recognition results for simple targets and detection results for complex targets are to be found in Fig. 17.14. These two sets of results were taken to be an indication that air/ground field detection of simple targets could be closely predicted from laboratory trials, if the

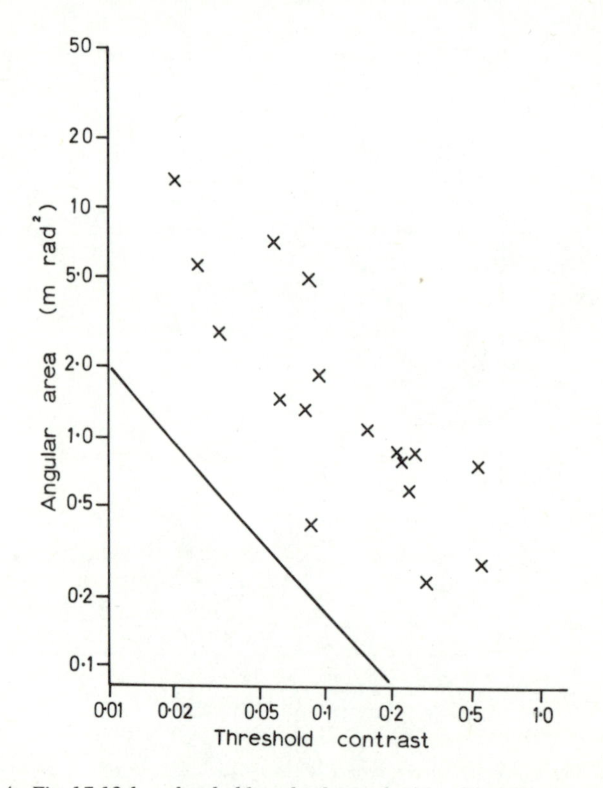

Fig. 17.13. As Fig. 17.12 but thresholds only shown for 'simple' targets (i.e. Targets where acquisition was believed to be a simple detection task).

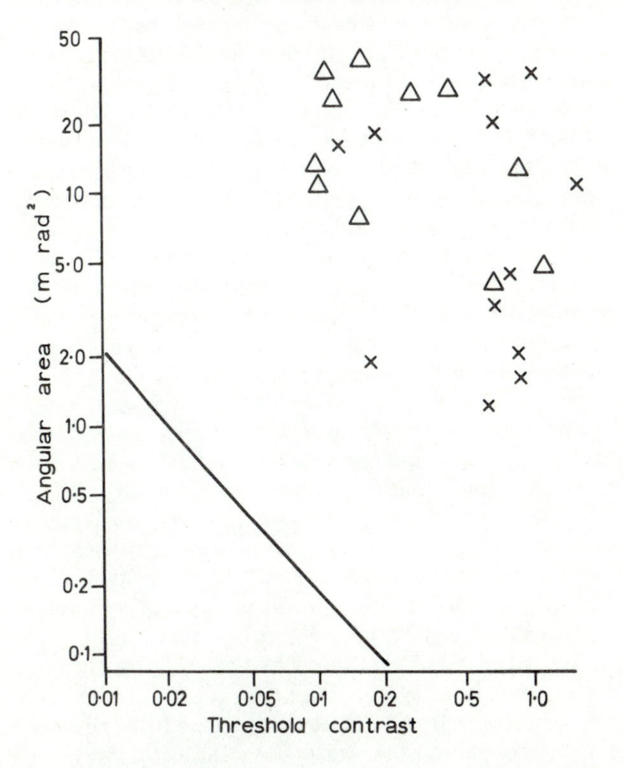

Fig. 17.14. Collected detection *data for complex targets compared with collected* recognition *data for simple targets in the field.*

appropriate surface properties and atmospherics were allowed for, whilst the 'detection' of complex targets in the field was really a recognition exercise (i.e. detection of detail) and should equally be predictable from appropriate laboratory trials. Subsequent work at BAC(GW) has served to strengthen this idea, since it has been realised that the 'field factor' necessary to 'degrade' Tiffany data to field detection for simple targets[22,23] is similar to the factor necessary to degrade forced choice laboratory thresholds to the free choice thresholds of Vos *et al*[24] (also see Fig. 4.1). Since field thresholds are, by nature, free choice they should be compared to Vos data rather than Tiffany data.

17.4 SIMULATION OF FIELD TRIALS

Since field trials are very expensive to run, difficult to control and instrument and at the mercy of the weather, it has been a major activity at BAC(GW) over many years to investigate the practicability of bringing the 'field trials' into the laboratory. At once it is important to note the differences between such simulated field trials and classical laboratory acquisition studies. Firstly the aim of simulation is to provide, in the laboratory, most, if not all, of the optical factors applicable to the field trials – scene structure and variability, atmospherics, target growth, etc.. Secondly some attempt is made to provide realism of situation, where possible. For instance, in an air/ground laboratory simulation, briefing by maps, as in flight, would be a prime requisite, whilst in ground/air simulation of acquisition using visual aids the field of view and aid/observer interface conditions would be of prime importance.

In recent years a variety of methods have been explored to provide at least some of the realism of field trials in the laboratory. Amongst those studied for air/ground simulation are cine photography (e.g. Murphy *et al*[25]), high quality still photography together with dynamic viewing facilities (e.g. Hobson[26]), high quality television displays in conjunction with photographic records, scale models of terrain or moving belts containing terrain detail (e.g. Erickson *et al*,[27] Shurmer[28;29]), photographic recording from scale models of terrain, and direct viewing of terrain models (e.g. Hilgendorf.[30] For ground to air simulation several 'dome' simulators have been developed, where an aircraft shape can be projected dynamically on a 'sky dome' and viewed from the centre through visual aids. Other simulators can provide realistic growth of size and contrast, whilst also providing search facilities and viewing through a visual aid (e.g. Kelly and Overington[31]). Two main drawbacks of such simulators are their inability to provide negative contrast targets (a very common condition for aircraft) and an inability to have target/background independence. Finally, an advanced concept being pursued at present permits the presentation, through a visual aid, of a very high quality target which is completely isolated from its background and which can be of either positive or negative contrast as required.[32]

In order to illustrate the practicability and limitations of simulation techniques, two extended studies on air/ground acquisition simulation will be discussed. The possibilities of adequate ground/air acquisition simulation will then be considered. An indication will be given of how modern methods of image evaluation are helping to isolate shortcomings of simulation media.

17.4.1 Air/ground Simulation by Cine Photography

An early attempt at laboratory simulation of the air/ground acquisition problem was carried out by BAC(GW) in conjunction with certain of the field trials reported by Murphy *et al.*[19] Cine films (16 mm) taken as a photometric record of the trials were carefully analysed and reprinted to provide a set of projection prints of controlled mean scene density and gamma, thus retaining a controlled relationship between the original scene and the projection image. These projection prints were then projected at a controlled scene luminance and size, and were viewed from such a distance that the angles subtended by ground details were as in the original field trial.[25] It was unrealistic to produce scene luminances equal to daylight using available projection equipment, so instead a normal cinema screen luminance was accepted whilst the overall commercial photographic process gamma of around 1.5 produced contrasts higher than the real scene. The edge contrast was known to be further enhanced by adjacency effects (Section 9.2.2). It was thus hoped to show that the overall effect (for air/ground approach situations) would be such that there was an acceptably constant ratio (preferably close to unity) between acquisition 'ranges' from projected 16 mm cine film and those in field trials.

As a comparison with field trials, the same aircrew who had flown the field trials were invited to BAC(GW) Filton to take part in the simulation exercise. Other aircrew, familiar with the type of field trial but not with the particular target approach runs, were also invited to take part. Since it was appreciated, throughout the exercise, that learning and memory would probably play a part (see Sections 3.3 and 4.15), three approach runs had been carried out in the field trials and three runs were also carried out in the simulation. Comparisons were made of various laboratory results with the three sets of field data. It was shown that the aircrew who did not know the target runs at all performed considerably worse than the aircrew who had flown the runs, but that there was similar learning from first to third run in both field and laboratory situations for the same aircrew. For full details and discussion of this work the reader is referred to Murphy *et al.*[25] We shall limit present discussion to the comparison of third run data in the field and in the laboratory by the same aircrew, it being assumed that the similar learning curves and the passage of time (some six months) between the field and laboratory trials has provided two sets of comparable, relatively uncontaminated data in these two cases. Figure 17.15 shows the set of results for this third run comparison. It will be seen that there is a very strong correlation between the two sets of data (a regression correlation of 0.91) and that there is no reason to suppose that a simple ratio may not be taken as adequate to describe the entire set of field trials in terms of laboratory simulation results. However, despite the high process gamma and adjacency effects, this ratio is well below unity (0.87 ± 0.11) and must be realised to be to some extent dependent on conditions.

It might be assumed that a larger cine format would yield a ratio closer to unity. No reliable comparative trials of such larger formats are known to the author, although BAC(GW) did use 16 mm reduction prints from 35 mm originals in a similar way to that described for the 16 mm cine records.[25] This resulted in a ratio in excess of unity (1.07 ± 0.11), but it is considered by the author that the compromises of gamma (contrast), luminance and adjacency

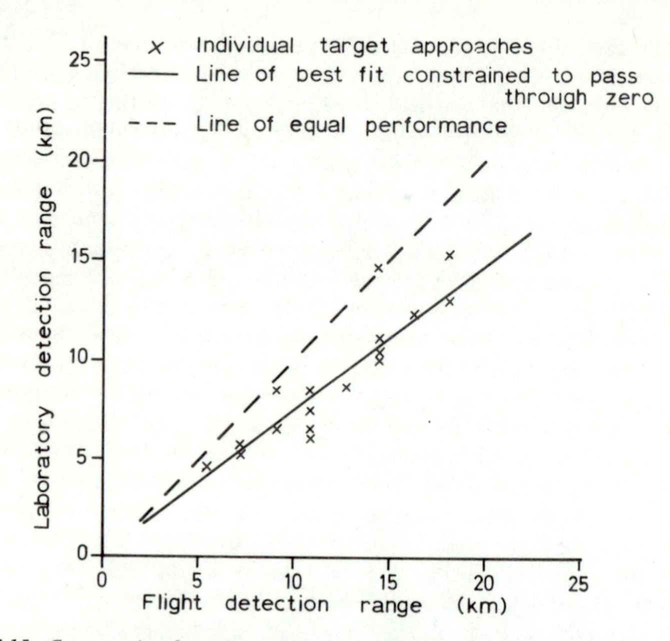

Fig. 17.15. Comparative detection ranges for a selection of simple targets in the field and in a cine simulation aimed at reproducing the field viewing conditions.

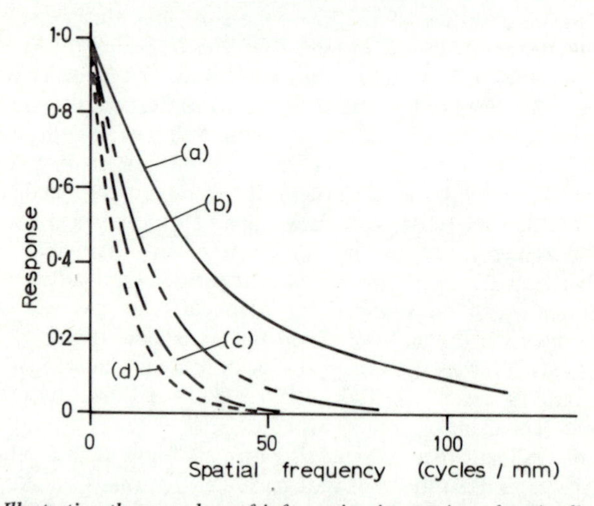

Fig. 17.16. Illustrating the gross loss of information in copying of optimally recorded fine grain cine films onto commercial print stock. Curve (a): MTF of typical fine grain film. Curve (b): Composite MTF due to copying onto similar film. Curve (c): Composite MTF due to two stages of copying. Curve (d): Composite MTF of a release print after three stages of copying. (Note:– In practice the losses are not quite so alarming since the original MTF must be degraded to some extent by the taking camera lens).

effects might well lead to a major variation of ratios, dependent on conditions.

As an aid to further understanding of the reasons for ratios being significantly different to unity, a major exercise was undertaken by BAC(GW), particularly concerned with the limitations of quality of 16 mm projection prints when produced by standard commercial processes.[33] It was found that if very high quality original negatives were produced then the high contrast resolution available was reduced markedly with each stage of copying. Since commercially the practice is usually to produce a duplicate negative for printing purposes, thus avoiding excessive use of the master original, no less than three copying stages are required in arriving at a standard release print. Starting with fine grain camera stock it was found possible to record high contrast periodic structure up to 120 line pairs/mm on the original negative, but that it was unlikely that resolutions better than 40 or 50 line pairs/mm could be achieved on the release print. The reason for this was found to be that the MTF's of commercial copying materials were of the same order as the original camera stock. Thus (see Chapter 10) the final release print quality could never be superior to that predicted by the product of the various stage MTF's as shown in Fig. 17.16, other than as a result of non-linear effects such as adjacency effects and high process gammas. In order to minimise the losses of quality in copying, and at the same time minimise distortion of local scene structure due to non-linearities, it has been found necessary to employ reversal films for camera use, thus providing a positive image in one stage whilst at the same time minimising adjacency effects.[33,34] The film is then copied, using a special high quality, ultra-fine grain film,[35] arranging that the final product gamma on the release print is approximately unity. In this way very nearly all the information contained in the original

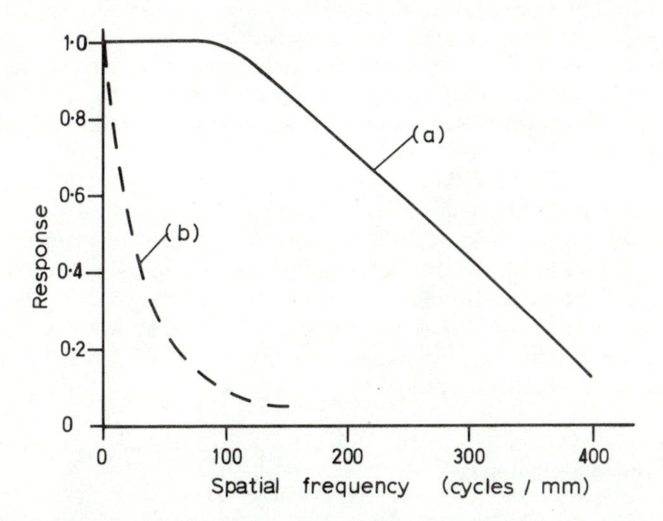

Fig. 17.17. Comparative MTF's of a typical fine grain camera emulsion and of special emulsion SO 156. Curve (a): SO 156. Curve (b): Fine grain camera emulsion.

camera record is retained in the projection print, as can be envisaged by looking at the comparative MTF's of a typical camera film stock and SO 156 (Fig. 17.17).

17.4.2 Air/ground Simulation with High Quality Still Photographs

In an endeavour to overcome some of the potential problems of simulation using cine film, BAC(GW) constructed a Still Target Acquisition Facility (STAF),[26] where a high quality transparency could be displayed in front of a back illuminated, diffuse screen at normal daylight levels. This transparency, originally taken at a given slant range, could be prepared to represent scene contrast at that slant range with a given atmospheric condition. It could then be viewed from up to 9m from a slowly moving, motorised chair. The viewing conditions were such that there should be no photographic distortion or limitation in the presented scene.

In order to test the ability of such a facility to yield field detection results for complex scenes, a number of scenes were selected where the field detection ranges coincided approximately with the taking range of a high quality still photograph (taken on the same flight). Transparencies were then prepared to simulate the measured (photometered) apparent contrast of the target at detection. Aircrew were presented with the task of detecting the targets from the moving chair, the target area thus growing approximately correctly in size but only being of the correct contrast at the field detection range. It was found that, in general, the detection 'ranges' obtained in the laboratory were much higher than the field detection ranges (of the order of twice as great). This was partially explained away as being due to the lack of contrast growth (at ranges much greater than the field detection the contrast would be much higher in the laboratory than in the field) but no entirely satisfactory explanation has been forthcoming other than the possibility that adjacency effects were still playing a significant part.

Between cine and still simulation we have shown it possible to produce simulated detection of a complex field situation where 'ranges' vary from around 0.85 of those in the field to double those in the field, dependent on the particular system used. Whilst such results are encouraging, a considerable amount of further study is required to determine exactly which of the various differences between field trials and the simulations really are the controlling factors. Image evaluation techniques now being refined (Chapter 10) may shortly help to answer this question.

17.4.3 Ground/air Simulation

The important points of a ground to air simulation are the provision of an adequate representation of an approaching aircraft, simulation of atmospherics, the provision of an adequate and appropriate search field and the provision of representative viewing optics where applicable. If the simulation is to represent viewing through visual aids, as is often the case, then it is of considerable importance to provide representative interface conditions with the observer[36,37] (see also Chapter 9.1). It is not enough to provide viewing optics with the same MTF as the aid to be simulated (or the same resolution). To date this is a very considerable problem in the design of ground/air simulations, since the only reliable way to ensure correct interface conditions is to provide, in the simulation, all components of the aid to be simulated which contribute significant aberrations. In practice this usually means use of the eye-piece assembly of a telescopic system, since most aberrations are produced there. It *may* then be permissible to replace the objective by a finite conjugate, high quality lens, subject to exploratory checks to confirm that both objective and simulation lens are nearly diffraction limited, or at least very much superior to the eyepiece of the system.

In the case of ground/air simulation, atmospheric attenuation may be simulated reasonably satisfactorily (but see Section 15.3.1) by means of a density wedge controlling the target luminance, since in many situations the background against which the target is viewed in the field is fairly constant. The other major atmospheric optical factor, turbulence, is usually ignored because of the difficulties of defining exactly what the optical effect of turbulence is in a given field situation (see Chapter 16 and Section 17.2.1). However, difficulties arise in provision of a growing aircraft shape with either positive or negative contrast against its background (typically aircraft present a negative contrast against their background), whilst providing adequate luminance, field of view and quality. Additional problems arise if the background contains broken cloud and the aircraft is required to cross the discontinuities. Once again these problems have been studied in some depth by BAC(GW) (e.g. Kelly and Overington,[31] Crowther,[38] Hawkins and Church,[39] Kelly and Overington[40]). Various possible solutions are as follows:

(a) A cine record of a particular aircraft approach track. This is the nearest which one can get to 'realism' in the general sense, but is limited very severely in practice by film quality, owing to the need for small aircraft angular subtense within a wide field of view. Typically a minimum aircraft *wingspan* of subtense around 1.5 mrad at the viewing optics is required in a field of view of many degrees, with aircraft detail subtending small fractions of a mrad. This form of simulation is also limited in respect of versatility, since each film specifically relates to one field situation. It is therefore expensive and difficult to provide input data for the simulation.

(b) A projection of a mechanically variable shape onto a background. This has been used extensively, both by BAC(GW) and others, but is limited in practice

to certain simple shapes – particularly circles and diamonds, which can be made to grow by adjustment of forms of iris.[31,41] It is also limited in that it is only possible to obtain positive contrast, thus raising the question of comparability of positive and negative contrast (see Section 4.13). Finally it is limited by target/background interaction when structured backgrounds are employed.

(c) A projection of an aircraft or other shape through a zoom lens onto a background. Here it is possible to overcome the problem of limitation to positive contrast by combination of a target projection beam and a secondary veiling beam, this latter being adjusted in luminance in opposition to the luminance of the target beam in order to retain the background level constant. Such a system can be envisaged for either plain background projection on a screen or structured backgrounds illuminated by transmission (see Fig. 17.18).

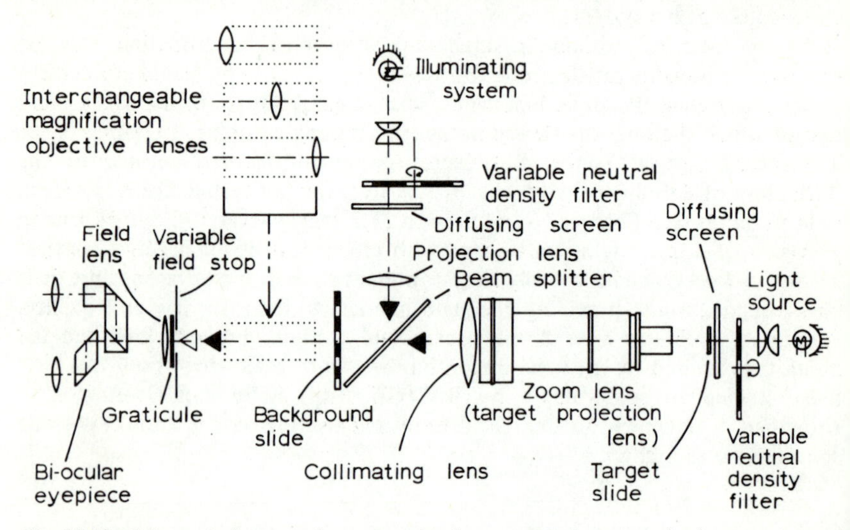

Fig. 17.18. Schematic layout of a zoom simulator permitting control of target size and target background contrast.

In the latter case there will inevitably be target/background interaction. This form of simulation depends greatly on zoom lens quality for its performance and it has been found in practice that currently available zoom lenses are other than entirely satisfactory for high quality target projection.[42]

(d) A double pass projection system, where a background is projected through a zoom lens onto a mirror or finely structured diffuse screen into which a target shape is etched.[32] The composite scene is then viewed through the same zoom

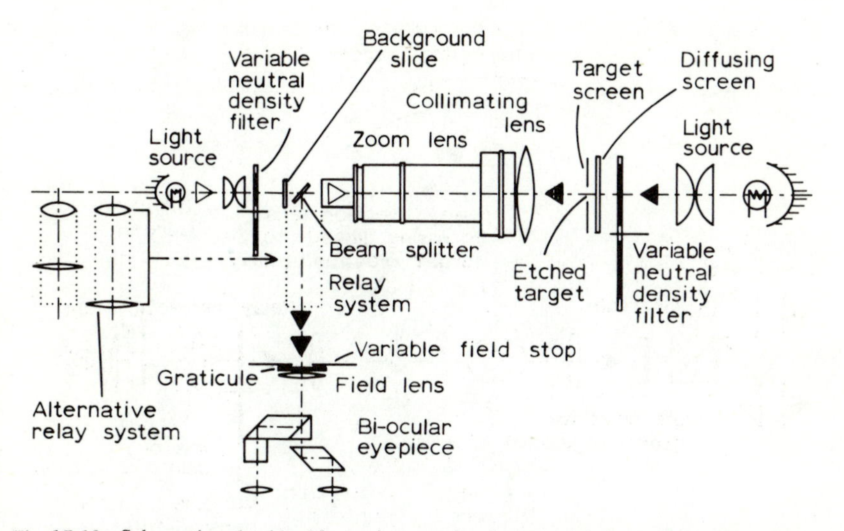

Fig. 17.19. Schematic of a 'double-pass' zoom simulator permitting completely independent control of target luminance and background parameters, whilst also permitting variation of target size.

lens (Fig. 17.19). The result is a simulation which can provide positive or negative contrast at will (by back illuminating the target) and has no target/background interaction. However, once again it is limited by zoom lens quality.

All forms of similation other than (a) have an additional common limitation that there is no ability to change target *shape* during a presentation. However, the zoom lens quality limitations of (c) and (d) can be largely overcome by use of a 'mechanical zoom' concept, a high quality process lens and the target mirror (or screen) being moved differentially to provide variable ratio finite conjugates. Such systems appear to hold great promise.[43]

It can be seen from the above that there appears to be no reasonable w ay to provide a *fully* versatile ground/air simulation. BAC(GW) have studied all the foregoing methods, and have recently completed a detail design study for a double pass zoom system which could be used as an exploratory tool for studying eyepiece/observer interactions.[44] In the meantime significant practical simulation at BAC(GW) is restricted to a series of limited experiments using a simulator of form (b). This simulator projected a 6.6:1 aspect ratio diamond onto a curved white screen, the diamond being changed in size by pulling the diamond leaves apart. The screen was flooded with white light and the target could be swept over the screen's surface by two mirrors. The contrast was controlled by an annular density wedge in the target beam and the whole screen was viewed via a finite conjugate objective coupled to a biocular eyepiece from a telescopic sight (Fig. 17.20).

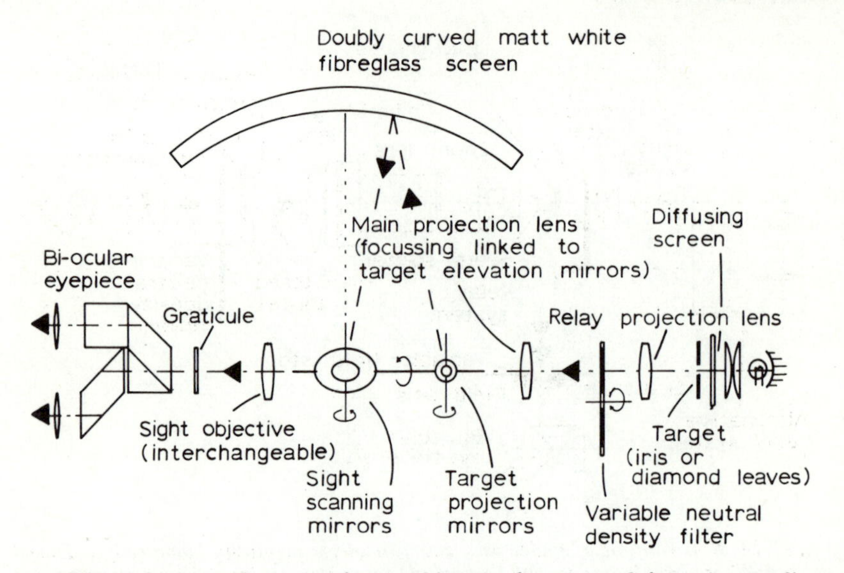

Fig. 17.20. Schematic of an aircraft acquisition simulator as used for various studies at B.A.C.(GW). Facilities include target growth (by iris adjustment), variable target contrast (positive only), two-dimensional target lateral motion and independent two-dimensional effective sight rotation.

Experiments were carried out with the foregoing simulator representing approaches of a typical aircraft (represented by equivalent presented area of the diamond) at various slant path visibilities. Comparisons were then attempted with the results of controlled ground/air field trials using a biocular sight similar to that from which the simulator eyepiece came. The comparisons appeared to show considerably degraded acquisition performance on the simulator as compared to the field trials for equivalent 'visibilities' and aircraft contrasts. At the time (some few years ago) this apparently degraded performance was attributed to the quality of the simulator. However, recent growing awareness of the important differences between slant path and horizontal visibilities (Section 17.2.1) and between high positive and negative contrasts of equal numerical value, together with findings that the effects of quality in a single glimpse viewing situation are not necessarily great[4 5] (see also Chapter 4) lead to speculation as to whether the results were really degraded if correct allowance is made for the contrast and visibility factors.

17.5 CONCLUSIONS

It must be concluded from the foregoing that any attempt to compare field trials acquisition results with laboratory thresholds is critically dependent on the ability to record the conditions under which the trials were carried out. Subject to adequate specification of field conditions, the limited data which it has been possible to check would suggest that there is little disagreement between field and laboratory results.

Attempts to simulate field trials in the laboratory are often thwarted by difficulties of instrumentation. Subject to adequate control of the optics and task specification of the simulation, limited evidence suggests that such simulations can often be considered reliable. However, it cannot be stressed too strongly how important it is to know that the optics are correct and that the task specified is adequately similar to the task in the field. Modern methods of optical assessment, as discussed in Chapter 10, are currently becoming sufficiently reliable and definitive to enable the optical engineer to specify the optical limitations of a simulation quantitatively.

REFERENCES

1. Erickson, R. A. and Gordon, J. I. (1970). 'Field Evaluation of a 1962 Vintage Visual Detection Model', NWC TP 5057, Naval Weapons Center, China Lake, California
2. Murphy, M. J., Overington, I. and Williams, D. G. (1965). 'Final Report on Visual Studies Contract', BAC(GW) Ref. R41S/11/VIS
3. Overington, I. and Duncan, I. (1968). 'Final Report on Film Library Contract', Sect. 3, BAC(GW) Ref. K03/20/PHY/125/897
4. Kelly, R. A. and Overington, I. (1972). 'Research into Factors affecting the Detection of Aircraft through Optical Sights – Final Report', Sect. 2.1, BAC(GW) Ref. L50/186/1449
5. Overington, I. (1972). 'Some Aspects of the Variation and Measurement of Contrast of Targets in Field Trials', in *Proceedings of the NATO/APOR Conference on Field Trials and Acquisition of Tactical Operational Data*, Vol. 1, p. 137
6. Overington, I., Duncan, I. and Brown, M. B. (1970). 'Final Report on Visual Studies II Contract' (App. 3.3.I), BAC(GW) Ref. L50/22/PHY/158/1164
7. Gordon, J. I. and Church, P. V. (1966). 'Overcast Sky Luminances and Directional Luminous Reflectances of Objects and Backgrounds under Overcast Skies', *Applied Optics*, 5, 919
8. Gordon, J. I. and Church, P. V. (1966). 'Sky Luminances and Directional Luminous Reflectances of Objects and Backgrounds for a Moderately High Sun', *Applied Optics*, 5, 793
9. Titmuss, G. E. (1972). 'Target Aircraft Contrast – Larkhill', Study Note No. 4 of 'A Second Phase of Research into Factors affecting the Detection of Aircraft through Optical Sights', BAC(GW) Rep. ST7951
10. McClatchey, R. A., Fenn, R. W., Selby, J. E. A., Garing, J. S. and Volz, F. E. (1970). 'Optical Properties of the Atmosphere', Air Force Cambridge Research Laboratories Rep. AFCRL – 70–0527
11. Gordon, J. I. (1969). 'Model for a Clear Atmosphere', *J. Opt. Soc. Am.*, **59**, 14
12. Duntley, S. Q., Johnson, R. W. and Gordon, J. I. (1972). 'Airborne Measurements of Optical Atmospheric Properties in Southern Germany', Visibility Laboratory, Scripps Institution of Oceanography, San Diego, Rep. SIO 72–64
13. Hoffman, H. E. (1972). 'The Visibility Range when observing an aircraft with and without Field–glasses', *Optica Acta*, **19**, 463
14. Overington, I. (1972). 'Modelling of Random Human Visual Search Performance based

on the Physical Properties of the Eye', in *AGARD Conference proceedings No. 100*, (Ed. H. F. Huddleston), London, p. B2–1
15. Murphy, M. J., Overington, I. and Williams, D. G. (1965). 'Final Report on Visual Studies Contract', Sect. 3.1, BAC(GW) Ref. R41S/11/VIS
16. Overington, I. and Duncan, I. (1968). 'Final Report on Film Library Contract', Sect. 2, BAC(GW) Ref. K03/20/PHY/125/897
17. Overington, I. (1967). 'Photographic Techniques for the Study of Atmospheric Attenuation in Air to Ground Viewing', *J. Phot. Sci.,* **15**, 164
18. Overington, I. (1967). 'The Measurement of Contrast and Range in Air to Ground Viewing', *J. Phot. Sci.,* **15**, 277
19. Murphy, M. J., Overington, I. and Williams, D. G. (1965). 'Final Report on Visual Studies Contract', (Sect. 3.2 & 3.4), BAC(GW) Ref. R41S/11/VIS
20. Overington, I. and Duncan, I. (1968). 'Final Report on Film Library Contract', BAC(GW) Ref. K03/20/PHY/125/897
21. Overington, I., Duncan, I. and Brown, M. B. (1970). 'Final Report on Visual Studies II Contract', App. 2.5.1, BAC(GW) Ref. L50/22/PHY/158/1164
22. Davies, E. B. (1968). 'Visual Theory in Target Acquisition', in *AGARD Conference Proceedings No. 41*, Paper A.1
23. Silverthorn, D. G. (1972). 'The 'K' Factor in Air-to-Ground Acquisition Modelling', in *AGARD Conference Proceedings No. 100*, (Ed. H. F. Huddleston), p. B3–1
24. Vos, J. J., Lazet, A. and Bouman, M. A. (1956). 'Visual Contrast Thresholds in Practical Problems', *J. Opt. Soc. Am.,* **46**, 1065
25. Murphy, M. J., Overington, I. and Williams, D. G. (1965). 'Final Report on Visual Studies Contract', Sect. 3.5, BAC(GW) Ref. R41S/11/VIS
26. Hobson, R. D. (1972). 'Measurement of Laboratory Thresholds', Annex B of 'Final Report of the Third Visual Studies Contract', BAC(GW) Ref. L50/196/1535
27. Erickson, R. A., Linton, P. M. and Hemingway, J. C. (1968). 'Human Factors Experiments with Television', NWC TP 4573, Naval Weapons Center, China Lake, California
28. Shurmer, C. R. (1968). Human Factors Study Note Series 4, No. 20, BAC(GW) Ref. R47/20/HMF/925
29. Shurmer, C. R. (1968). 'Description of the 3000:1 Terrain Belt', Human Factors Study Note Series 4, No. 31, BAC(GW) Ref. R47/20/HMF/1810
30. Hilgendorf, R. L. (1972). 'Air-to-Ground Target Acquisition with Flare Illumination', in *AGARD Conference Proceedings No. 100*, (Ed. H. F. Huddleston), p. B9–1
31. Kelly, R. A. and Overington, I. (1972). 'Research into Factors affecting the Detection of Aircraft through Optical Sights – Final Report', App. IV.2, BAC(GW) Ref. L50/186/1449
32. Kelly, R. A. and Overington, I. (1972). 'Research into Factors affecting the Detection of Aircraft through Optical Sights – Final Report', App. IV.4. BAC(GW) Ref. L50/186/1449
33. Overington, I. and Duncan, I. (1968). 'Final Report on Film Library Contract', Sect. 2.1, BAC(GW) Ref. K03/20/PHY/125/897
34. Overington, I. (1967). 'The Stability of the Photographic Process for Absolute Measurement of Radiation', *J. Phot. Sci.,* **15**, 11
35. Overington, I., Duncan, I. and Brown, M. B. (1970). 'Final Report on Visual Studies II Contract', Sect. 4 BAC(GW) Ref. L50/22/PHY/158/1164
36. Overington, I. and Gullick, S. A. (1973). 'Evaluation of a Total System – Optics plus Operator', *Optica Acta,* **20**, 49
37. Overington, I. (1973). 'The Importance of Coherence of Coupling when viewing through Visual Aids', *Optics and Laser Technology,* **5**, 216
38. Crowther, A. G. (1971). 'Contrast Attenuation by the Atmosphere along a Slant Path', BAC(GW) Ref. L50/22/PHY/186/1197

39. Hawkins, K. and Church, N. T. (1969). 'Contrast Sign Dependence', Study Note No. 2 of 'Research into Factors affecting the Detection of Aircraft through Optical Sights', BAC(GW) Ref. L50/20/PHY/186/1059
40. Kelly, R. A. and Overington, I. (1972). 'Research into Factors affecting the Detection of Aircraft through Optical Sights – Final Report', BAC(GW) Ref. L50/186/1449
41. Smith, E. S. (1966). 'Visual Search for Simulated Approaching Aircraft Targets', RRE Memorandum 2259
42. Overington, I. (1973). 'A Feasibility Study for a Double Pass Zoom Simulator', Study Note No. 6 of 'A Second Phase of Research into Factors affecting the Detection of Aircraft through Optical Sights', BAC(GW) Rep. ST10240
43. Overington, I. (1973). 'Design Considerations for a Mechanical Zoom Simulator', Study Note No. 7 of 'A Second Phase of Research into Factors affecting the Detection of Aircraft through Optical Sights', BAC(GW) Rep. ST10344
44. Crowther, A. G. and Overington, I. (1975). 'Factors affecting Aircraft Detection through Optical Sights – Final Report', BAC(GW) Rep. ST13527
45. Crowther, A. G. and Overington, I. (1975), 'Experiments on the Detection of Blurred Targets', BAC(GW) Rep. No. ST10840

Symbols and Units

Owing to the broad scope and interdisciplinary nature of this book it is inevitable that symbols and units used by various authors whose work is cited conflict to a considerable extent. In the interests of uniformity the symbols used herein have been changed where necessary and possible. It is hoped that this procedure will aid the reader. Likewise SI units have been used where possible. For the reader's convenience there follows a list of the more commonly used symbols, a list of units used and a conversion chart from SI units to those commonly found in past literature.

COMMONLY USED SYMBOLS

$a\ A$	area (various)
b	atmospheric scattering coefficient (also special constants)
B	stimulus luminance
B'	background luminance
C_μ	refractive index structure function
C	psychometric contrast ($\equiv (B - B')/B'$)
d	displacement, distance of separation
$D_{(p)}$	wave structure function
E	illuminance
E_R	retinal illuminance
f_c	critical fusion frequency
f_t	temporal frequency
f_s	spatial frequency
F	flux
g	retinal point
$h\ H$	height
I	intensity
k_λ	wavenumber
l_0	inner scale of turbulence
L_0	outer scale of turbulence
m	number of glimpses
M	magnification
$M(\ )$	modulation transfer function with respect to ( )
n	number of retinal receptors
N_p	number of rare events
N_q	number of quanta
N_e	noise equivalent passband

N	number of particles, noise
p	probability
P	power
q	$(1 - p)$
Q	neural or sensory response
r	radius
R	range
s_R	receptor spacing
S	signal
$S(\phi)$	scattering coefficient
t	time
T	transmission
u	visual acuity
v	velocity
V	potential difference
Z	function of contrast (specific)
α	angular diameter
$\beta'(\phi)$	volume scattering coefficient
δ	vision modelling constant (specific)
$\theta\ \phi\ \psi$	angles
Φ	cumulative probability
λ	wavelength
η	efficiency
τ	time constant
$\Delta\tau$	volume element of the atmosphere
Π	product
Σ	sum
ω	angular velocity
Ω	solid angle
ρ	reflectance
ϵ	50% threshold contrast
σ	standard deviation (normal distribution)
σ_e	atmospheric extinction coefficient
κ	threshold signal/noise
$\mathcal{R}$	resolution

UNITS USED

Physical quantity	*Name of unit*	*Symbol*
Length	micrometre	μm
	millimetre	mm
	metre	m
	kilometre	km
Time	second	s
	minute	min
Luminous intensity	candela	cd
Luminance	candelas/metre^2	cd/m^2
Retinal illuminance		
photopic	trolland	trol
scotopic	scotopic trolland	scot trol
Angle	milli radian	mrad
	radian	rad
Spatial frequency	cycles/milliradian	c/mrad
Temporal frequency	cycles/second (≡ Hertz)	Hz

COMMONLY USED CONVERSIONS

1 millimetre	≡ 0.0394 inches
1 metre	≡ 3.28 feet
1 kilometre	≡ 3280 feet ≡ 0.6214 miles
1 candela/metre^2	≡ 0.292 foot lamberts ≡ 0.314 millilamberts
1 milliradian	≡ 3.42 minutes of arc
1 radian	≡ 57.3 degrees
1 cycle/milliradian {	≡ 0.292 cycles/minute of arc
	≡ 17.4 cycles/degree

Bibliography

Should the reader wish to pursue any of the topics discussed in this work beyond the references given in the appropriate chapters, the following literature is suggested for additional reading.

1 PHYSIOLOGY AND NEUROPHYSIOLOGY

Alpern, M. (1971). 'Rhodopsin Kinetics in the Human Eye', *J. Physiol.*, **217**, 447

Alpern, M., Rushton, W. A. H. and Torrii, S. (1970). 'The Size of Rod Signals', *J. Physiol.*, **206**, 193

Alpern, M., Rushton, W. A. H. and Torrii, S. (1970). 'The attenuation of Rod Signals by Backgrounds', *J. Physiol*, **206**, 209

Aquilar, M. and Stiles, W. S. (1954). 'Saturation of the Rod Mechanism of the Retina at High Levels of Stimulation', *Optica Acta*, **1**, 59

Bagrash, F. M., Thomas, J. P. and Shimamura, K. K. (1974). 'Size-Tuned Mechanisms: Correlation of Data on Detection and Apparent Size', *Vision Research*, **14**, 937

Baker, H. (1973). 'Area Effects and the Rapid Threshold Decrease in Early Dark Adaptation', *J. Opt. Soc. Am.*, **63**, 749

Bartley, S. H. (1951). 'The Physiology of Vision', in *Handbook of Experimental Psychology*, (Ed. S. S. Stevens) Wiley, pp. 921–984

Berny, F. (1969). 'Study of the Formation of Retinal Images and Determination of the Spherical Aberration of the Human Eye' (in French), *Vision Research*, **9**, 977

Blake, R., Cool, S. J. and Crawford, M. L. J. (1974). 'Visual Resolution in the Cat', *Vision Research*, **14**, 1211

Blakemore, C. and Sutton, P. (1969). 'Size Adaptation: a New After-effect', *Science, N.Y.* **166**, 245

Brindley, G. S. (1962). 'Beats produced by Simultaneous Stimulation of the Human Eye with Intermittent Light and Intermittent or Alternating Electric Current', *J. Physiol.*, **164**, 157

Brink, G. Van den (1962). 'Measurements of the Geometrical Aberration of the Eye', *Vision Research*, **2**, 233

Brown, J. L., Metz, J. W. and Yohman, J. R. (1969). 'Test of Scotopic Suppression of the Photopic Process', *J. Opt. Soc. Am.*, **59**, 1677

Campbell, F. W. (1960). 'Correlation of Accommodation between the Two Eye's,' *J. Opt. Soc. Am.*, **50**, 738

Campbell, F. W. and Gregory, A. H. (1960). 'The Spatial Resolving Power of the Human Retina with Oblique Incidence', *J. Opt. Soc. Am.*, **50**, 831

Campbell, F. W. and Kulilowski, J. J. (1971). 'An Electrophysiological Measure of the Psychophysical Contrast Threshold', *J. Physiol.*, **217**, 54

Campbell, F. W. and Kulikowski, J. J. (1972). 'The Visual Evoked Potential as a Function of Contrast of a Grating Pattern', *J. Physiol.*, **222**, 345

Campbell, F. W. and Maffei, L. (1971). 'The Tilt After-effect: A Fresh Look', *Vision Research*, **11**, 833

Campbell, F. W., Maffei, L. and Piccolino, M. (1973). 'The Contrast Sensitivity of the Cat', *J. Physiol.*, **229**, 719

Campbell, F. W., Robson, J. G. and Westheimer, G. (1959). 'Fluctuations of Accommodation under Steady Viewing Conditions', *J. Physiol.*, **145**, 579

Campbell, F. W. and Westheimer, G. (1959). 'Factors influencing Accommodation Responses of the Human Eye', *J. Opt. Soc. Am.*, **49**, 568

Campbell, F. W. and Westheimer, G. (1960). 'Dynamics of Accommodation Responses of the Human Eye', *J. Physiol.*, **151**, 285

Chase, R. and Kalil, R. E. (1972). 'Suppression of Visual Evoked Responses to Flashes and Pattern Shifts during Voluntary Saccades', *Vision Research*, **12**, 215

Clarke, F. J. J. (1957). 'Rapid Light Adaptation of Localised Areas of the Extra-Foveal Retina', *Optica Acta*, **4**, 69

Cleland, B. G., Levick, W. R. and Sanderson, K. J. (1973). 'Properties of Sustained and Transient Ganglion Cells in the Cat Retina', *J. Physiol.*, **228**, 649

Cobb, P. W. (1915). 'The Influence of Pupilliary Diameter on Visual Acuity', *Am. J. Physiol.*, **36**, 335

Crawford, B. H. (1947). 'Visual Adaptation in Relation to Brief Conditioning Stimuli', *Proc. Roy. Soc. B.*, **134**, 283

Crozier, W. J. and Holway, A. H. (1939). 'Theory and Measurement of Visual Mechanisms', *J. Gen. Physiol.*, **22**, 341

Dowling, J. E. and Werblin, F. S. (1969). 'Organisation of the Retina of the Mud-puppy, *Necturus Maculosus*. I. Synaptic Structure', *J. Neurophysiol*, **32**, 315

Enoch, J. M. (1960). 'Response of a Model Retinal Receptor as a Function of Wavelength', *J. Opt. Soc. Am.*, **50**, 315

Enoch, J. M. (1960). 'Waveguide Modes: Are they present, and what is their Role in the Visual Mechanism?', *J. Opt. Soc. Am.*, **50**, 1025

Enoch, J. M. (1964). 'Physical Properties of the Retinal Receptors and Response of the Retinal Receptors', *Psych. Bull.*, **614**, 242

Enoch, J. M. and Fry, G. A. (1958). 'Characteristics of a Model Retinal Receptor studied at Microwave Frequencies', *J. Opt. Soc. Am.*, **48**, 899

Fender, D. H. and Nye, P. W. (1961). 'An Investigation of the Mechanisms of Eye Movement Control', *Kybernetik*, **1**, 81

Foster, D. H. (1973). 'A Note on whether a non-linearity precedes the De Lange Filter in the Human Visual System', *Optica Acta*, **20**, 325

Fry, G. A. (1963). 'Retinal Image Formation: Review, Summary and Discussion', *J. Opt. Soc. Am.*, **53**, 94

Fry, G. A. and Alpern, M. (1953). 'The Effect on Foveal Vision produced by a Spot of Light on the Sclera near the Margin of the Retina', *J. Opt. Soc. Am.*, **43**, 187

Fry, G. A. and Bartley, S. H. (1935). 'The Relation of Stray Light in the Eye to the Retinal Action Potential', *Am. J. Physiol.*, **111**, 335

Green, D. G. (1967). 'Visual Resolution when Light enters the Eye through Different Parts of the Pupil', *J. Physiol.*, **190**, 583

Gubisch, R. W. (1967). 'Optical Performance of the Human Eye', *J. Opt. Soc. Am.*, **57**, 407

Hartridge, H. (1950). *Recent Advances in the Physiology of Vision*, The Blakiston Co., Philadelphia

Hubel, D. H. and Wiesel, T. N. (1970). 'Cells Sensitive to Binocular Depth in Area 18 of the Macaque Monkey Cortex', *Nature*, **225**, 41

Ivanoff, A. (1953). 'The Aberrations of the Eye', (in French), *Ann. Opt. oculaire*, **2**, 97

Johnson, E. P. (1958). 'The Character of the b-wave in the Human ERG', *Arch. Ophth.*, **60**, 565

Julesz, B. (1971). *Foundations of Cyclopean Perception*, University of Chicago Press

Koenderink, J. J. (1972). 'Contrast Enhancement and the Negative After-image', *J. Opt. Soc. Am.*, **62**, 685

Krauskopf, J. (1962). 'Light Distribution in Human Retinal Images', *J. Opt. Soc. Am.*, **52**, 1046

Krueger, H. (1973). 'An Apparatus for Continuous Objective Measurement of Refraction of the Human Eye', *Optica Acta*, **20**, 277

Lange, H. de (1958). 'Research into the Dynamic Nature of the Human Fovea-Cortex Systems with Intermittent and Modulated Light. I. Attenuation Characteristics with White and Colored Light', *J. Opt. Soc. Am.*, **48**, 777

Lange, H. de (1958). 'Research into the Dynamic Nature of the Human Fovea-Cortex Systems with Intermittent and Modulated Light. II. Phase Shift in Brightness and Delay in Colour Perception', *J. Opt. Soc. Am.*, **48**, 784

Lipetz, L. E. (1961). 'Mechanisms of Light Adaptation', *Science*, **133**, 639

Lotmar, W. (1971). 'Theoretical Eye Model with Aspherics', *J. Opt. Soc. Am.*, **61**, 1522

Lotmar, W. and Lotmar, T. (1974). 'Peripheral Astigmatism in the Human Eye: Experimental Data and Theoretical Model Predictions', *J. Opt. Soc. Am.*, **64**, 510

Lowenstein, O. and Loewenfeld, I. E. (1958). 'Electronic Pupillography', *AMA. Arch. Ophthalmol*, **59**, 352

Maffei, L. and Fiorentini, A. (1972). 'Processes of Synthesis in Visual Perception', *Nature*, **240**, 479

Makous, W. and Boothe, R. (1974). 'Cones block Signals from Rods', *Vision Research*, **14**, 285

Mansfield, R. J. W. (1973). 'Brightness Function: Effect of Area and Duration', *J. Opt. Soc. Am.*, **63**, 913

Marks, E. L. (1973). 'Brightness and Equivalent intensity of intrinsic light', *Vision Research*, **13**, 371

Marr, D. (1974). 'The Computation of Lightness by the Primate Retina', *Vision Research*, **14**, 1377

Meeteren, A. Van (1974). 'Calculations on the Optical Modulation Transfer Function of the Human Eye for White Light', *Optica Acta*, **21**, 395

Mollon, J. (1974). 'After-effects and the Brain', *New Scientist*, 21st. February

Mote, F. A., Riupelle, A. J. and Meyer, D. R. (1950). 'The Effect of Intermittent Preadapting Light upon Subsequent Dark Adaptation in the Human Eye', *J. Opt. Soc. Am.*, **40**, 584

Murray, M. J. (1974). 'A Remark on Sensory Encoding of Two-Parameter Signals', *IEEE Transactions on Biomedical Engineering*, **BME–21**, 501

Nadell, M. C. and Knoll, H. A. (1956). 'The Effect of Luminance, Target Configuration and Lenses upon Refractive State of the Eye', *Am. J. Optom*, **33**, 24

Nesteruk, V. F. and Porfirieva, N. N. (1974). 'Concerning the Law of Visual Light Sensation', *Vision Research*, **14**, 899

Normann, R. A. and Werblin, F. S. (1974). 'Control of Retinal Sensitivity I. Light and Dark Adaptation of Vertebrate Rods and Cones', *J. Gen. Physiol.*, **63**, 37

Ohzu, H. (1966). 'Comments on the Application of Fiber Optics to Retinal Studies – Part I. Image Processing by Retinal Receptors', *J. clin. Ophthal.*, Japan, **20**, 1031

Pettigrew, J. D. (1972). 'Development and Neurophysiological Basis of Stereoscopic Vision', presented at the topical meeting on *Design and Visual Interface of Biocular Systems*, Annapolis, Maryland, May

Pi, H. T. (1925). 'The Total Peripheral Aberration of the Eye', *Trans. Ophthtal. Soc.*, UK., **45**, 393

Polack, A. (1923). 'The Chromatism of the Eye', (in French), in *Bull. Soc. Ophthalmol.*, France, No. 9 bir.

Regan, D. and Richards, W. (1973). 'Brightness Contrast and Evoked Potentials', *J. Opt. Soc. Am.*, **63**, 607

Richards, W. (1969). 'Saccadic Suppression', *J. Opt. Soc. Am.*, **59**, 617

Riggs, L. A. (1969). 'Progress in the Recording of Human Retinal and Occipital Potentials', *J. Opt. Soc. Am.*, **59**, 1558

Rodieck, R. W. (1965). 'Quantitative Analysis of Cat Retinal Ganglion Cell Response to Visual Stimuli', *Vision Research*, **5**, 583

Rodieck, R. W. and Stone, J. (1965). 'Analysis of Receptive Fields of Cat Retinal Ganglion Cells', *J. Neurophysiol.*, **28**, 833

Ronchi, L. and Moreland, J. D. (1957). 'The Effect on the Human Electroretinogram of the Distribution of Flux in a Light Stimulus of Finite Duration', *Optica Acta*, **4**, 31

Ronchi, L. and Mori, G. F. (1959). 'On the Factors which affect the Contrast Enhancement in a Figure with 'Quasi Perceptive Contours' and a Practical Application of such a Figure', *Atti della Fondazione Giorgio Ronchi*, **14**, 495

Rushton, W. A. H. (1965). 'Bleached Rhodopsin and Visual Adaptation', *J. Physiol.*, **181**, 645

Schweitzer, N. M. J. and Troelstra, A. (1964). 'On the Relationship between the Single-flash ERG and the ERG elicited by more Complex Stimuli', in *Flicker* (Junk, The Hague) 114

Sekuler, R. W., Rubin. E. L. and Cushman, W. H. (1968). 'Selectivities of Human Visual Mechanisms for Direction of Movement and Contour Orientation', *J. Opt. Soc. Am.*, **58**, 1146

Simon, J. F. and Denieul, P. M. (1973). 'Influence of the Size of Test Field employed in Measurement of Modulation Transfer Function of the Eye', *J. Opt. Soc. Am.*, **63**, 894

Spekreuse, H., Van der Tweel, L. H. and Zuidema, Th. (1973). 'Contrast Evoked Responses in Man', *Vision Research*, **13**, 1577

Stanioch, W. (1973). 'Relation of the Electrical Activity to the Spectral Response of the Retina', in *Colour 73*, Adam Hilger, London, 263

Stark, L. (1959). 'Stability, Oscillations, and Noise in the Human Pupil Servomechanism', *Proc. I.R.E.*, **47**, 1925

Stark, L. (1968). *Neurological Control Systems*, Sect. 3, Plenum Press

Valois, R. L. De (1966). 'Neural Processing of Visual Information', in *Frontiers in Physiological Psychology*, (Ed. R. W. Russell) Academic Press

Walls, G. L. (1942). *The Vertebrate Eye*, Cranbrook Institute

Werblin, F. S. (1972). 'Functional Organisation of a Vertebrate Retina: Sharpening up in Space and Intensity', in *Annals N.Y. Acad. Sci.: Patterns of Integration from Biochemical to Behavioural Processes, (Ed. G. Hadu)*, **193**, 75

Werblin, F. S. (1974). 'Control of Retinal Sensitivity. II Lateral Interconnections at the Outer Plexiform Layer', *J. Gen. Physiol.*, **63**, 62

Werblin, F. S. and Copenhagen, D. R. (1974). 'Control of Retinal Sensitivity. III Lateral Interconnections at the Inner Plexiform Layer', *J. Gen. Physiol.*, **63**, 88

Weisel T. N. (1960). 'Receptive Fields of Ganglion Cells in the Cat's Retina', *J. Physiol.*, **153**, 583

2 VISUAL PERFORMANCE

Baker, C. A. and Steedman, W. C. (1961). 'Perceived Movement in Depth as Function of Object Luminance', *Science* (Washington), **133**, 1356

Baker, H. D. (1953). 'The Instantaneous Threshold and Early Dark Adaptation', *J. Opt. Soc. Am.*, **43**, 798

Barlow, H. B. (1956). 'Retinal Noise and Absolute Threshold', *J. Opt. Soc. Am.*, **46**, 634

Baron, W. S. and Westheimer, G. (1973). 'Visual Acuity as a Function of Exposure Duration', *J. Opt. Soc. Am.*, **63**, 212

Békésy, G. V. (1972). 'Mach Bands measured by a Compensation Method', *Vision Research*, **12**, 1485

Blackwell, H. R. and Smith, S. W. (1959). 'The Effects of Target Size and Shape on Visual Detection: II Continuous Foveal Targets at Zero Background Luminance', University of Michigan, Eng. Research Inst. Rep. 2144–334–T

Bloomfield, J. R. (1972). 'Peripheral Acuity with Complex Stimuli at Two Viewing Distances', *AGARD Conference Proceedings No. 100*, (Ed. H. F. Huddleston) London, p. B6–1

Bouman, M. A. and Van den Brink, G. (1952). 'On the Integrate Capacity in Time and Space of the Human Peripheral Retina', *J. Opt. Soc. Am.*, **42**, 617

Boynton, R. M. (1973). 'Implications of the Minimally Distinct Border', *J. Opt. Soc. Am.*, **63**, 1037

Brown, B. (1972). 'Resolution Thresholds for Moving Targets at the Fovea and in the Peripheral Retina', *Vision Research*, **12**, 293

Brown, B. (1972). 'Dynamic Visual Acuity, Eye Movements and Peripheral Acuity for Moving Targets', *Vision Research*, **12**, 305

Brown, J. L., Graham, C. H., Leibowitz, H. and Ranken, H. B. (1953). 'Luminance Thresholds for Resolution of Visual Detail during Dark Adaptation', *J. Opt. Soc. Am.*, **43**, 197

Campbell, F. W. and Gregory, A. H. (1960). 'Effect of Size of Pupil on Visual Acuity', *Nature*, **187**, 1121

Campbell, F. W. and Westheimer, G. (1958). 'Sensitivity of the Eye to Difference in Focus', *J. Physiol.*, **143**, 18P

Clark, W. C. (1958). 'Relation between the Threshold for Single and Multiple Light Pulses', Doctorial Dissertation, University of Michigan

Crawford, W. A. (1960). 'The Perception of Moving Objects. IV. Accuracy of Fixation Required in the Perception of Detail in Moving Objects', Flying Personnel Research Committee (GB) Memo. No. 150d October

Crawford, W. A. (1960). 'The Perception of Moving Objects, V. The Moment of Perception', Flying Personnel Research Committee (GB) Memo. No. 150e September

Dudley, L. P. (1965). 'Stereoscopy', Chapter 2 in *Applied Optics and Optical Engineering*, (Ed. R. Kingslake) Vol. 2, Academic Press

Fankhauser, F. and Enoch, J. M. (1962). 'The Effects of Blur upon Perimetric Thresholds', *A.M.A. Arch. of Ophthal.*, **86**, 240

Fry, G. A. (1947). 'The Relation of the Configuration of a Brightness Contrast Border to its Visibility', *J. Opt. Soc. Am.*, **37**, 166

Gilmour, J. D. (1973). 'A Systematic Approach for Prediction and Improvement of Target Acquisition Performance', in *Rep. OA 6201, Vol. I: A Collection of Unclassified Technical Papers on Target Acquisition*, Martin Marietta Aerospace, Orlando, Florida, p. 179

Gottsdanker, R., Frick, T. W., and Lockard, R. B. (1961). 'Identifying the Acceleration of Visual Targets', *Brit. J. Psychol.*, **52**, 31

Hay, G. A. and Chesters, M. S. (1972). 'Signal-transfer Functions in Threshold and Suprathreshold Vision', *J. Opt. Soc. Am.*, **62**, 990

Hills, B. L. (1968). 'Pattern Detection and Recognition in the Human Visual System', Ph.D. Thesis, Dept. of Electrical and Electronic Engineering, University of Nottingham

Holladay, L. L. (1926). 'The Fundamentals of Glare and Visibility', *J. Opt. Soc. Am.*, **12**, 271.

Holladay L. L. (1927). 'Action of a Light Source in the Field of View in Lowering Visibility', *J. Opt. Soc. Am.*, **14**, 1

Keesey, U. T. (1960). 'Effects of Involuntary Eye Movements on Visual Acuity', *J. Opt. Soc. Am.*, **50**, 769

Kelly, D. H. (1959). 'Effects of Sharp Edges in a Flickering Field', *J. Opt. Soc. Am.*, **49**, 730

Krauskopf, J. (1957). 'Effect of Retinal Motion on Contrast Thresholds for Maintained Vision', *J. Opt. Soc. Am.*, **47**, 740

Kristofferson, A. B. (1957). 'Visual Detection as influenced by Target Form', in *Form Discrimination as related to Military Problems*, (J. W. Wulfeck and J. H. Taylor, Eds.) Nat. Acad. Sci. NRC, Washington DC 109

Latour, P. L. (1962). 'Visual Thresholds during Eye Movements', *Vision Research*, **2**, 261

Leibowitz, J. (1952). 'The Effect of Pupil Size on Visual Acuity for Photometrically Equated Test Fields at Various Levels of Luminance', *J. Opt. Soc. Am.*, **42**, 416

Meeteren, A. Van and Vos, J. J. (1972). 'Resolution and Contrast Sensitivity at Low Luminances', *Vision Research*, **12**, 825

Mitrani, L., Yakimoff, N. and Mateeff, St. (1973). 'Saccadic Suppression in the Presence of Structural Backgrounds', *Vision Research*, **13**, 517

Otero, J. M. (1951). 'Influence of the State of Accommodation on the Visual Performance of the Human Eye', *J. Opt. Soc. Am.*, **41**, 942

Ronchi, L. (1972). *An Annotated Bibliography on Variability and Periodicities of Visual Responsiveness*, Fondazione 'Giorgio Ronchi', XVII, Firenze

Ronchi, L. and Barca, L. (1972). 'On the Influence of Eye Version on the Visibility of Small Targets', *Atti della Fondazione Giorgio Ronchi*, **27**, 79

Salvi, G. and Innocenti, F. B. (1971). 'On the Relative Visibility of Single Break and Double Break Rings', *Atti della Fondazione Giorgio Ronchi*, **26**, 109

Steinman, R. M. (1965). 'Effect of Target Size, Luminance and Color on Monocular Fixation', *J. Opt. Soc. Am.*, **55**, 1158

Stigmar, G. (1971). 'Blurred Visual Stimuli', *Acta Ophthalmol.*, **49**, 364

Stiles, W. S. and Crawford, B. H. (1937). 'The Effect of a Glaring Light Source on Extrafoveal Vision', *Proc. R. Soc. B.*, **122**, 255

Thorn, F. and Boynton, R. M. (1974). 'Human Binocular Summation at Absolute Threshold', *Vision Research*, **14**, 445

Yarbus, A. L. (1967) *Eye Movements and Vision*, Plenum Press

3 RECOGNITION

Alluisi, E. A., Hawkes, G. R. and Hall, T. J. (1964). 'Effects of Distortion on the Identification of Visual Forms under Two Levels of Multiple Task Performance', *J. Eng. Psychol.*, **3**, 29

Anderson, N. S. (1957). 'Pattern Recognition: A Probability Approach', in *Form Discrimination as related to Military Problems*, (Eds. J. W. Wulfeck and J. H. Taylor), Nat. Acad. Sci. NRC, Washington DC pp. 45–49

Attneave, F. (1951). 'The Relative Importance of Parts of a Contour', Research Note P & MS: 51–8, Human Resources Center, San Antonio, Texas

Bailey, H. H. (1973). 'Target Acquisition through Visual Recognition: an Early Model', in *Rep. OA6201, Vol. I: A Collection of Unclassified Papers on Target Acquisition*, Martin Marietta Aerospace, Orlando, Florida, p. 115

Borg, G. (1964). 'Studies of Visual Gestalt Strength', Report, Dept. Education, Umea University, Sweden

Clark, H. J. and Knoll, R. L. (1968). 'Variables underlying the Recognition of Random Shapes', in *AGARD Conference Proceedings No. 41*, Paper A.3

Erickson, R. A. and Hemingway, J. C. (1970). 'Image Identification on Television', NWC TP 5025, Naval Weapons Center, China Lake, California

Forsyth, G. A. and Brown, D. R. (1967). 'Recognition-discrimination Performance as a Function of Stimulus Characteristics to which a Subject attends', Paper read at Midwestern Psychological Association, Chicago, May

French, R. S. (1954). 'Identification of Dot Patterns from Memory as a Function of Complexity', *J. Exp. Psychol.*, **47**, 22

Glezer, V. D., Leushina, L. I., Nevskaya, A. A. and Prazdnikova, N. V. (1974), 'Studies on Visual Pattern Recognition in Man and Animals', *Vision Research*, **14**, 555

Kause, R. H. (1965). 'Interpretation of Complex Images – Literature Survey', GER–10830 REV.A. Akron, Ohio, February

Østerberg, H. and Smith, L. W. (1964). 'Resolution of Shape of Self-radiant Line Elements', *J. Opt. Soc. Am.*, **54**, 599

Rock, I. (1974). 'The Perception of Disoriented Figures', *Scientific American*, 78
Rusis, G. and Snyder, H. L. (1965). 'Laboratory Studies in Air-to-Ground Target Recognition: II. The Effect of TV Camera Field of View', Autonetics Report T5–133/3111
Rusis, G. and Calhoun, R. L. (1965). 'Laboratory Studies in Air-to-Ground Target Recognition: III. The Effects of Aircraft Speed and Time-to-go Information'. Autonetics Report T5–134/3111
Vanderplas, J. M. and Garvin, E. A. (1959). 'Complexity, Association Value and Practice as Factors in Shape Recognition following Paired Associates Training', *J. Exp. Psychol.*, **57**, 155

4 COLOUR VISION

Benham, C. E. (1894). 'Artificial Spectrum Top', *Nature*, **51**, 113
Bidwell, S. (1897). 'On Subjective Colour Phenomena attending Sudden Changes of Illumination', *Proc. R. Soc.*, **50**, 368
Boynton, R. M. (1971). 'Colour Vision' in *Experimental Psychology*, (Ed. L. A. Riggs & J. Kling) Holt, Rinehart and Winston, pp. 315–368
Brindley, G. S., Croz, J. J. du and Rushton, W. A. H. (1966). 'The Flicker Fusion Frequency of the Blue-sensitive Mechanism of Colour Vision', *J. Physiol.*, **183**, 497
Clarke, F. J. J. (1973). 'Needs and Prospects for a Tetrachromatic System of Large Field Colorimetry', in *Colour 73*, Adam Hilger, London, 319
Coates, E., Kiszka, R. C., Provost, J. R. and Rigg, B. (1973). 'The Accuracy of Colour-difference Equations in relation to Perceived Colour Differences', in *Colour 73*, Adam Hilger, London, 300
Granger, E. M. and Heurtley, J. C. (1973). 'Visual Chromaticity-modulation Transfer Function', *J. Opt. Soc. Am.*, **63**, 1173
Halsey, R. M. and Chapanis, A. (1954). 'Chromaticity-Confusion Contours in a Complex Viewing Situation', *J. Opt. Soc. Am.*, **44**, 442
Horst, G. J. C. Van der (1969). 'Fourier Analysis and Colour Discrimination', *J. Opt. Soc. Am.*, **59**, 1670
Kilmer, E. and Kilmer, W. (1968). 'Temporal Reversal of Land Effect Colour Rules', *Nature*, **218**, 883
Kuehni, R. G. (1973). 'Is there a Special Significance to Colours of Dominant Wavelength 495 nm and 570 nm in Colour Vision?, in *Colour 73*, Adam Hilger, London, 286
Lange, H. de (1958). 'Research into the Dynamic Nature of the Human Fovea-Cortex Systems with Intermittent and Modulated Light. I. Attenuation Characteristics with White and Coloured Light', *J. Opt. Soc. Am.*, **48**, 777
Lange, H. de (1958). 'Research into the Dynamic Nature of the Human Fovea-Cortex Systems with intermittent and Modulated Light. II. Phase Shift in Brightness and Delay in Colour Perception', *J. Opt. Soc. Am.*, **48**, 784
MacNichol, E. F. (Jr.), Feinberg, R. and Hárosi, F. I. (1973). 'Colour Discrimination Processes in the Retina', in *Colour 73*, Adam Hilger, London, 191
Mollon, J. D. and Krauskopf, J. (1973). 'Reaction Time as a Measure of the Temporal Response Properties of Individual Colour Mechanisms', *Vision Research*, **13**, 27
Parsons, Sir J. H. (1924). *Colour Vision*, Cambridge University Press
Pointer, M. R. (1973). 'The Effect of White Light Adaptation on Colour Discrimination', in *Colour 73*, Adam Hilger, London, 283
Regan, D. and Tyler, C. W. (1971). 'Temporal Summation and its Limit for Wavelength Changes: An Analog of Bloch's Law for Colour Vision', *J. Opt. Soc. Am.*, **61**, 1414
Rentschler, I. (1973). 'Colour Discrimination at a Blurred Border', in *Colour 73*, Adam Hilger, London, 273

Rubinstein, C. B. and Limb, J. O. (1973). 'Colour Border Sharpness', in *Colour 73*, Adam Hilger, London, 377

Sperling, H. G. and Lewis, W. G. (1959). 'Some Comparisons between Foveal Spectral Sensitivity Data obtained at High Brightness and Absolute Threshold', *J. Opt. Soc. Am.*, **49**, 983

Stanioch, W. (1973). 'Relation of the Electrical Activity to the Spectral Response of the Retina', in *Colour 73*, Adam Hilger, London, 263

Uttal, W. R. (1973). 'Chromatic and Intensive Effects in Dot-pattern Masking: Evidence for Different Time Constants in Colour Vision', *J. Opt. Soc. Am.*, **63**, 1490

Yates, J. T. (1974). 'Chromatic Information Processing in the Foveal Projection (Area Striata) of Unanaesthetized Primate', *Vision Research*, **14**, 163

5 STABILISED VISION

Bossler, F. B. (1968). 'Visual Image Stabilisation Measurements and Specifications', *Applied Optics.*, **7**, 1155

Campbell, F. W. and Robson, J. G. (1961). 'A Fresh Approach to Stabilised Retinal Images', *J. Physiol.*, **158**, 1

Ditchburn, R. W., Fender D. H. and Mayne, S. (1959). 'Vision with Controlled Movements of the Retinal Image', *J. Physiol.*, **145**, 98

Evans, C. R. and Piggins, D. J. (1963). 'A Comparison of the Behaviour of Geometrical Shapes when viewed under Conditions of Steady Fixation, and with Apparatus for Producing a Stabilised Retinal Image', *B.J. of Phys. Opt.*, **20**, 1

Gerrits, H. J. M. and Vendrik, A. J. H. (1970). 'Artificial Movements of a Stabilised Image', *Vision Research*, **10**, 1443

Keesey, U. T. (1969). 'Visibility of a Stabilised Target as a Function of Frequency and Amplitude of Luminance Variation', *J. Opt. Soc. Am.*, **59**, 604

Ratliff, F. (1958). 'A Stationary Retinal Image requiring no Attachments to the Eye', *J. Opt. Soc. Am.*, **48**, 274

Tulunay–Keesy, U. (1973). 'Stabilised Target Visibility as a Function of Contrast and Flicker Frequency', *Vision Research*, **13**, 1367

6 STATISTICS, PROBABILITY AND OTHER MATHEMATICAL PROCESSES

Aitken, A. C. (1962). *Statistical Mathematics*, Oliver and Boyd

Bendat, J. S. and Piersol, A. G. (1971). *Random Data: Analysis and Measurement Procedures*, Wiley-Interscience

Dirac, P. A. M. (1958). *The Principles of Quantum Mechanics*, 4th Edn., Clarendon Press

Finney, D. J. (1947). *Probit Analysis*, Cambridge University Press

Green, D. M. and Swets, J. A. (1966). *Signal Detection Theory and Psychophysics*, Wiley

Jennison, R. C. (1961). *Fourier Transforms and Convolutions for the Experimentalist*, Pergamon Press

Johnson, C. B. (1973). 'Point-spread Functions, Line-spread Functions and Edge-response Functions associated with MTF's of the Form $\exp[-(\omega/\omega_c)^n]$', *Applied Optics*, **12**, 1031

Marchand, E. W. (1964). 'Derivation of the Point Spread Function from the Line Spread Function', *J. Opt. Soc. Am.*, **54**, 915

Marchand, E. W. (1965). 'From Line to Point Spread Function: The General Case', *J. Opt. Soc. Am.*, **55**, 352

Miller, G. A. and Frick, F. C. (1949). 'Statistical Behavioristics and Sequences of Responses', *Psychol. Review.*, **56**, 311
Peterson, W. W. and Birdsall, T. G. (1953). 'The Theory of Signal Detectability', Electronic Defense Group, University of Michigan, Tech. Rep. No. 13
Raynor, A. J. (1972). 'The Use of Kelly's Repertory Grid Technique for assessing Subjective Estimates of Important Parameters for Target Acquisition', *AGARD Conference Proceedings No. 100*, (Ed. H. F. Huddleston), London, p. B14–1
Rosenfeld, A. and Thurston, M. (1970). 'Edge and Curve Detection for Visual Scene Analysis', Tech. Report No. 70–128, Computer Science Center, University of Maryland, AFCRL–70–0488
Shannon, C. E. (1948). 'A Mathematical Theory of Communication', *Bell Systems Tech. J.*, **27**, 379 and 623
Shannon, C. E. and Weaver, W. (1964). *The Mathematical Theory of Communication*, University of Illinois Press
Shaw, R. (1962). 'The Application of Fourier Techniques and Information Theory to the Assessment of Photographic Image Quality', *Phot. Sci. & Eng.*, **6**, 281
Torgerson, W. S. (1967). *Theory and Methods of Scaling*, Wiley
Winer, B. J. (1962). *Statistical Principles in Experimental Design*, McGraw-Hill
Woodward, P. M. (1953). *Probability and Information Theory with Applications to Radar*, Pergamon Press

7 VISION MODELLING

Bliss, J. C. and MacCurdy, W. B. (1961). 'Linear Models for Contrast Phenomena', *J. Opt. Soc. Am.*, **51**, 1373
Bouman, M. A. (1953). 'Visual Thresholds for Line-Shaped Targets', *J. Opt. Soc. Am.*, **43**, 209
Bouman, M. A. and Velden, H. A. van der (1947). 'The Two-Quanta Explanation of the Dependence of the Threshold Values and Visual Acuity on the Visual Angle and Time of Observation', *J. Opt. Soc. Am.*, **37**, 908
Bouman, M. A. and Velden, H. A. van der (1948). 'The Two-Quanta Hypothesis as a General Explanation for the Behaviour of Threshold Values and Visual Acuity for the Several Receptors of the Human Eye', *J. Opt. Soc. Am.*, **38**, 570
Bouman, M. A., Vos, J. J. and Walraven, P. L. (1963). 'Fluctuation Theory of Luminance and Chromaticity Discrimination', *J. Opt. Soc. Am.*, **53**, 121
Breitmeyer, B. G. (1973). 'A Relationship between the Detection of Size, Rate, Orientation and Direction in the Human Visual System', *Vision Research*, **13**, 41
Bruscaglioni, R. (1952). 'The Equation of Visual Adaptation and the Enhancement of the Sensitivity to Contrast and of the Perception Threshold for Varying Luminance. (in Italian) *Atti della Fondazione Giorgio Ronchi*, 7, 1
Bryngdahl, O. (1964). 'Visual Transfer Characteristics from Mach Band Measurements', *Kybernetik*, **2**, 71
Crandall, W. E. (1973). 'Digital Retinal Vision Theory', in *Colour 73*, Adam Hilger, London, 265
Davies, E. B. (1971). 'The Effect of Length/Breadth Ratio on Thresholds for Visual Detection', RAE Tech. Memo. WE1359
Greening, C. P. and Wyman, M. J. (1970). 'Experimental Evaluation of a Visual Detection Model', *Human Factors*, **12**, 435
Harris, J. L. (Sr.) (1968). 'Image Processing as it relates to the Human System', in *Current Developments in Optics and Vision*, Meeting of Committee on Vision, 1967, 78–88. Ac. Sci., Washington DC
Kornfeld, G. H. and Lawson, W. R. (1971). 'Visual Perception Models', *J. Opt. Soc. Am.*, **61**, 811

Matin, L. (1968). 'Critical Duration, the Differential Luminance Threshold, Critical Flicker Frequency and Visual Adaptation: a Theoretical Treatment', *J. Opt. Soc. Am.*, **58**, 404

Merchant, J. (1965). 'Sampling Theory for the Human Visual Sense', *J. Opt. Soc. Am.*, **55**, 1291

Pinegin, N. I. and Travnikova, N. P. (1971). 'The Probability of Visual Detection of Objects as a Function of their Angular Size, Contrast and Search Time', *Optical Technol.*, **38**, 257

Ronchi, L. (1968). 'Bruscaglioni's Equation of Vision (1939–1941) versus Rose's Model (1946–1948).' *Atti della Fondazione Giorgio Ronchi*, **23**, 782

Schiffman, H. and Crovitz, H. F. (1972). 'A Two-stage Model of Brightness', *Vision Research*, **12**, 2121

Smith, L. J. (1966), 'Theoretical Visual and Televisual Detection Ranges based on Target Size and Contrast', RAE Tech. Rep. No. 66157

Sperling, G. and Sondhi, M. M. (1968). 'Model for Luminance Discrimination and Flicker Detection', *J. Opt. Soc. Am.*, **58**, 1133

Sutherland, N. S. (1968). 'Outlines of a Theory of Visual Pattern Recognition in Animals and Man', *Proc. R. Soc. B.*, **171**, 297

Walraven, A. L. (1973). 'Theoretical Models of the Colour Vision Network', in *Colour 73*, Adam Hilger, London, 11

8 VISUAL SEARCH

Davies, E. B. (1965). 'Contrast Thresholds for Air to Ground Vision', RAE Tech. Rep. 65089

Enoch, J. M. and Fry, G. A. (1958). 'Visual Search of a Complex Display: A Summary Report', MCRL TP No. (696)–17–282, Ohio State University, Columbus, Ohio

Erickson, R. A. (1964). 'Visual Search for Targets: Laboratory Experiments', NOTS TP 3328, Naval Ordnance Test Station, China Lake, California

Greening, C. P. (1972). 'The Likelihood of Looking at a Target', in *AGARD Conference Proceedings No. 100*, (Ed. H. F. Huddleston), London, p. B1–1

Howarth, C. I. and Bloomfield, J. R. (1968). 'Towards a Theory of Visual Search', in *AGARD Conference Proceedings No. 41*, Paper A.2

Koopman, B. O. (1956). 'The Theory of Search, I. Kinematic Bases', *Operations Research*, **4**, 324

Koopman, B. O. (1957). 'The Theory of Search, III. Optimum Distribution of Searching Effort', *Operations Research*, **5**, 613

Pinegin, N. I. and Trasnikova, N. P. (1971). 'The Probability of Visual Detection of Objects as a Function of their Angular Size, Contrast and Search Time', *Optical Technol.*, **38**, 257

Smith, L. J. (1968). 'The Application of Visual Lobe Search Theory to Air to Ground Target Detection', RAE Tech. Rep. 68253

Teichner, W. H. and Krebs, M. J. (1973). 'Visual Search for Symbolically-coded Targets', in Report OA 6201, Volume I: A Collection of Unclassified Technical Papers on Target Acquisition. Martin Marietta Aerospace, Orlando, Florida, p. 149

Williams, L. G. (1966). 'A Study of Visual Search using Eye Movement Recordings', Honeywell Inc. Rep. No. 12009–IR1

9 VISUAL AIDS

Blackwell, H. R. (1968). 'Visual Factors related to the Design and Use of Direct-view Electro-optical Devices', in *Current Developments in Optics and Vision*, Meeting of Committee on Vision, 1967, 93–108. Ac. Sc. Washington DC

Carvennec, F. le (1969). 'Photo-electronic Image Devices', in *Advances in Electronic and Electron Physics,* (Eds. J. D. McGee, D. McMullen, E. Kaham and B. L. Morgan). Academic Press 28A, 265

Catchpole, C. E. (1971). 'The Channel Image Intensifier', Chap. 8 of *Photoelectronic Imaging Devices*, Volume 2, (Eds. L. M. Biberman and S. Nudelman), Plenum

Coleman, H. S. (1947). 'Stray Light in Optical Systems', *J. Opt. Soc. Am.*, 37, 434

Desvignes, F., Revuz, J. and Zeida, R. (1969). 'Photoelectric Solid-State Devices and the Perception of Images in the Infra-red', *Philips Tech. Review,* 30, 264

Fink, D. G. (1957). *Television Engineering Handbook*, McGraw-Hill

Hall, J. A. (1971). 'Evaluation of Signal-generating Image Tubes', Chapter 4 of *Photoelectronic Imaging Devices*, Vol. 2, (Eds. L. M. Biberman and S. Nudelman), Plenum

Harker, G. S. (1972). 'Vision: Monocular, Bi-ocular, Binocular', USAMRL Rep. No. 984

Keesee, R. L. (1973). 'Detectability Thresholds for Line-scan Displays', in *Report OA 6201, Volume I: A Collection of Unclassified Technical Papers on Target Acquisition*, Martin Marietta Aerospace, Orlando, Florida, p. 89

Lawson, W. R. (1971). 'Electro-optical System Evaluation', in *Photoelectronic Imaging Devices*, Vol. 1, p. 375, L. M. Biberman and S. Nudelman (Eds.) Plenum

Mees, C. E. K. and James, T. H. (1966). *The Theory of the Photographic Process*, MacMillan Co. New York

Meeteran, A. van and Boogaard, J. (1972). 'Visual Contrast Sensitivity with Ideal Image Intensifiers', Institute for Perception RVO–TNO, Report IZF1972–18

Richards, W. (1969). 'Saccadic Suppression', *J. Opt. Soc. Am.*, **59**, 617

Rosell, F. A. (1971). 'Television Camera Tube Performance and Data', in *Low-light-level Devices: a Designer's Manual*, IDA Rep. R169, 175

Schade, O. H. (1973). 'Image Reproduction by a Line Raster Process', in *Perception of Displayed Information*, (L. M. Biberman, Ed.), Plenum

Snyder, H. L. (1972). 'A Unitary Measure of Video System Image Quality' in *A Collection of Unclassified Technical Papers on Target Acquisition, Rep. OA 6201*, Vol. I, Office of Naval Research, Orlando Florida

10 OPTICAL QUALITY

Anon. (1971). 'Recommendations for Measurement of the Optical Transfer Function of Optical Devices', *British Standard BS 4779*

Aznarez, J., Corno, J., Lamare, M. and Simon, J. (1974). 'Contribution to the Determination of Transfer Functions by the Edge Trace Method', *Optica Acta.*, 2, 809

Barakat, R. (1965). 'Determination of the Optical Transfer Function directly from the Edge Spread Function', *J. Opt. Soc. Am.*, **55**, 1217

Barnard, T. W. (1972). 'Image Evaluation by means of Target Recognition', *Phot. Sci. and Eng.*, **16**, 144

Biberman, L. M. (1973). 'Image Quality', in *Perception of Displayed Information*, (L. M. Biberman, Ed.) Plenum

Brock, G. C. (1968). 'A Review of Current Image Evaluation Techniques', *J. Phot. Sci.*, **16**, 241

Coleman, H. S. (1947). 'Stray Light in Optical Systems', *J. Opt. Soc. Am.*, **37**, 434

Coleman, H. S. (1947). 'The Reduction in Image Contrast caused by the Aberrations in Telescopic Systems', *J. Opt. Soc. Am.*, **37**, 684

De Velis, J. B. (1965). 'Comparison of Methods for Image Evaluation', *J. Opt. Soc. Am.*, **55**, 165

Dixon, F. A. (1961). 'Optical Frequency Response in relation to other Methods of Image Assessment', Weapons Research Establishment Tech. Note OID19, Australia

Dow Smith, F. (1963). 'Optical Image Evaluation and the Transfer Function', *Applied Optics*, **2**, 335

Dubenskov, V. P., Tybkina, A. I. and Marinchenko, YuM. (1972). 'The Effect of Vibration on the Visual Resolution of a Telescope', *Optical Technol.*, **39**, 522

Gilmore, H. R. (1967). 'Models of the Point Spread Function of Photographic Emulsions based on a Simplified Diffusion Calculation', *J. Opt. Soc. Am.*, **57**, 75

Gullick, S. A. (1970). 'A Survey of Methods of Evaluation of the Imaging Performance of Optical Systems', Study Note No. 10 of Ministry of Technology Contract KV/B/813/CB6 4B, BAC (GW) Ref. L50/22/PHY/186/1153

Harris, J. L. (1964). 'Resolving Power and Decision Theory', *J. Opt. Soc. Am.*, **54**, 606

Hopkins, H. H. (1966). 'The use of Diffraction Based Criteria of Image Quality in Automatic Optical Design', *Optica Acta*, **13**, 343

Hufnagel, R. E. (1965). 'Search for a Summary Measure of Image Quality – Part II', *J. Opt. Soc. Am.*, **55**, 1564

Kelsall, D. (1973). 'Rapid Interferometric Technique for MTF Measurements in the Visible or Infra-red Region', *Applied Optics*, **12**, 1398

Niederpruem, C. J., Nelson, C. N. and Yule, J. A. C. (1966). 'Contrast Index', *Phot. Sci. & Eng.*, **10**, 35

Perrin, F. H. and Altman, J. H. (1951). 'Photographic Sharpness and Resolving Power. II. The Resolving-power Cameras in the Kodak Research Laboratory' *J. Opt. Soc. Am.*, **41**, 265

Preston, K. (1963). 'Modulation Transfer Function Instrumentation', Perkin Elmer Symposium on Practical Application of Modulation Transfer Functions, 6–1

Roetling, P. G. (1970). 'Image Enhancement by Noise Suppression', *J. Opt. Soc. Am.*, **60**, 867

Rosell, F. A. (1969). 'Limiting Resolution of Low-light-level Imaging Sensors', *J. Opt. Soc. Am.*, **59**, 539

Rosenau, M. D. (1963). 'Image-motion Modulation Transfer Functions', Perkin Elmer Symposium on Practical Applications of Modulation Transfer Functions, 5–1

Rosenhauer, K. and Rosenbruch, K. J. (1967/8). 'On the Influence of Stray Light on the OTF', *Optik*, **26**

Schade, O. H. (1971). 'Resolving Power Functions and Integrals of High-definition Television and Photographic Cameras – a New Concept in Image Evaluation', *RCA Review*, December

Scott, F. (1963). 'Film Modulation Transfer Functions', *Perkin Elmer Symposium on Practical Application of Modulation Transfer Functions*, 4–1

Scott, F. (1968). 'The Search for a Summary Measure of Image Quality – A Progress Report', *Phot. Sci. and Eng.*, **12**, 154

Shaw, R. (1962). 'The Application of Fourier Techniques and Information Theory to the Assessment of Photographic Image Quality', *Phot. Sci. and Eng.*, **6**, 281

Simon, J. F. and Denieul, P. M. (1973). 'Influence of the Size of Test Field employed in Measurement of Modulation Transfer Function of the Eye', *J. Opt. Soc. Am.*, **63**, 894

Snyder, H. L. (1972). 'A Unitary Measure of Video System Image Quality' in *A Collection of Unclassified Technical Papers on Target Acquisition, Rep. OA 6201*, Vol. I, Office of Naval Research, Orlando, Florida

Williams, T. L., Leach, B. A. and Biddles, B. J. (1972). 'A Workshop Instrument for Testing Binocular and Other Sights using the MTF Criterion', *Opt. and Laser Technol.*, **4**, 115

11 SPATIAL FREQUENCY RESPONSE

Abadi, R. V. and Kulikowski, J. J. (1973). 'Linear Summation of Spatial Harmonics in Human Vision', *Vision Research*, **13**, 1625

Atkinson, J. and Campbell, F. W. (1974). 'The Effect of Phase on the Perception of Compound Gratings', *Vision Research*, **14**, 159

Blakemore, C. and Campbell, F. W. (1968). 'Adaptation to Spatial Stimuli', *J. Physiol.,* **200**, 11P

Blakemore, C., Nachmias, J. and Sutton, P. (1970). 'Perceived Spatial Frequency Shift: Evidence of Frequency Selective Neurons in the Human Brain', *J. Physiol.,* **210**, 727

Burton, G. J. (1973). 'Evidence for Non-linear Response Processes in the Human Visual System from Measurements on the Thresholds of Spatial Beat Frequencies', *Vision Research,* **13**, 1211

Campbell, F. W., Cooper, G. F., Robson, J. G. and Sachs, M. B. (1969). 'The Spatial Selectivity of Visual Cells of the Cat and the Squirrel Monkey', *J. Physiol.,* **204**, 120

Campbell, F. W. and Maffei, L. (1974). 'Contrast and Spatial Frequency', *Scientific American,* 106

Campbell, F. W., Nachmias, J. and Jukes, J. (1970). 'Spatial-frequency Discrimination in Human Vision', *J. Opt. Soc. Am.,* **60**, 555

Foster, D. H. and Idris, I. I. M. (1974). 'Spatio-temporal Interaction between Visual Colour Mechanisms', *Vision Research,* **14**, 35

Fry, G. A. (1969). 'Visibility of Sine-wave Gratings', *J. Opt. Soc. Am.,* **59**, 610

Graham, N. (1972). 'Spatial Frequency Channels in Human Vision: Effects of Luminance and Pattern Drift Rate', *Vision Research,* **12**, 53

Lohmann, A. W. (1968), 'Experiments in Spatial Filtering', in *Current Developments in Optics and Vision*, Meeting of Committee on Vision, 1967, 89–92, Ac. Sci., Washington DC

Lowry, E. M. and DePalma, J. J. (1961). 'Sine-wave Response of the Visual System. I. The Mach Phenomenon', *J. Opt. Soc. Am.,* **51**, 740

Macleod. I. D. G. and Rosenfeld A. (1974). 'The Visibility of Gratings: Spatial Frequency Channels or Bar-detecting Units? *Vision Research,* **14**, 909

Maffei, L. and Fiorentini, A. (1973) 'The Visual Cortex as a Spatial Frequency Analyser', *Vision Research,* **13**, 1255

Meeteren, A. van and Boogaard, J. (1972). 'Visual Contrast Sensitivity with Ideal Image Intensifiers', Institute for Perception RVO–TNO, Rep. IZF1972–18

Meeteren, A. van and Vos, J. J. (1972). 'Resolution and Contrast Sensitivity at Low Luminances', *Vision Research,* **12**, 825

Nachmias, J. and Sansbury, R. V. (1974). 'Grating Contrast: Discrimination may be better than Detection', *Vision Research,* **14**, 1039

Nachmias, J., Sansbury, R., Vassilev, A. and Weber, A. (1973). 'Adaptation to Square-wave Gratings in Search of the Elusive Third harmonic', *Vision Research,* **13**, 1335

Nes, F. L., Koenderink, J. J., Nas, H. and Bouman, M. A. (1967). 'Spatio-temperal Modulation Transfer in the Human Eye', *J. Opt. Soc. Am.,* **57**, 1082

Stecher, S., Sigel, C. and Lange, R. V. (1973). 'Spatial Frequency Channels in Human Vision and the Threshold for Adaptation', *Vision Research,* **13**, 1691

Westheimer, G. (1960). 'Modulation Thresholds for Sinusoidal Light Distribution on the Retina', *J. Physiol.,* **152**, 67

12 RECEPTIVE FIELD STUDIES

Barlow, H. B. (1958). 'Temporal and Spatial Summation in Human Vision at Different Background Intensities', *J. Physiol.,* **141**, 337

Békésy, G. Von. (1968). 'Mach and Hering-type Lateral Inhibition in Vision', *Vision Research,* **8**, 1483

Brown, D. R., Schmidt, M. J., Fulgham, D. D. and Cosgrove, M. P. (1973). 'Human Receptive Field Characterisitics: Probe Analysis of Stabilised Images', *Vision Research,* **13**, 231

Cavonius, C. R. and Hilz, R. (1973). 'Invariance of Visual Receptive-field Size and Visual Acuity with Viewing Distance', *J. Opt. Soc. Am.,* **63**, 929

Fiorentini, A. and Maffei, L. (1973). 'Contrast in Night Vision', *Vision Research*, **13**, 73

Fiorentini, A. and Zoli, M. T. (1966). 'Detection of a Target superimposed to a Step Pattern of Illumination', *Atti della Fondazione Giorgio Ronchi*, **21**, 338

Fry, G. A. and Bartley, S. H. (1935). 'The Effect of one Border in the Visual Field upon the Threshold of Another', *Am. J. Physiol.*, **112**, 414

Hartline, H. K., Ratliff, F. and Miller, W. H. (1961). 'Inhibitory Interaction in the Retina and its Significance in Vision', in *Nervous Inhibition*, (E. Florey, Ed.), Pergamon Press, 241–284

Luria, S. M. and Ryan, A. (1973). 'Adaptation to a Masking Stimulus', *J. Opt. Soc. Am.*, **63**, 201

Maffei, L. (1968). 'Inhibitory and Facilitatory Spatial Interactions in the Retinal Receptive Fields', *Vision Research*, 8, 1187

Saunders, R. McD. (1974). 'The Contribution of Spatial and Border Interactions to the "Westheimer Effect",' *Vision Research*, **14**, 379

Thomas, J. P. (1968). 'Linearity of Spatial Integrations involving Inhibitory Interactions', *Vision Research*, **8**, 49

13 NOISE

Benton, S. A. (1971). 'Properties of Granularity Wiener Spectra', *J. Opt. Soc. Am.*, **61**, 524

Berwart, L. (1969). 'Wiener Spectrum of Experimental Emulsions with Cubic Homogeneous Grains, Comparison of the Spectra with the Wiener Spectra of Commercial Emulsions', *J. Phot. Sci.*, **17**, 41

Coltman, J. W. and Anderson, A. E. (1960). 'Noise Limitations to Resolving Power in Electronic Imaging', *Proc. I.R.E.*, **48**, 858

Davies, E. B. and Alpin, J. E. (1965). 'Preliminary Visual Experiments on Background and Noise Effects using a Television Display', RAE Tech. Memo. WE1169(E)

Doerner, E. C. (1962). 'Wiener-Spectrum Analysis of Photographic Granularity., *J. Opt. Soc. Am.*, **52**, 669

Doerner, E. C. (1965). 'The Use of the Wiener Spectrum (Power–Spectrum) for studying the Granularity of Photographic Systems', Paper presented at the Congrès Internationale de Science Photographique, Paris

Erikson, R. A. and Hemingway, J. C. (1970). 'Visibility of Raster Lines in a Television Display', *J. Opt. Soc. Am.*, **60**, 700

Gorokhovskiy, Yu. N. and Filimonov, R. P. (1973). 'Some Measurements of Wiener Spectra of Photographic Noise', *Optical Technol.*, **40**, 653

Roetling, P. G. (1970). 'Image Enhancement by Noise Suppression', *J. Opt. Soc. Am.*, **60**, 867

Rosell, F. A. and Willson, R. H. (1973). 'Recent Psychophysical Experiments and the Display Signal-to-noise Ratio Concept', in *Perception of Displayed Information*, (L. M. Biberman, Ed.) Plenum

Shaw, R. (1972). 'Photon Fluctuations and Photographic Noise', *J. Phot. Sci.*, **20**, 64

Shaw, R. (1972). 'The Photographic Process as a Photon Counting Device', *J. Phot. Sci.*, **20**, 174

Shaw, R. and Shipman, A. (1969). 'Practical Factors influencing the Signal-to-noise Ratio of Photographic Images', *J. Phot. Sci.*, **17**, 205

Weinberg, H. and Cooper, R. (1972). 'The Recognition Index: A Pattern Recognition Techique for Noisy Signals', *Electroenceph. Clin. Neurophysiol.*, **33**, 608

Yoneyama, M. (1973). 'Spatial Filtering for Noise Reduction in a Multirecording Telecine System', *Applied Optics*, **12**, 2721

14 ATMOSPHERICS

Clifford, S. F. (1971). 'Temporal-frequency Spectra for Spherical Wave Propagating through Atmospheric Turbulence', *J. Opt. Soc. Am.*, **61**, 1285

Consortini, A., Ronchi, L. and Moroder, E. (1973). 'Role of the Outer Scale of Turbulence in Atmospheric Degradation of Optical Images', *J. Opt. Soc. Am.*, **63**, 1246

Coulman, C. E. and Hall, D. N. B. (1967). 'Optical Effects of Thermal Structure in the Lower Atmosphere', *Applied Optics*, **6**, 497

Duntley, S. Q. (1948). 'The Visibility of Distant Objects', *J. Opt. Soc. Am.*, **38**, 237

Gambling, D. J. and Billard, B. (1967). 'A Study of the Polarisation of Skylight', *Aust. J. Phys.*, **20**, 675

Hampton, W. M. (1933). 'The Visibility of Objects in a Searchlight Beam', *Proc. Phys. Soc.*, London, **45**, 663

Harger, R. O. (1967). 'On Processing Optical Images Propagated through the Atmosphere', *IEEE Trans. on Aerospace and Electronic Systems*, AES3, 819

Hufnagel, R. E. (1963). 'Random Wavefront Effects', *Perkin Elmer Symposium on Practical Application of Modulation Transfer Functions*, 3–1

Korff, D. (1973). 'Analysis of a Method for obtaining Near-diffraction-limited Information in the Presence of Atmospheric Turbulence', *J. Opt. Soc. Am.*, **63**, 971

Lahart, M. J. (1974). 'Maximum-likelihood Restoration of Non-stationary Imagery', *J. Opt. Soc. Am.*, **64**, 17

Lutomirski, R. F. and Yura, H. T. (1974). 'Imaging of Extended Objects through a Turbulent Atmosphere', *Applied Optics*, **13**, 431

Rosenau, M. D. (Jnr) (1962). 'The Alteration of Object Modulation by Real Atmospheres as it affects Aerial Photography', *Phot. Sci. and Eng.*, **6**, 265

Wagner, H. F. and Bartsch, G. (1973). 'A Field Apparatus for Scanning Skylight Intensities', *J. Phys. E.*, **6**, 1084

15 FIELD TRIALS

Bryson, M. R. (1973). 'Air-to-Ground and Ground-to-Air Detection Experiments', in *Rep. OA 6201, Vol. I: A Collection of Unclassified Technical Papers on Target Acquisition*, Martin Marietta Aerospace, Orlando, Florida, P. 17

Rosenau, M. D. (Jnr) (1962). 'The Alteration of Object Modulation by Real Atmospheres as it affects Aerial Photography', *Phot. Sci. and Eng.*, **6**, 265

Index